Historic Watermills of the South

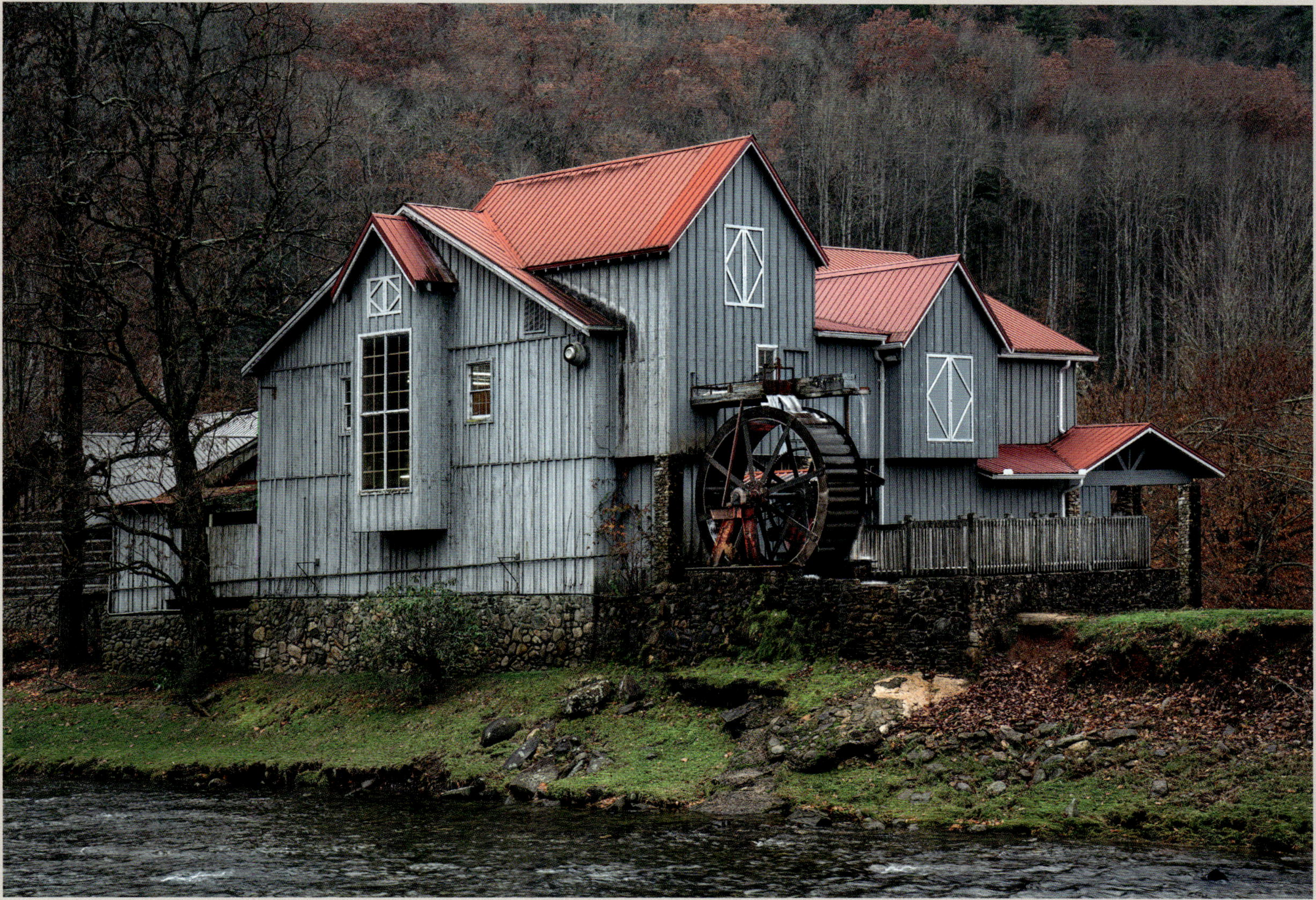

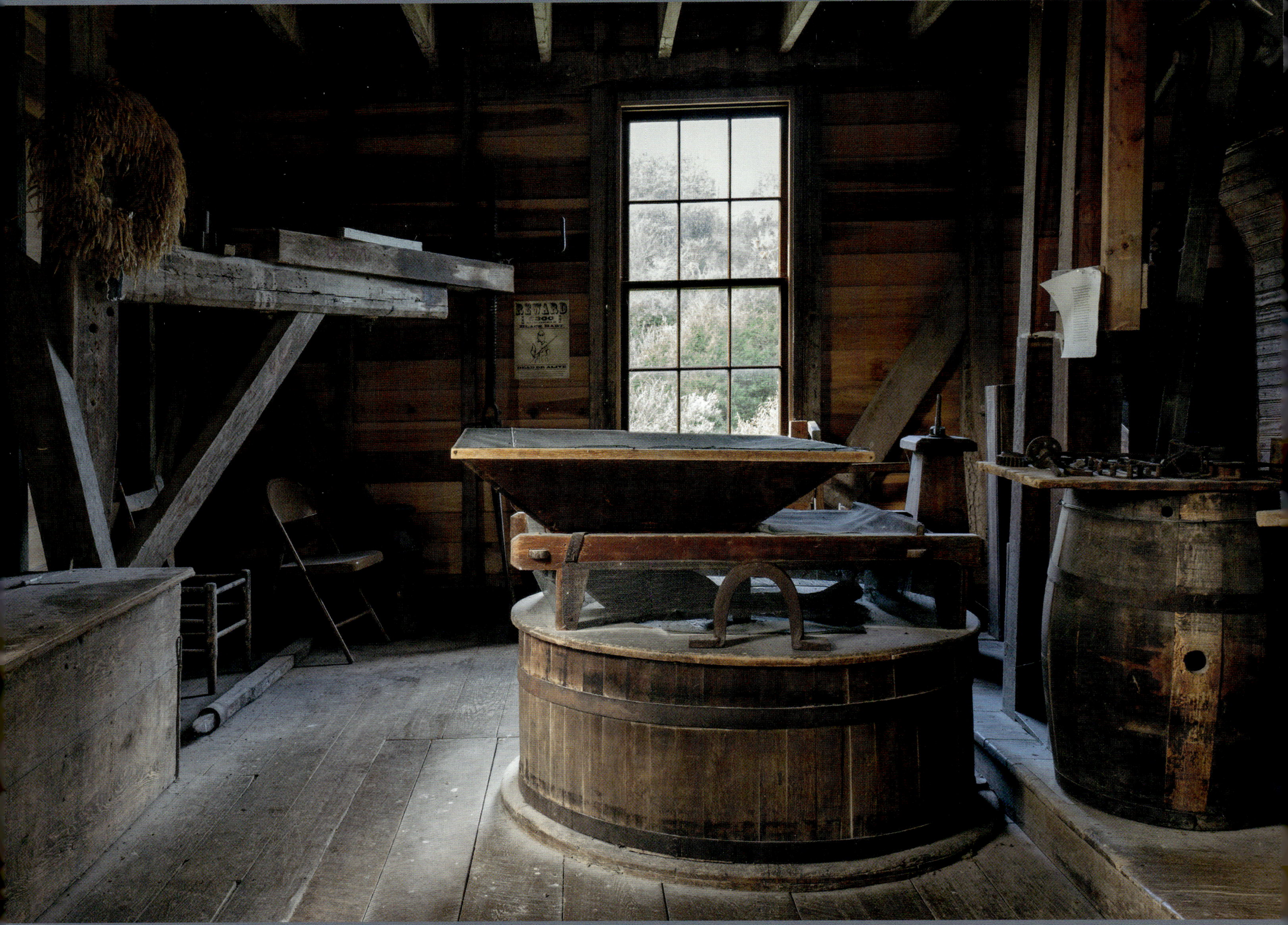
REWARD
300

Historic Watermills of the South

Enduring Legacy and Timeless Charm

KEN BOYD

THE UNIVERSITY OF GEORGIA PRESS ATHENS

A Wormsloe
FOUNDATION
PUBLICATION

Published by the University of Georgia Press
Athens, Georgia 30602
www.ugapress.org

Designed by Erin Kirk
Set in Arno Pro
Printed and bound by Martin Book Management
The paper in this book meets the guidelines for permanence and durability of the Committee on Production Guidelines for Book Longevity of the Council on Library Resources.

Printed in the United States of America
30 29 28 27 26 C 5 4 3 2 1

EU Authorized Representative
Easy Access System Europe—Mustamäe tee 50, 10621 Tallinn, Estonia,
gpsr.requests@easproject.com

Library of Congress Cataloging-in-Publication Data
Names: Boyd, Ken, 1952– author photographer
Title: Historic watermills of the South : enduring legacy and timeless charm / Ken Boyd.
Description: Athens : The University of Georgia Press, [2026] | Series: A Wormsloe Foundation publication | Includes bibliographical references and index.
Identifiers: LCCN 2026006172 | ISBN 9780820376882 hardback
Subjects: LCSH: Water mills—Southern States | LCGFT: Illustrated works
Classification: LCC TJ859 .B695 2026
LC record available at https://lccn.loc.gov/2026006172

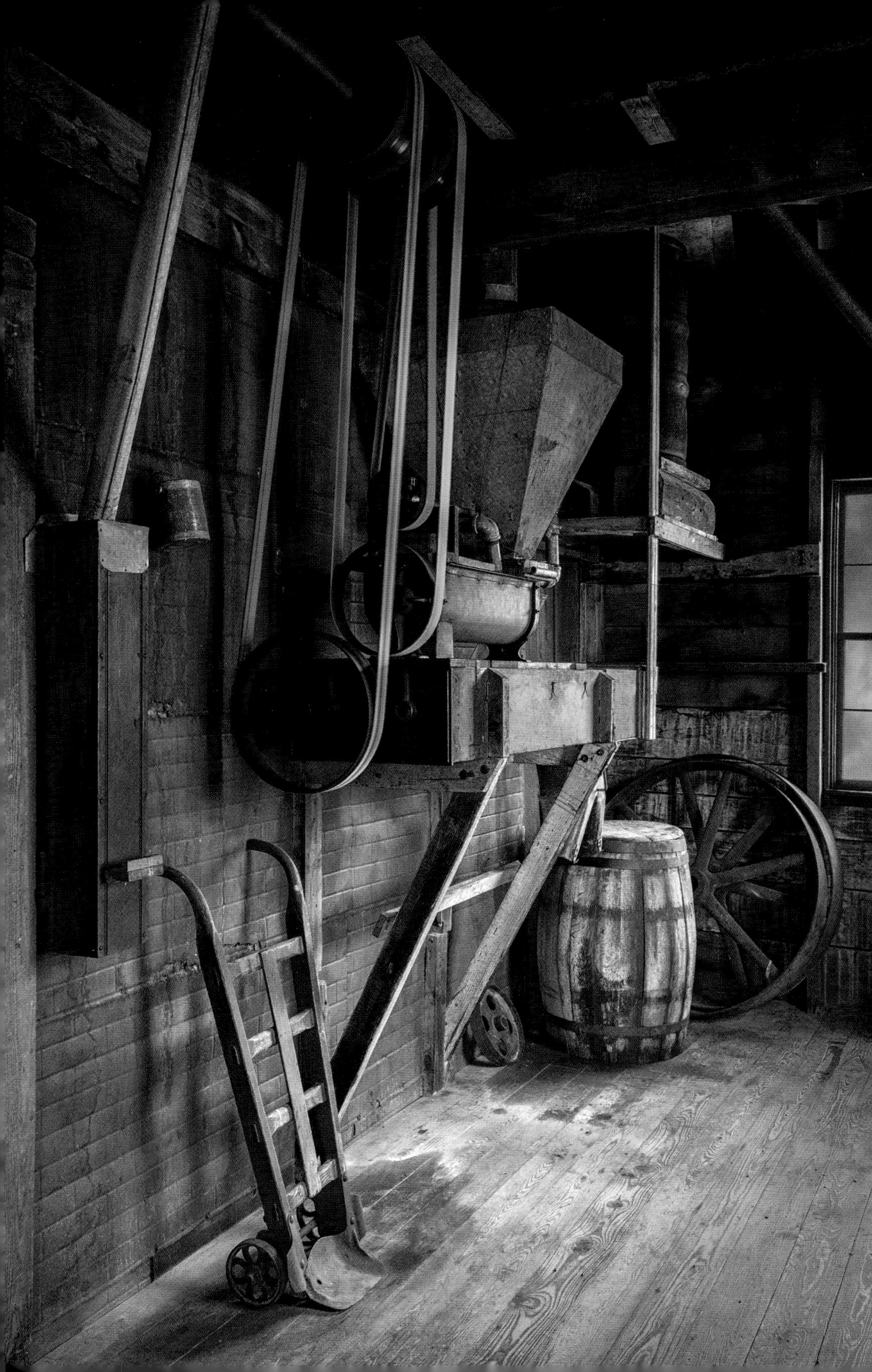

This book is dedicated to the Society for the Preservation of Old Mills (SPOOM). Chartered in 1972, this international organization has worked tirelessly to foster an awareness of and appreciation for the heritage and preservation of old mills.

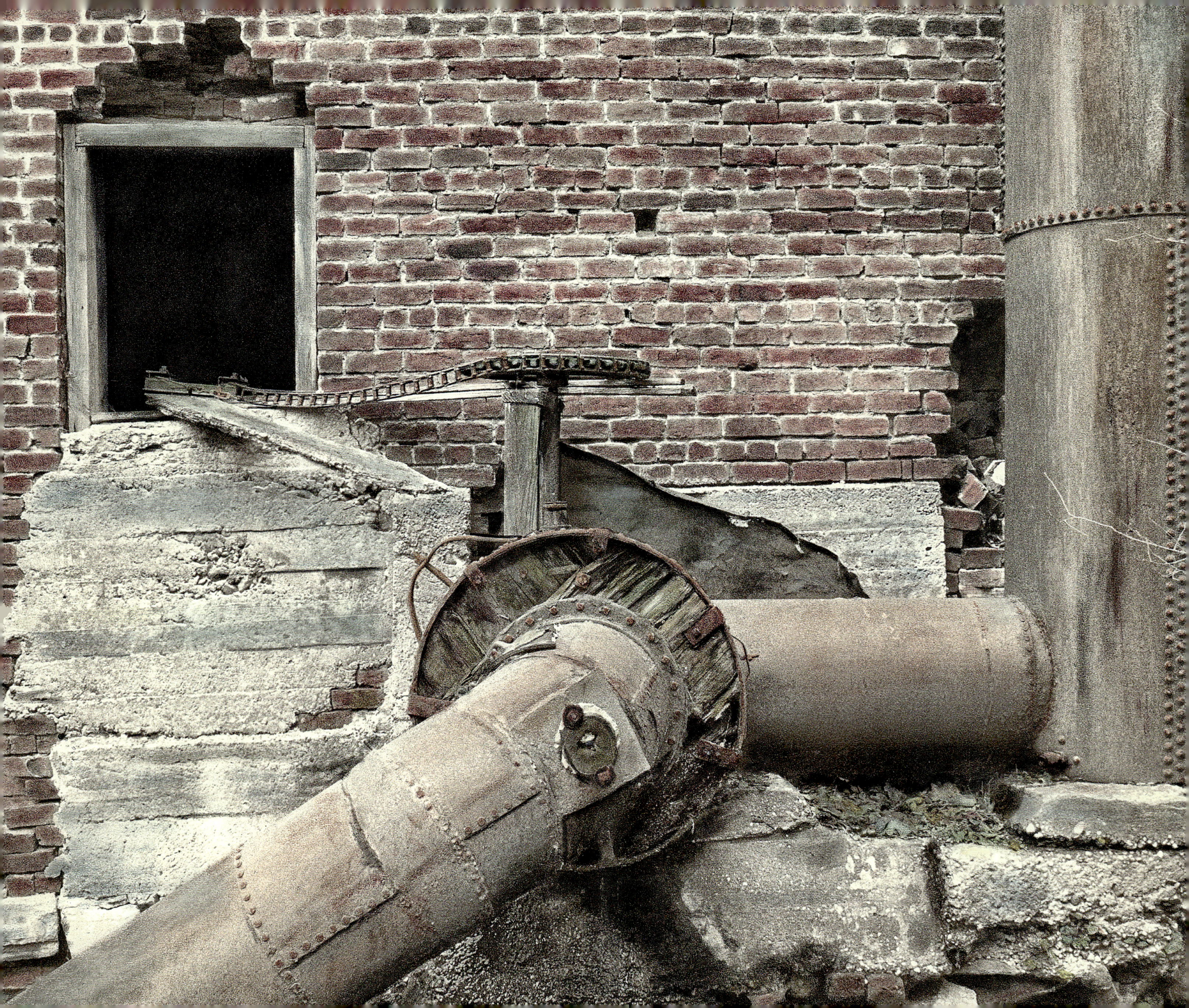

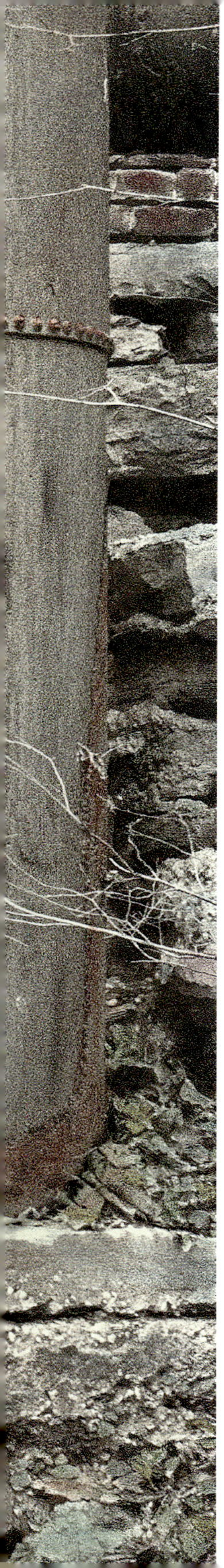

Contents

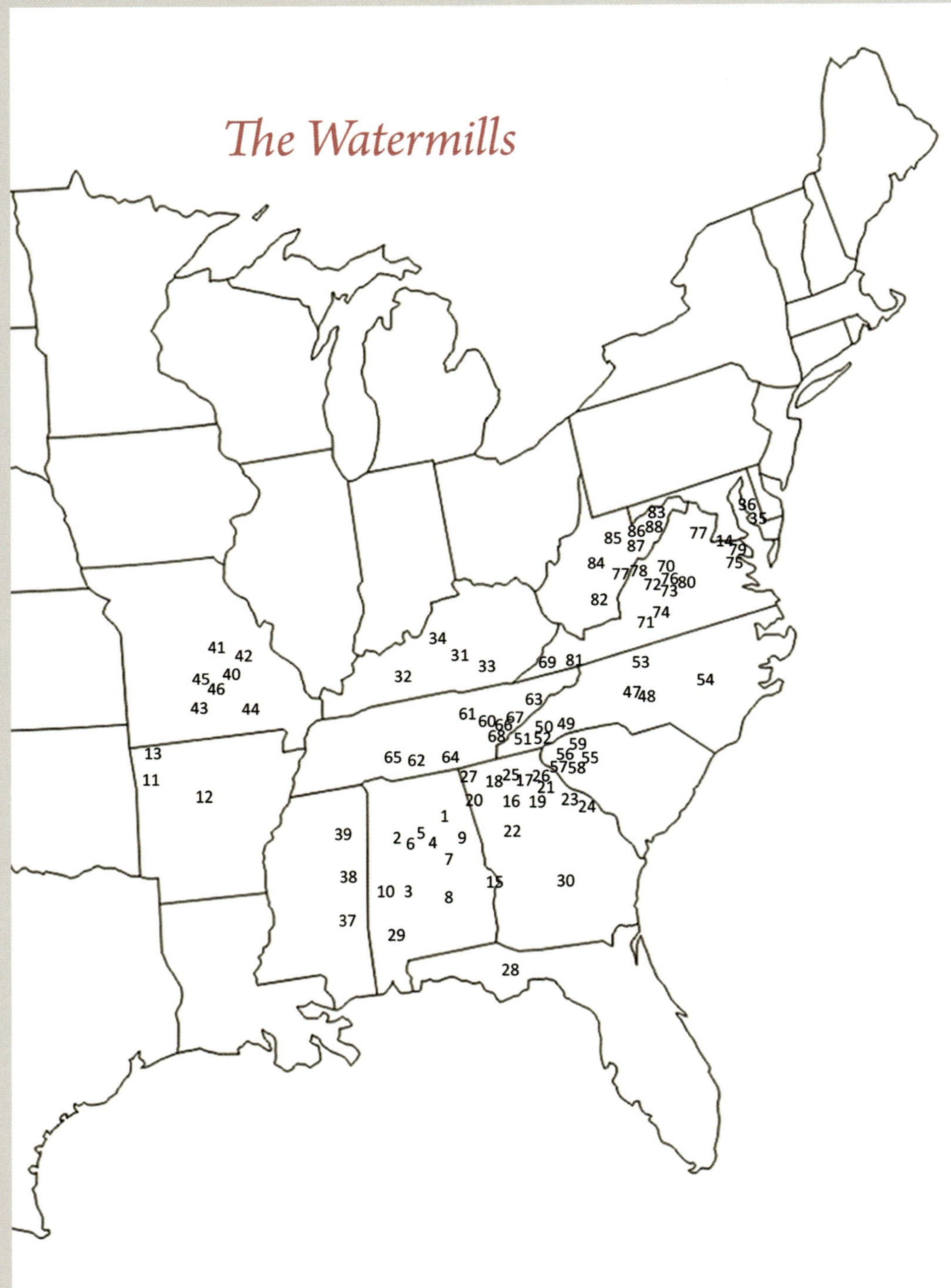

The Watermills

(Key to Location Map)

ALABAMA

1 Aderholdt's Mill
2 John Wesley Hall Grist Mill and Cotton Gin and the Roupes Valley blower house
3 Pratt Cotton Gin Mill
4 Swann Millhouse
5 Robert Jemison millhouse
6 Sweetwater Falls Millhouse
7 Kymulga Grist Mill
8 Tallassee Mills ruins
9 Wehadkee Yarn Mills
10 Kenan's Mill

ARKANSAS

11 Johnson Mill
12 Pugh's Mill (the *Gone with the Wind* mill)
13 War Eagle Mill

DISTRICT OF COLUMBIA

14 Peirce Mill

GEORGIA

15 Columbus Historic Riverfront Industrial District
16 Johnson Grist Mill
17 Loudermilk Mill
18 Nora Mills Granary
19 Stone Mountain Grist Mill
20 The Old Mill at Berry College
21 Ragsdale Mill

22 Starr's Mill
23 Sibley Mill / Confederate Powder Works
24 King Mill
25 Pardue Mill
26 Sells Mill
27 Prater's Mill

GULF COASTAL PLAIN

28 Shepard's Mill
29 Rikard's Mill
30 Parrish Mill

KENTUCKY

31 Mill Springs Mill
32 Bardstown Mill
33 McHargue's Mill
34 Weisenberger Mill

MARYLAND

35 Linchester Mill
36 Old Wye Mill

MISSISSIPPI

37 Lake Bounds Mill
38 Carroll-Richardson Grist Mill
39 Sciple's Water Mill

MISSOURI: OZARK MOUNTAIN SPRINGS

40 Alley Spring Mill
41 Reed Spring Mill
42 Markham Springs Mill
43 Hodgson Water Mill
44 Falling Spring Mill
45 Rockbridge Mill
46 Topaz Mill

NORTH CAROLINA

47 Allison Woods Upper Mill
48 Allison Woods Lower Mill
49 Murray's Mill
50 Francis Grist Mill
51 Mingus Mill
52 Saunooke's Mill
53 The Old Mill of Guilford (Bailes Old Mill)
54 Yates Mill

SOUTH CAROLINA

55 Anderson Mill
56 Gilreath's Mill
57 Hagood Mill
58 Suber's Corn Mill
59 Ballenger Mill

TENNESSEE

60 The Old Mill in Pigeon Forge
61 Hacker Martin Grist Mill
62 Falls Mill
63 K. C. French Mill
64 Ketner's Mill
65 G. T. Wilburn & Son Grist Mill
66 Noah Ogle Tub Mill
67 Alfred Reagan Tub Mill
68 Cable Mill

VIRGINIA

69 Bush Mill
70 Meadow Run Mill
71 Mabry Mill
72 McCormick's Grist Mill
73 Osceola Mill
74 Pine Creek Mill
75 Lee's Mill at Stratford Hall
76 Kennedy-Wade Mill
77 Colvin Run Mill
78 Tingler's Mill
79 George Washington's Gristmill
80 Woodson's Mill
81 White's Mill

WEST VIRGINIA

82 Cook's Old Mill
83 Fitz-Tuscarora Iron Works
84 Glade Creek Grist Mill
85 Jackson's Mill
86 Yellow Spring Mill
87 McClung's Mill
88 Bunker Hill Flour Mill

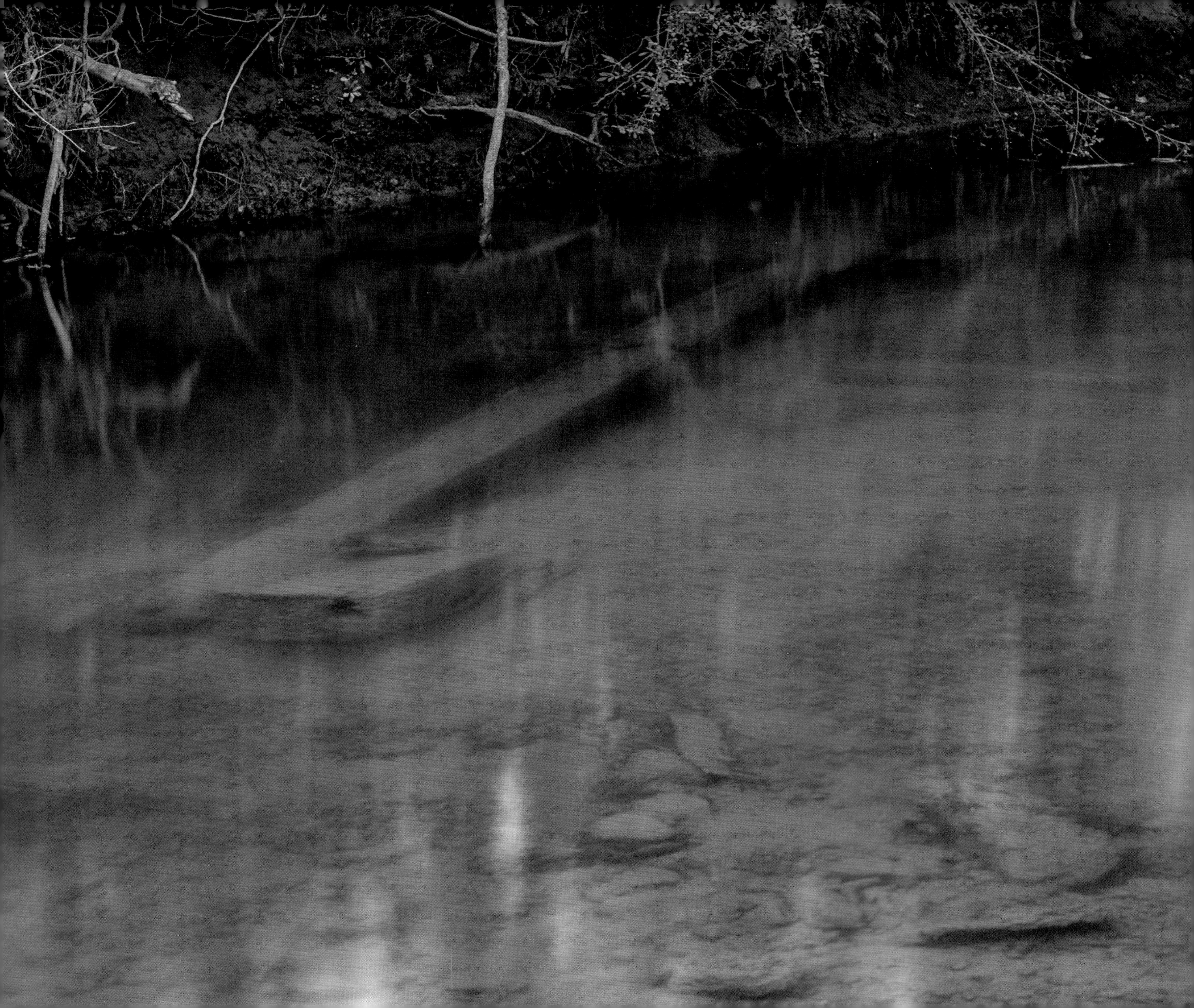

Prologue

NOT SO LONG AGO, watermills turned on creeks, streams, and rivers all across the South. Thousands were in operation, from small farm mills to large commercial enterprises. However, as the twentieth century opened, the South's watermills began to disappear from the landscape. Steam and electricity proved to be better, more reliable, and more economical sources of power, and the expansion of transportation networks rendered the old watermills obsolete.

I grew up in rural northeast Alabama in the 1950s and 1960s. Early in my life I developed a fascination with and later a passion for these old and dying water-powered structures that for several centuries had been so fundamental and vital to life throughout the South.

As a child I heard the stories, especially from my grandfather, of the old mills on Ohatchee Creek near the family farm. I could see the remains of the mills in the creek, which by that time were little more than foundations, millstones, and a few metal relics, mostly submerged in the water. This all left a profound and lasting impression on me.

My interest in watermills continued to grow, and as a young adult in the 1970s I began to travel around the South to see what remained of the region's watermill heritage. Most of the sites I visited were abandoned, decaying edifices that only hinted at what they once had been, but to me they were still beautiful, charming, and, as often described, "magical" in some captivating way. They drew me in to experience what they had been not so many years ago.

Maybe I was enchanted by the splashing clear water, or the old stones and weathered wood, or the quiet pastoral setting. But I know now that beyond what I could see, my captivation also had a lot to do with personal reflections on a time when much of life throughout the South had revolved around the grinding of grain, sawing of wood, forging of iron, and ginning and weaving of cotton. For me, a visit to a mill site was an emotional transport back to that earlier, now vanished time.

I imagined that I could see the waterwheel turning, the mill running, and people going about their everyday lives inside and around the mill site. All of that was now silent and lost to us with the passage of time. Perhaps my photographs could in a small way capture and preserve some of whatever had been left for me to experience.

To be clear, I was in no way first in this pursuit. Artists of all media have been drawn to watermills for centuries. Painters and sketchers were pulled from their studios into the countryside to experience and hopefully capture this "magic." Poets and authors would sit for hours, days, and seasons exploring and contemplating choice words to describe their impressions of the mill.

Through the 1970s, 1980s, and 1990s, I visited, studied, and photographed scores of watermills, from the Delmarva Peninsula along the Chesapeake Bay, south to Florida, west across the Appalachian Mountains, down to the Gulf of Mexico, over the Mississippi River, through the Ozarks, and just beyond the Mississippi Delta. This region we think of as the South is physically quite diverse, with several distinct geological areas ranging from coastal plains and river basins to plateaus and mountainous terrain. Yet many of these locations were well suited for watermills of some type.

A part of the intrigue for me is that no two watermills were ever created alike. There was never any standard design, and thus each watermill was unique. The way a mill looked and the approach to its design were functions of the terrain, the available flow of water, the local building materials, the purpose of the mill,

OPPOSITE: The remains of an old mill can be seen beneath the waters of Ohatchee Creek near my home.

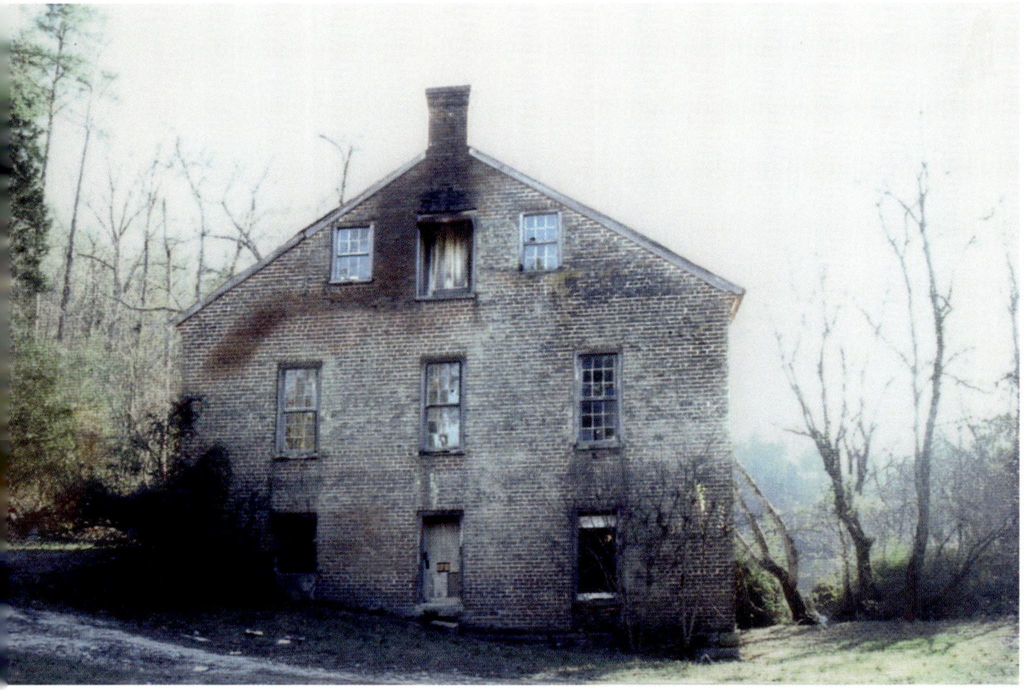

This was my first photograph of a watermill, taken at Aderholdt's Mill during a visit there with my father in about 1969 or 1970.

the construction budget, the size of the endeavor, and the creative vision of the millwrights and millers who did the work.

As my appreciation for watermills grew and my skills as a photographer expanded, my objectives also evolved. My goal became not just to document each mill with my camera but to produce images that would stand as a testament to the mill. I sought to create "works of art" and to bring back and share the "magic."

In striving to produce art images of our Southern watermills, I have attempted to visually illustrate the incredible character, detail, and intricacy of mill designs, as well as the strikingly beautiful landscapes around the mills. The structures clearly hark back to an earlier era of Southern history. Without question, this was a crucial period that has much to do with who we are today as a people. To more fully understand who we are, we need to appreciate where we came from and how that has shaped our lives today.

Many of my images were taken on traditional, high-quality black-and-white films and with the large cameras of the time. More recent renderings were made with twenty-first-century equipment. In this book I have included several types of images that span the technologies of my lifetime.

Some of the watermills in this book are captured in highly detailed, traditional black-and-white photographs. These images are complemented by a number of hand-tinted images. In the early 1980s, I began to experiment with colorizing enlarged black-and-white prints using transparent oils. This technique may be more familiar from colorized portraits produced in the early and mid-twentieth century. With transparent oils, pastel colors can be added without obscuring the delicate details of an image.

Over the years, I have hand-colored prints of more than eighty watermill scenes as fitting representations of such historic Southern subjects. Sadly, detailed hand-tinting of black-and-white photographs is almost a lost art now. Modern digital colorization rarely produces a satisfactory result.

Finally, to complete the portfolio, some mills were photographed using modern digital media, but even these images have been post-processed with an awareness of the historic character of the subject matter.

I have also included images of details such as waterwheels, millstones, saw blades, clever wooden contraptions, and assorted old machines. My intent is to promote a greater appreciation of the invention that literally revolutionized colonial life, the early frontier, and rural regions of the South.

Through the details of mill interiors, the true skills and art of the South's master craftspeople are revealed. Stepping inside a watermill is transformative, and akin to walking into a giant clockworks. Every cog, pin, pulley, and gear has been meticulously carved and fitted to move with precision and to last for generations. This is the true hallmark of watermill design and construction. Each mill is indeed a work of art!

As you view these photographs, please understand that some of them were taken almost fifty years ago, and much has changed over that time. Some of the watermills have been fairly well preserved since then, others have been operationally restored, and some have been repurposed as businesses, restaurants, and even private residences. Sadly, a number of the mills I visited no longer exist.

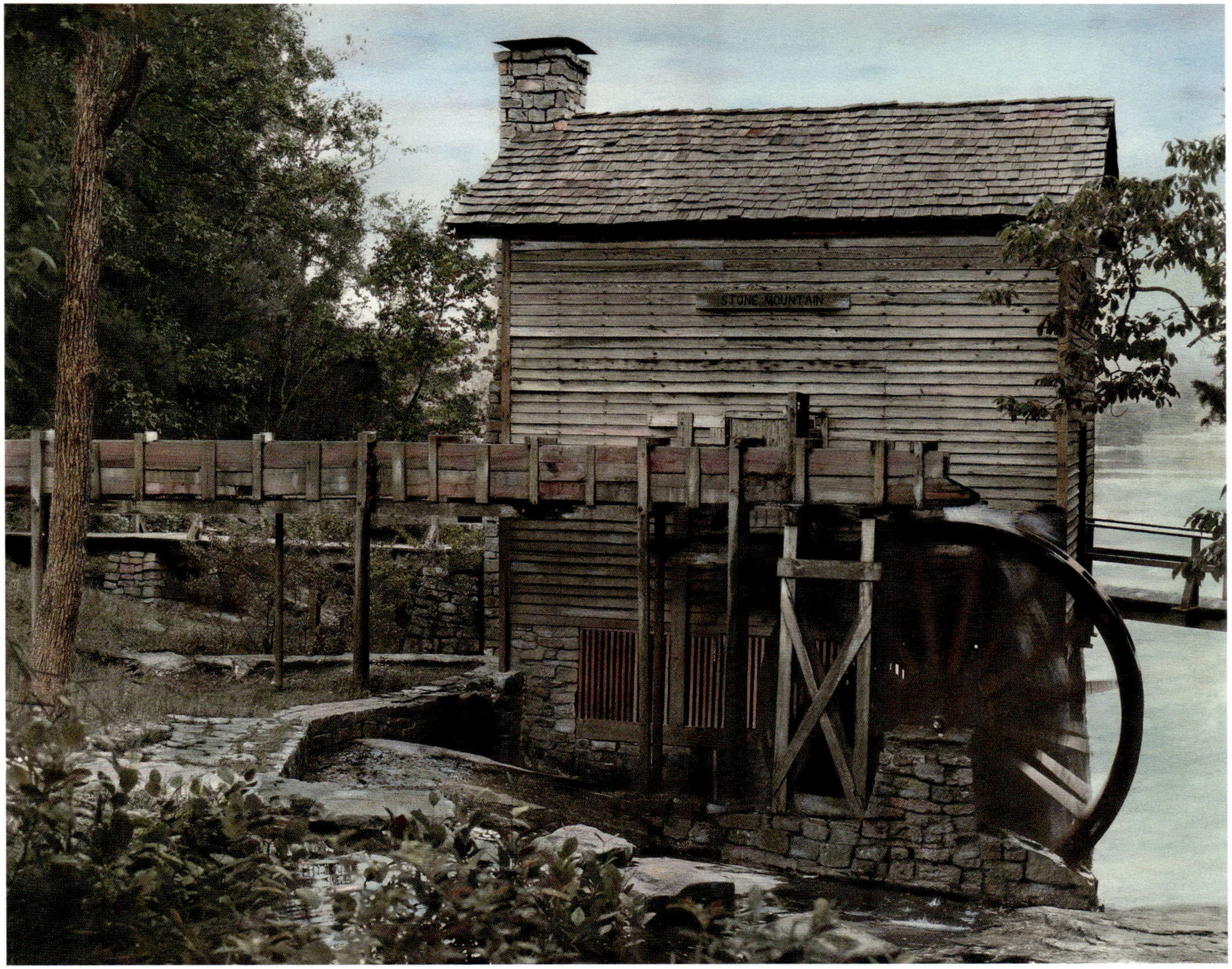

This photograph of the old watermill at Stone Mountain is a good example of my early hand-tinted art. The waterwheel was turning when I took the original black-and-white photo, so movement is apparent in the image.

While I could have replaced watermill images from my early decades with current photographs, I chose not to do so. My vintage expressions better preserve the visual and historical qualities of these ancient watermills, at least from my perspective.

My grandfather, whose stories helped spark my deep and abiding interest in watermills, was born in 1880. In the twenty-first century, we cannot fully grasp much of what he told me about watermills along Ohatchee Creek, but through these images we can at least in some ways better appreciate this fascinating aspect of our Southern heritage.

Historic Watermills of the South

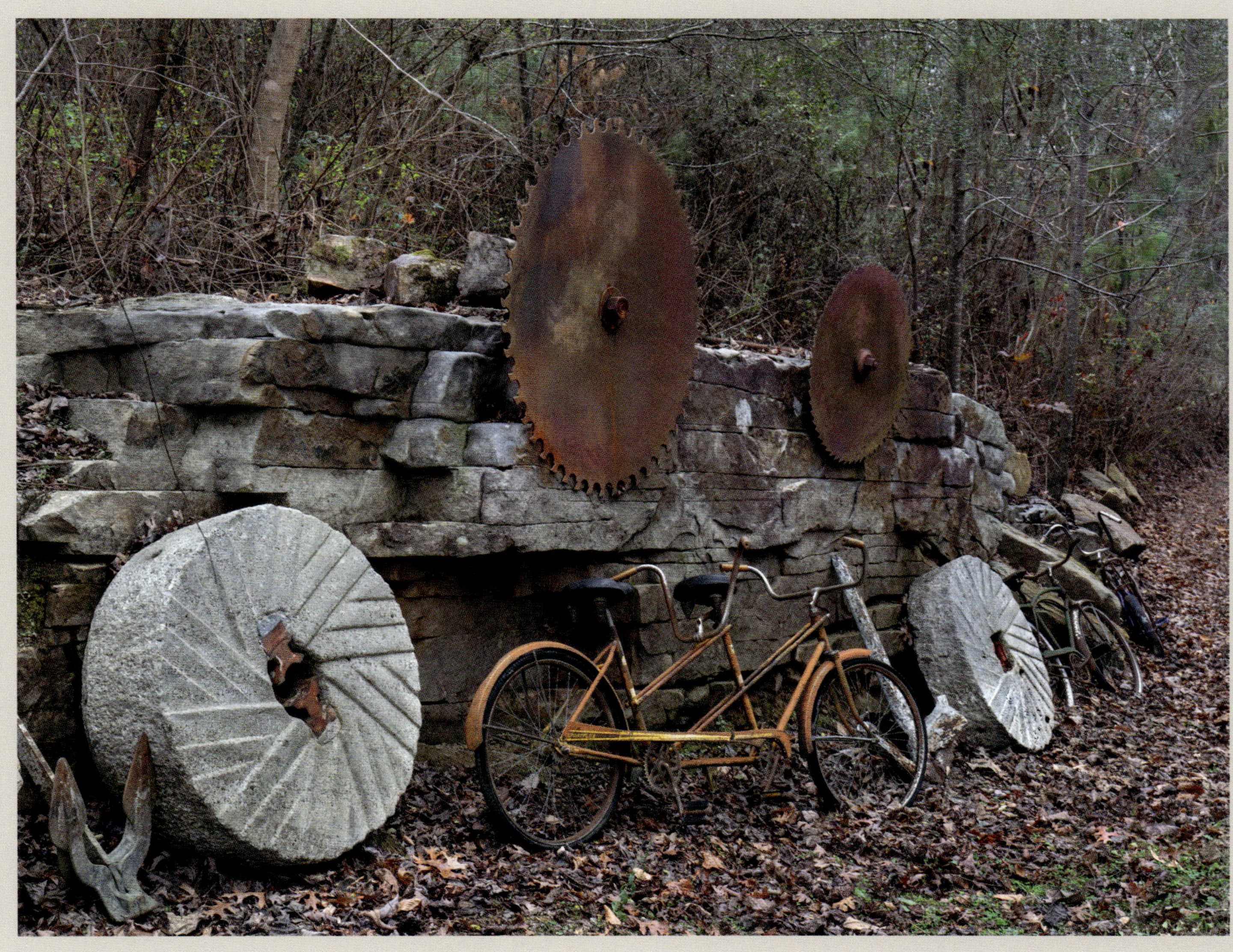

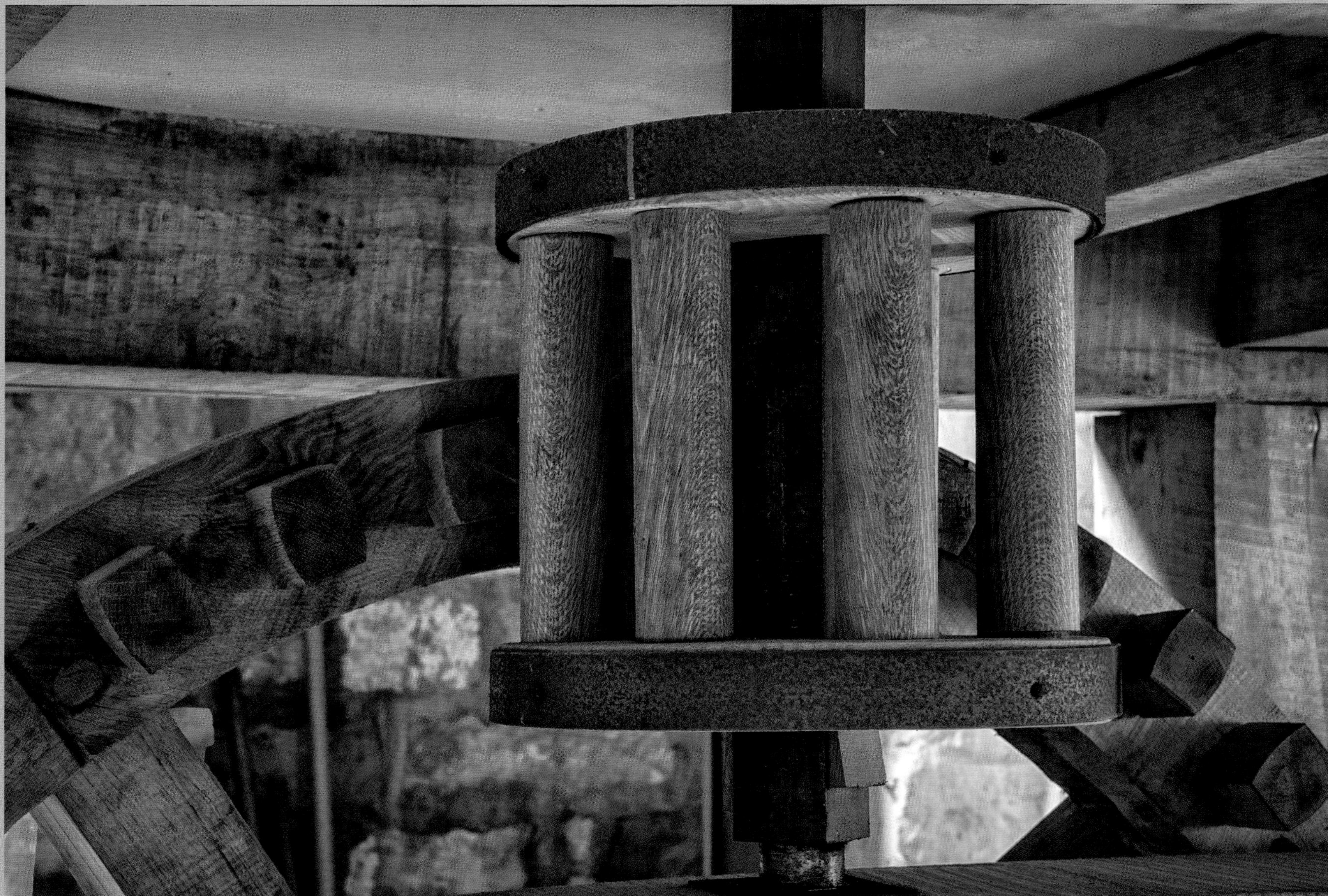

Introduction *The Story of the Watermill in the South*

AS THE YEARS PASS in the twenty-first century, increasingly few people can still recall actual experiences with authentic, working Southern watermills. Other individuals have more recently visited watermills at historic parks and tourist attractions where demonstrations and tours are conducted. Many more of today's Southerners have seen watermills only in photos, paintings, or movies, if at all. Few are aware of the significance of these venerable structures and their importance to life across the South from colonial times until they were displaced by steam engines and electricity around the turn of the twentieth century.

What is a watermill, and how does it work? Simply put, a classic watermill harnesses the power of water as it flows in a stream and falls from a higher level to a lower one. The linear motion and vertical weight of the water under gravitational pull are transferred into rotational energy by a waterwheel for operation of machinery in a mill. A waterwheel may weigh several thousand pounds; once rotating, it can produce an astonishing measure of mechanical energy for driving machinery.

Historically, watermills churned on almost every suitable stream in the South to grind grain for food and to saw timber for building materials. Watermills also powered textile machinery, drove ironworks, pumped water, crushed ore, processed paper, and pressed fruits and vegetables. In all, as many as forty applications for watermills have been documented, with working examples of many of these mill types in the South.

Wherever watermills were constructed, the technology was revolutionary. These were the world's first genuine machines. Before the watermill, essentially all work was done manually or with the help of animals. Without mills, grain was laboriously ground into flour using the prehistoric methods of pounding and rolling the grain with rocks or hard wood. Timber was cut with primitive saws and hewn into lumber using simple tools. For those charged with performing these exhausting tasks, the watermill provided a much better quality of life.

Although watermills may at first look somewhat simple and primitive from today's perspective, for folks living in earlier times, they provided an almost quantum leap forward in the way everyday tasks were performed. Today, many people consider watermills primarily for their aesthetics, with little to no understanding or awareness of their historical significance.

Conceptually, watermills were also the first notable example of applying one function—the flow of water—to perform an unrelated task, like grinding grain. Clearly, the social and economic implications brought on by the watermill rival the impact of many of the most important inventions throughout world history.

With that said, the watermill was not a new invention when it arrived in the South. The first watermills can be traced to the Middle East in about the second or third century BCE, with likely parallel development in China and India. First the Greeks and then the Romans refined the machines and processes and spread the technology across Europe.

The first known watermills in the southern portion of North America were built in the early seventeenth century by English settlers along the Atlantic coast, in what would become Virginia. These were tidal mills. Unlike the more familiar concept of a watermill built along a flowing stream, a tidal mill is a unique type of mill that operates on the difference between high and low tide to power a waterwheel; water at high tide is trapped in a millpond, and as the tide turns, the water is released to flow over and rotate a waterwheel. Water in an elevated "high tide" pond might fall or drop several feet over a waterwheel and into a "low tide" water elevation beyond the mill before flowing out to sea. Most of the earliest Southern tidal mills were sawmills rather than mills for grinding grains.

In the succeeding decades, and continuing into the early years of the eighteenth century, settlers began to clear farmland in inland areas, away from the tidewater regions near the coast, and constructed watermills along local streams and creeks to grind grains. Where waterfalls existed, mills could be sited to use the falling water as a source of energy. Steep, rushing streams—especially streams with rapids—and even gushing mountain springs could also rotate waterwheels. Elsewhere, dams and impoundments became necessary to elevate the water level. Water channeled from a millpond behind a dam could produce the power needed to rotate a waterwheel that was in turn connected to machinery for grinding grain, cutting lumber, or performing other tasks.

In addition to the mills in Virginia, which date to the Jamestown era, watermills began operating in Maryland around 1634. The Swedish also built watermills and windmills in the Delmarva area (comprising most of Delaware and the Eastern Shore of Maryland and Virginia) in the 1640s. The Spanish constructed mills further south during this same period, first in Florida and then along the Gulf Coast, though less is known about these mills.

As settlers moved further west into the interior of North America, they found abundant flowing streams for siting watermills in the rugged foothills of the Appalachians and along the geological fall line that extends as far south and west as what is today the Alabama/Mississippi border. By the early 1700s, colonists were opening up new frontiers throughout western Virginia and the Carolinas and into Georgia.

Almost all frontier watermills were gristmills, often built to grind corn or maize, a practice that European settlers adopted from Indigenous Americans living nearby. Sawmills soon followed to provide lumber for building.

Why were gristmills so important? For the early settlers, bread was the staple of life and a foundation of their diet. Hunting and fishing might not always provide daily meat for the table, but bread could be counted on for each meal. Large quantities of grain were required for every family and community.

Most people today envision a "classic" watermill with a large vertical rotating wheel and buckets or paddles spinning in splashing water. In fact, as seventeenth- and eighteenth-century pioneers moved inland from the coastal areas, they constructed their watermills along the frontier in a much more rudimentary fashion—with small horizontal waterwheels mounted near ground level. The arrangement might look more like a wagon wheel fitted with blades laying on its side. The blades were spun by a rushing stream of water, and the wheel was connected directly to a simple vertical shaft and crude rotating stones. The materials available for building mills and the skill of the builders varied by location but were universally basic. Most of the mills from this era were of common log construction and used stones from nearby streams for grinding.

Some folks referred to these early mills as "tub mills" because the horizontal wheel was often partially enclosed in a wooden tub or barrel-like vessel for better water control. Similar mills had been used much earlier in Europe, the Middle East, and Asia to grind grain for a single family or an extended family group. Those mills were sometimes known as "Norse mills."

As larger mills came to be needed, the expertise of a millwright was sought for initial design and construction, and a dedicated miller was necessary for mill operation. Construction was still primarily of locally available materials, with wooden siding and framing, stone and masonry foundations, and seasoned hardwood for mechanical and stressed components. Millstones were brought in from specialized quarries where stonecutters shaped matched pairs of stones. The finest stones were shipped in from Europe, especially France, with some coming from Germany.

A pair of millstones for grinding corn or wheat consisted of a stationary "base" stone that did not move and a rotating upper millstone, known as a "runner" stone. The stones were round in shape, averaged three to six feet in diameter, and weighed up to several thousand pounds. Millwrights set and aligned the stones upon initial installation, and millers subsequently maintained and sharpened the stones as necessary. Great care was taken in their alignment: the millstone spacing needed to be close enough that the stones could shear a kernel of corn or a grain of wheat without touching each other, but stone-to-stone contact could damage the stones, break mill linkage, or even spark a fire. Larger mills often operated with two or more sets of stones—one set for large corn kernels and a second set for smaller grains, such as wheat.

Such massive stones demanded powerful waterwheels for operation. The small horizontal waterwheels of the pioneer tub mills were replaced by the more familiar large vertical waterwheels. Larger waterwheels also required more waterpower, dependable streams, and substantial millponds. A typical waterwheel might operate at about 10 to 12 revolutions per minute, which meant that gearing was necessary to achieve 100 to 120 revolutions per minute at the millstones. Large-diameter wheels turned more slowly, completing only a few revolutions per minute.

Early colonial mills had often depended on a design in which a vertical waterwheel simply paddled in a rapidly moving stream. In the first half of the eighteenth century, many Southern mills evolved toward designs in which water flowed over the top of a vertical wheel or hit the wheel partway up its height, releasing a tremendous amount of mechanical energy; such arrangements were known as "overshot" and "breastshot" wheels, respectively. Waterwheels could be tall and thin or low and wide, or almost any dimensions in between. The terrain, volume of water available, and milling requirements dictated the configuration of each waterwheel and the actual watermill structure. Each site was unique.

As mentioned already, sites for large mills almost invariably required the construction of a dam to store water in an elevated millpond, from which energy could be released to turn the waterwheel. A trough, pipe, or channel, often referred to as a "flume," "penstock," or "mill race," directed the water from the millpond to the waterwheel. This infrastructure required constant attention and ongoing maintenance.

A sawmill might operate on the same stream and even the same millpond as a gristmill. For years, sawmills used straight blades that moved up and down on a crank. The blade motion was like that made by a two-man saw. An ingenious moving frame inched the timber forward and into the blade as it was cut. Circular saw blades came later, as more substantial and durable iron and steel materials became available. These circular blades, which might be half an inch or more thick, significantly accelerated cutting speed relative to that of a straight blade and greatly increased lumber production.

Sawmills could cause considerable environmental harm, as logs were floated in streams and wood waste was released into the water downstream of the mill and dam. Some streams became clogged with debris.

By the time of the American Revolution, certain aspects of the gristmilling process were beginning to be mechanized, but other steps still involved challenging manual labor, such as moving sacks and barrels of grain and bulky equipment. In 1790, Oliver Evans, an inventor and engineer, applied for a patent that described what essentially became a fully automated milling system. His system incorporated water-powered pulleys, belts, and lift technologies to make milling less physically demanding and more manageable for the operators. Evans's system or something similar was installed in many of the South's larger gristmills.

An automatic mill required a highly skilled millwright for its design and construction. When a millwright arrived at the planned site for an automatic mill or even a large but simpler mill, families from across the area often worked together to help erect the mill and then bring in a miller and his family to operate it. Every rural area or community wanted a mill and a competent and experienced miller, and the competition for millers could be intense. Various incentives were offered to attract millers. Over the years, a miller might come to be respected in their community, or they could become known as a cheat and a scoundrel.

Fascinating stories about crooked and deceitful millers were common across the South. The typical Southern miller held a monopoly over a farming area, and because customers were often uneducated and trusting, the temptation to steal was too much for some dishonest millers. A bad miller might substitute old meal, gypsum, or chalk for a farmer's fresh meal, or they might adjust the scales to their advantage or use their thumb as they weighed out their toll. Even more common was for a miller to simply take more than the agreed-on percentage for grinding, a practice recounted in Chaucer's *Canterbury Tales*.

Despite the occasional unscrupulous operator, watermills were destined to function as community gathering places and as a key to pioneering social structure in the rural South. A mill might typically serve a radius of five to ten miles or more by the mid-1800s, if the condition of roads and trails could support wagon travel. During peak harvest seasons, farmers would line up with their wagons, carts, and sacks for a turn at the millstones.

While the farmers waited, sometimes accompanied by other family members, they would camp, trade, visit, and catch up on news. They participated in sharpshooter competitions, baking contests, and other enjoyable events to pass the time. These gatherings and experiences enriched Southern culture. Over time, a small hamlet and then a town might spring up around a mill. A blacksmith shop and a small general store would open nearby, followed by a church, a school, and even a post office.

During the Revolutionary War, watermills not only provided flour and meal for the Continental army but also produced black powder for muskets and cannons. Initially, black powder had been ground using a simple mortar and pestle technique. However, producing the quantities of powder needed to conduct a war required watermills with large grinding stones. The components—soft charcoal, saltpeter (potassium nitrate), and sulfur—could be pulverized to fine powders and blended using waterpower. Alcohol was often added to dampen the powder and reduce the risk of a spark and explosion in the mill.

Although most black powder was produced in Pennsylvania and Massachusetts during the Revolutionary War, significant quantities were delivered from Maryland and Delaware. The black powder mills in Delaware would soon become the largest in North America. (Powder mills operating in Augusta, Georgia, during the American Civil War were said to have been the second largest in the world at that time.)

In the decades prior to the Revolutionary War, ironworks were established in Maryland and Virginia. These ironworks came to rely on a watermill with a robust waterwheel to drive large bellows that forced "blasts" of air into the furnace—thus the term "blast furnace." The oxygen in the blasts of air raised the furnace temperature to produce a higher-quality iron product.

The blast furnace was not a new watermill technology; it had been developed in Europe and China several centuries earlier. Iron ore furnaces came to the South from New England after reserves of iron ore were discovered in the wetlands and swamps around the Chesapeake Bay. Known as "bog iron," this low-grade ore had been soaked up from streams and stormwater runoff and had concentrated in sediments over many years.

These blast furnaces produced crude iron ingots (pig iron) for shipment to Great Britain. The British Iron Act of 1750 restricted the production of refined and finished iron products in the colonies so as to preserve this lucrative consumer market for furnaces in the mother country. By 1760, almost two hundred tons of iron were being shipped annually from the colonies to England and Scotland. Some of the most productive Southern ironworks were the Fredericksville and Marlboro Furnaces in Virginia and the Keep Tryst Furnace in West Virginia. As war approached in the early 1770s, iron cannons were being forged for the Continental army.

Further inland, and further from British control, early iron deposits were discovered in East Tennessee. However, the South's largest production of iron ore emerged in the central Alabama region during the 1830s. By the time of the Civil War, the water-powered furnaces at Russellville, Tannehill, Oxmoor Valley, and Chalybeate Mountain were providing iron for armories across the South. By the late 1800s, the area had come to be known as the Birmingham District, and it was the third-largest producer of pig iron in the world.

Another important application for the Southern watermill was the cotton gin, along with related textile milling machines to spin and weave fabrics. Eli Whitney invented a cotton gin in 1793 that could separate the sticky seeds from the cotton fibers via a rotating mechanism, eliminating a task that was grueling work when done by hand. Initially, cotton gins were small hand-operated devices; these were followed later by larger models powered by horses. Within a few decades, the full potential of Whitney's invention was realized, and large water-powered gins were built across the South.

Over the years, as more settlers arrived and others pushed along wagon trails and further into the Southern frontier, still larger and more efficient watermills were built. According to the U.S. census of 1840, 10,612 gristmills and 6,568 sawmills were operating in the South, along with scores of ironworks and cotton gins. Many additional watermills, especially small farm mills and tub mills, were likely still in operation during this era.

The construction and operation of so many watermills required a large labor force. Before the Civil War, much of this work, especially the most difficult and physically demanding tasks, was done by enslaved laborers. Enslaved people also operated some of the watermills, especially mills that were owned by the wealthy. After the war, skilled Black millers worked alongside White millers into the twentieth century.

During the era of the early to mid-1800s, wealthy Southern plantation owners introduced a new dynamic to milling—the plantation mill. As plantations grew in economic, political, and social power and influence, the plantation owners constructed mills on their land and required everyone associated with the plantations to use the mills and pay a toll. In addition, some larger plantations also took on a commercial dimension, providing fine, high-quality flour to paying customers while producing a coarse, lower-quality flour to be consumed by enslaved people, poor White laborers, sharecroppers, and most everyone else on the plantation.

In the years leading up to the American Civil War, Southern plantations, together with the water-powered textile mills in New England, were producing two-thirds of the world's cotton supply. As the Civil War broke out, these watermills were ginning, spinning, and weaving incredible quantities of textile fabrics, but with a huge human toll in the miserable cotton fields of the South and in the brutal textile mills across New England.

Although some popular accounts of watermills from the mid-1800s limit themselves primarily to discussion of antebellum plantation mills, many Southern watermills did not have anything to do with the plantation system or even with the cotton and tobacco economies. Nonplantation watermills existed to support other aspects of the Southern economy, including industry and manufacturing, as well as grinding grain for the populations living and working on small farms and in the towns and cities across the region.

An important difference between Southern and Northern mills in the mid-1800s was their respective markets. As industrial centers, Southern mills were almost invariably community based. Northern mills were larger industrial production operations. With the onset of the Civil War, the watermill-based industry of the North was in many respects a major advantage for the Union side. Small Southern mills in 1860 were limited in their capability to contribute to the war effort.

As the war progressed, mills across the South were burned or destroyed as part of the Northern strategy to deprive the Southern population of its food supply and limited industrial base. In the closing years of the war, Southern iron furnaces and other industrial watermills were widely demolished. Some were even set afire weeks after General Grant and General Lee had signed the terms of surrender at Appomattox Court House, Virginia.

By the end of the war, much of the South, including many of the region's watermills, was devastated. During the difficult Reconstruction era of the late 1860s and 1870s, watermills were rebuilt, and many were larger and more productive than their predecessors. In fact, some mills that were not damaged or destroyed during the war were also replaced with new and more efficient designs. The postbellum era in the South was a time of rebuilding and adapting to changing technologies, cultures, and lifestyles, which contributed to the desire and need for larger and more productive mills and other industries.

All across the South, the vertical waterwheel remained the basis for regional postbellum industries, and the overshot wheel design was almost universal across the region. Some mills even had multiple waterwheels, with a few featuring as many as four wheels churning the water.

Throughout this period of Southern history, watermills were also important for crushing, pressing, and rolling operations. Water-powered hammers could pound ore, squeeze fruits and vegetables, and roll iron and other materials into sheets, wire, and rods. There also were mills for producing plaster, tobacco snuff, paper, flax, cider, mustard, bone meal, axes, barrel staves, linseed oil, and granulated salt. Waterwheels equipped with the appropriate buckets, cables, and linkage could remove water from underground mines and other subterranean operations, deliver water to elevated water tanks, and meet the water needs for agricultural and industrial applications.

As transportation improved, larger and more centrally located mills thrived. In some locations, the railroads facilitated transportation to and from the mills, but other mills were bypassed by the rail lines and typically did not fare as well. Although the South remained largely agricultural until after World War II, mining, lumbering, and iron and steel mills expanded, and a few industrial centers emerged. In the late nineteenth century, textile mills began their relocation from New England to the Southern Piedmont region.

In the decades following the Civil War, the water turbine was introduced across the South as an upgrade and a technological advancement relative to the classic waterwheel. Water turbines provided more power than waterwheels while requiring less maintenance. Turbines were fully enclosed and operated out of sight underneath the mill, and they were oriented horizontally, much like the arrangement in a modern hydroelectric facility.

The familiar old waterwheels were removed. Existing mills were repowered with turbines, and new mills were designed and constructed around this favored technology. For this reason, some mills that survive to the present day seem to be missing their waterwheels, but many of them never had a waterwheel to begin with. At other mills, wooden waterwheels were updated with metal wheels, which were now generally available and required less maintenance. For a time, some mills even featured waterwheels *and* turbines.

So how did the rotational power of a waterwheel or turbine reach the mill machinery inside? At a traditional and simple country gristmill or sawmill, linkage to a pair of millstones or a saw blade might be as straightforward as a driveshaft extending from the waterwheel to a few gears or belts and pulleys.

For larger, automated, and more commercial-scale mills, the linkage reflected state-of-the-art feats of engineering. The mill machinery could unload grain, deliver it to cleaners, regulate the rate of grinding, sift and separate the milled grain into several products (such as grits, cornmeal, fine white flour, middlings, and bran), and package the final products into barrels or bags. A textile mill could card, spin, and weave cloth at speeds unrivaled by early manual processes.

To perform all these tasks with a single waterwheel or turbine required an inspired and complex series of wooden or metal cogs, gear shafts, belts, and linkages, as well as regulated orientation and speed of rotation and other parameters for operation of the entire facility. Rotating power from a waterwheel could literally be directed to machinery on multiple floors throughout a mill. Today, a visit to an operating or restored mill can become a mesmerizing adventure into the brilliance and ingenuity of this early industrial technology!

Despite this water-powered evolution in technology, by the early twentieth century traditional watermills were disappearing from use all across the South. Alternative energy sources such as steam, fossil fuels, and electricity proved to be much simpler to generate or harness, more predictable, and less costly.

Continuing improvements in transportation, especially those brought on by steam locomotives and railroads, also facilitated larger and more centrally located milling operations.

In the mid-to-late 1800s, as many as 100,000 watermills turned across the United States, with thousands more operating in Canada. Over the first few decades of the twentieth century, alternative sources of energy in larger mills and factories replaced most of these watermills. The transition was fairly rapid.

In most locations, traditional watermills could not compete with alternative energy sources. The classic country mill endured mostly in rural settings where dedicated millers and loyal customers were determined to carry on the old ways. In the South, these mills remained in operation longer than in some other regions of the country because many Southerners, both Black and White, continued to stay on farms well into the twentieth century. Some current research points to the rural watermill as contributing to the survival of poor subsistence communities in certain parts of the South.

Many now-familiar changes in culture and lifestyle did not come to the South until after the Great Depression and World War II. The New Deal programs were in part designed to bring the South into the national culture and economy, where the region had lagged since the antebellum and Reconstruction eras. For example, the Tennessee Valley Authority provided low-cost electricity and created jobs in seven Southern states, while incidentally displacing, flooding, and destroying many watermills in valleys and along creeks. Military bases set up to support the World War II effort also brought better-paying work and more industry to the South.

Through these transitions, the watermills continued to lose importance. Farming to grow corn, raise chickens and pigs, and barter products transitioned to town and factory work, with regular paychecks followed by purchases at general merchandise stores. The mule and wagon duo were replaced by a truck or car, and the trip to the watermill with a cotton sack of shelled corn was supplanted by a stop at the market or grocery store.

As watermills went out of business, they were closed up and basically abandoned year by year through much of the twentieth century. The structures, machinery, and sites were most often just left to deteriorate and crumble. The Southern practice was not to demolish old structures but just to leave them and walk away, with rarely fulfilled intentions of getting back to them later. And so, forsaken to the elements, rot, insects, vandals, and natural events such as floods and fires, these abandoned mills surrendered themselves to their surroundings.

The mill owners would sometimes scrap the metal equipment, especially during the world wars. Occasionally, machinery was sold to markets in the developing world. As towns grew up, mills that had been sited at choice geographic locations fell victim to commercial site development; ironically, the structures were removed to make way for new developments in communities that had built up around the watermills.

Hydrologically, creeks and rivers often changed course over time. An energetic stream could gradually or even suddenly erode a creek bank and claim the land where the mill structure stood. At some old mill locations, the mill foundation can be seen in the water at the bottom of the creek or in the mud along the bank.

Today it is estimated that fewer than four thousand watermills remain in the United States, and most are in peril or in ruins. Perhaps one-fourth or more of these mills are in the South, with many located in the mid-Atlantic Southern states. Although preservation and restoration work is ongoing at some sites, such efforts are costly and labor intensive.

For much of the twentieth century, Americans were understandably preoccupied by the two world wars, the Great Depression, and the Cold War, but during the last quarter of the century, there was a general awakening in the South to the preservation of old buildings, including structures such as watermills. More recently, a number of watermills have been resurrected and preserved. Some have been restored to operation, others have been cosmetically refurbished, and still others have been repurposed as stores, antique shops, restaurants, homes, and even bed-and-breakfast destinations. These mills receive varying levels of care, maintenance, and authentic preservation.

A few select and "fortunate" mills have been adopted by federal, state, or local park services, historical groups, or individuals with a calling to preserve a site. These mills are generally located in areas in which awareness and appreciation of their value exists. Area affluence and access to financial resources are fundamental to site restoration and maintenance. Thanks to the ongoing care of preservationists, these mills should be with us for years to come.

As it happens, the recent popular interest in natural and organic foods has given new life to a handful of old watermills! Those who run these mills have targeted and fostered markets of restaurants, retailers, and individuals who are willing to pay a premium for stone-ground grains from water-powered mills. In the twenty-first century, this market seems to be growing.

Most of the preserved watermills in the South and across the continent are gristmills. Sawmills were often visually less attractive and were sometimes little more than sheds. Although there are a few exceptions, textile mills and other larger industrial mills are generally beyond the scope of almost any budget to restore and preserve, unless they can be totally repurposed as commercial or residential properties.

Understandably, watermills remain fascinating to many people across the region, even to those with little understanding of their history, technical design, and operation. No two watermills are alike, as terrain, available construction materials, and the skills and innovative genius of the builders determined the layout of each mill. This wide variation in appearance and design adds to their historical and sometimes mystical appeal.

In the twenty-first century, dilapidated old Southern mills and abandoned mill sites offer an enduring legacy and timeless charm to advocates, historians, archaeologists, and artists. Watermills in almost any condition can be inherently beautiful and picturesque. Old structures in ruin are sometimes said to "have character." Artists, photographers, painters, and writers may feel a special call to depict and share the plight of mills and to capture their essence and patina before they are all gone. As images and descriptions of old mills circulate, awareness of this important dimension of Southern history has been further realized and highlighted.

As another matter of concern, the survival and restoration of watermills in very rural, isolated, and economically depressed areas faces additional challenges. Some of the South's most historic, charming, and magnificent watermills sit neglected along almost-forgotten stretches of the region's more difficult and less traveled routes. The prospects for attracting interest to these mills in the face of so many competing and more visible historic projects are discouraging and disheartening.

What more can and should be done? Today's greater appreciation for the preservation of old and historic buildings has certainly helped. While some watermills are beyond hope of salvation, others can still be reclaimed. An assessment and evaluation, followed by stabilization and historic restoration where possible, can preserve a substantial number of remaining mills.

The goal should be to save as many of these structures as possible. Where they cannot be saved, communities, municipalities, and other groups should make the effort to document, photograph, and describe whatever remains. This work is vital to preserving the history of the South.

Thus, in many ways the story of the watermill in the South parallels the region's colorful history. For some four hundred years, watermills were vital to survival and to the region's way of life. These old mills tell us much about the resourcefulness, determination, and character of a people who arrived as colonists, settlers, and pioneers and developed a great region.

Not to be forgotten or overlooked, the invaluable work of thousands of enslaved people was fundamental to the construction, development, and operation of Southern watermills. Finally, no history of the South is complete without acknowledging the central role of Indigenous Southerners who introduced corn and cotton and brought other important cultural dimensions to the region.

Now, through all this history and weathered wood, cracked masonry, and flaking iron, please join me in a visual celebration of the historic watermills of the South.

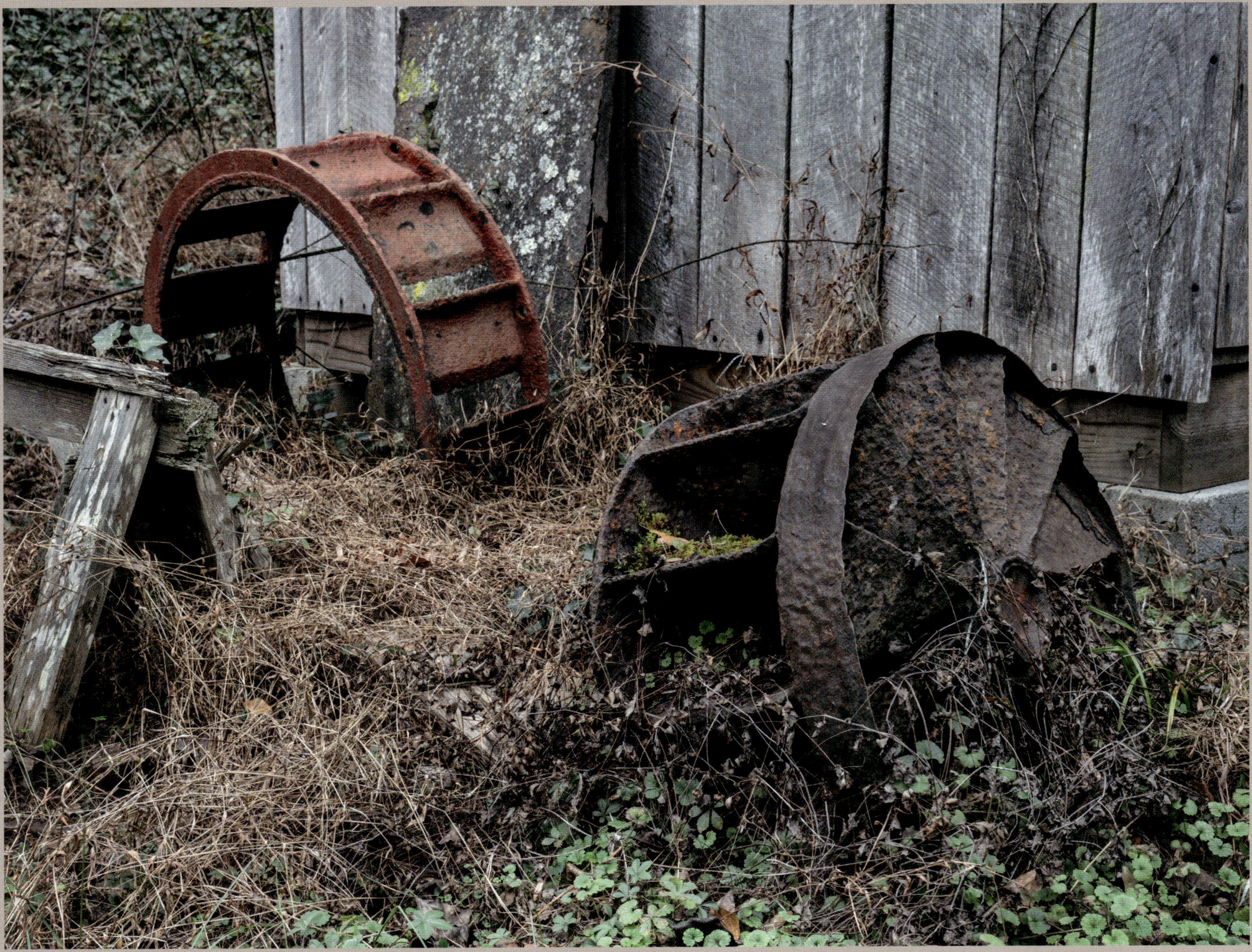

Lost, Unknown, and Imperiled Watermills of the South

DURING THE TWENTIETH CENTURY, much of our rural Southern history and heritage was lost to time. To some extent this is understandable, as generations were consumed with World War I, the Great Depression, World War II, the Cold War, and the threat of nuclear destruction. The century was also a time of incredible technological change, and of mass migration from an agrarian lifestyle to urban living all across the South.

Watermills are a classic example of something that had been a vital Southern institution at the beginning of the century but was antiquated, outmoded, and almost forgotten by century's end. Watermills were simply disregarded by a society preoccupied with war, economic upheaval, and constant adaptation to change. After all, the power that had come from a familiar, nearby rotating waterwheel was now supplied by a few massive, centralized, and visually remote industrial sources, and energy, food, and other essentials were available at retail outlets or could be delivered directly to any home, business, or other destination.

Depending on their location, watermills throughout the South were locked up, abandoned, and left to rot or later pillaged for building materials. Some were eventually repurposed for a different use, such as a hardware store, restaurant, or private residence. If their location was desirable for commercial development, mill sites were bulldozed and replaced with parking lots, roadways, high-rise buildings, or almost anything profitable.

As a result, many of the thousands of watermills that had been so important to generations of earlier Southerners are unfortunately now gone—lost and even unknown. It is impossible to know the overall number of watermills in the South at any point in time, but as noted earlier, more than 10,000 gristmills and in excess of 6,500 sawmills were operating in the South at the time of the 1840 census.

Information from the late 1800s about watermills, agriculture, and economies suggests that perhaps 70,000 to 100,000 watermills may have been in operation across the United States at the peak of their usage. Probably a third or a fourth of these mills were in the South. (Remember that vast regions of the country were too dry or flat or were otherwise ill-suited for siting watermills.)

Some states did maintain limited, partial records on the number of mills within their borders. For example, Georgia records indicate that about 1,200 watermills were operating in the state in the late 1800s. This estimate is based on tax records, business licenses, surveys, and other assorted but generally incomplete information.

A comprehensive historical listing or total of watermills in the South or the United States does not currently exist and would be difficult to put together. The lists that have been developed usually include not only water-powered mills but also mills powered by wind or animals, or even by humans. Further complicating any attempt to inventory watermills was the introduction of steam and internal combustion engines in the late 1800s and early 1900s. Significant numbers of mills were converted to, supplemented by, or built for operation with steam engines and later with internal combustion engines before the transition to electric motors began in the twentieth century. These non-water-powered mill operations are also often included in historical lists of mills.

The best source for data about watermills today is the Society for the Preservation of Old Mills (SPOOM). For decades, this organization has worked to develop state-by-state and county-by-county information on mills. As of 2024, SPOOM lists a total of about 1,400 known mills across fourteen Southern and border states. SPOOM's listings include gristmills, textile mills, windmills, sawmills, factories, and ruins, as well as gift shops, restaurants, and other repurposed mills.

The SPOOM data are excellent and provide an invaluable service. Not only are the society's state mill rosters the best available resource, but they also include the only information on some known mills that, practically speaking, no longer exist. A number of the entries on SPOOM's lists are for sites that are now little more than a partial foundation or the remains of a dam in a creek. Other mills are lost each year to flood, fire, wind, neglect, or demolition without those tragedies being reported to SPOOM or to anyone involved in tracking the remaining watermills. Without SPOOM, there likely would be no record for future generations that many of these mills ever existed.

Furthermore, thousands of Southern watermills operated for decades and then disappeared before the invention of photography and in areas with little written historical documentation. By contrast, artists in Europe produced paintings and drawings of early watermills in the era before photography. This type of artistic documentation was rare in rural and frontier America.

Tragically, even many watermills that survived into the twentieth century before vanishing were never seriously photographed. When those last few individuals who can recall these mills pass from this world, any visual remembrances will be gone forever. The images shown here capture just a few examples of literally thousands of similar sites that are now gone.

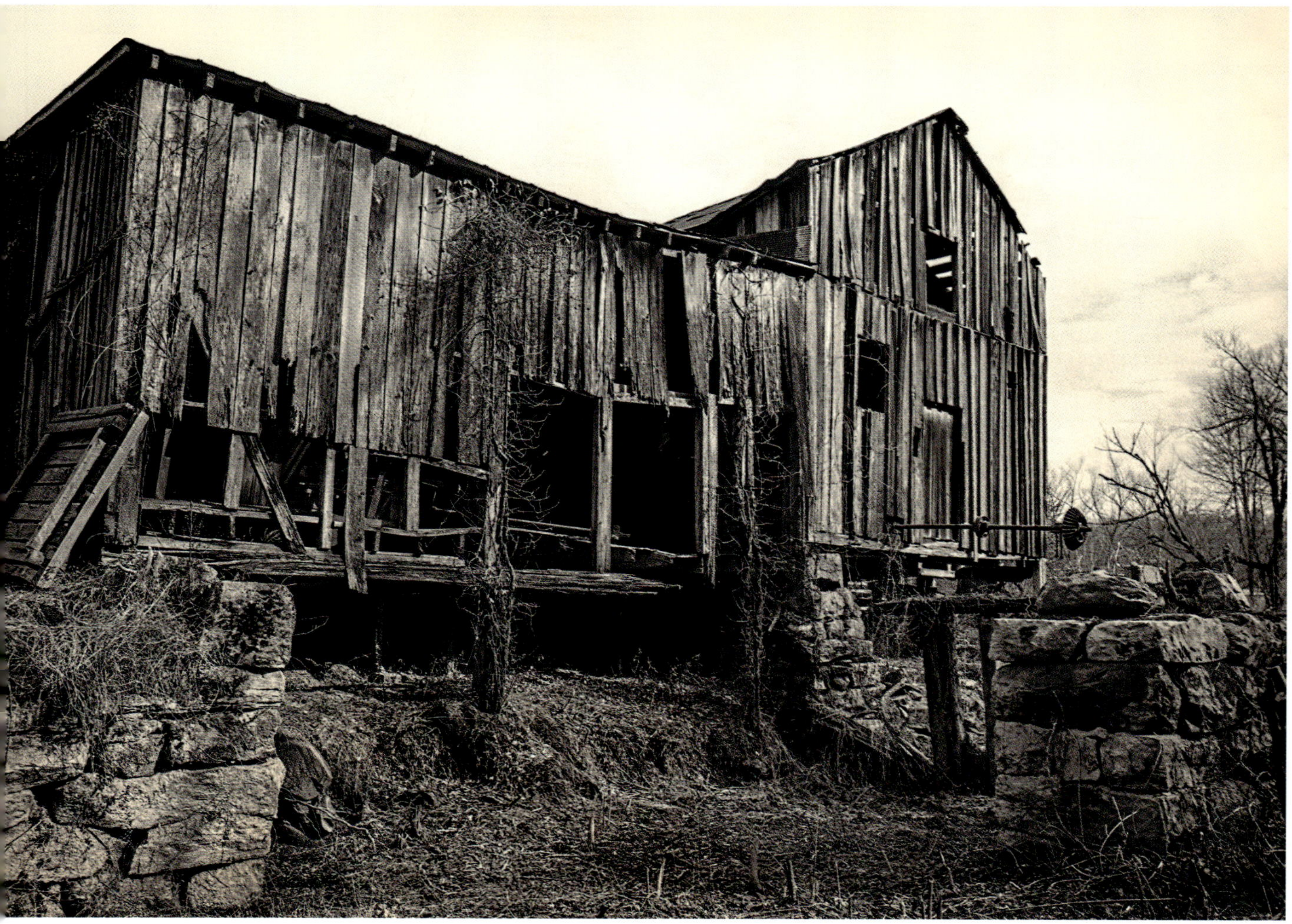

This image of an abandoned (and now vanished) mill is fifty years old. The water raceway in the lower foreground appears to have been stone-lined at some point in the past. A metal driveshaft with a beveled gear still extends from the side of the structure at right, but the waterwheel had been gone for years by the time the author took this photograph. This mill is thought to have been somewhere in southwest Virginia.

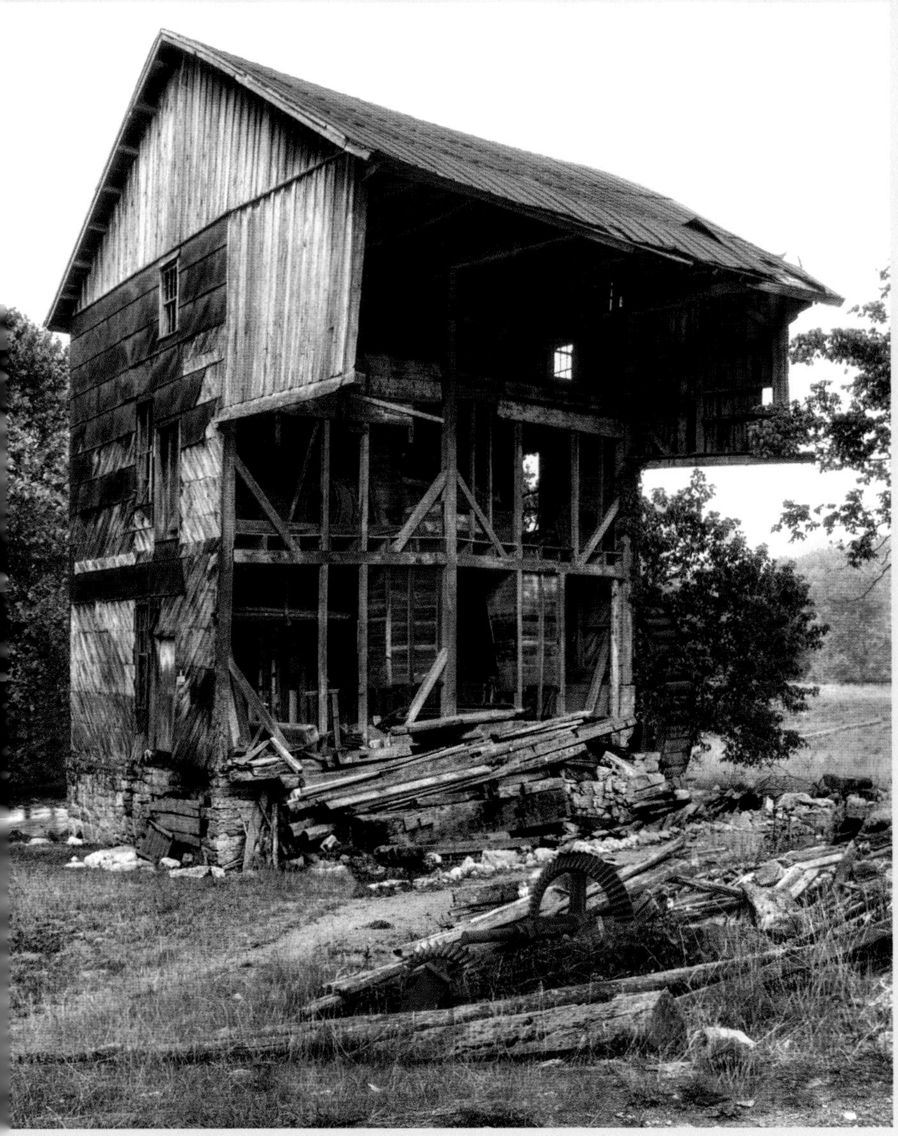

This is an example of an old watermill that has been scavenged for construction materials. The waterwheel is still attached at the right end of the structure, and assorted metal gears and linkages are scattered about the property.

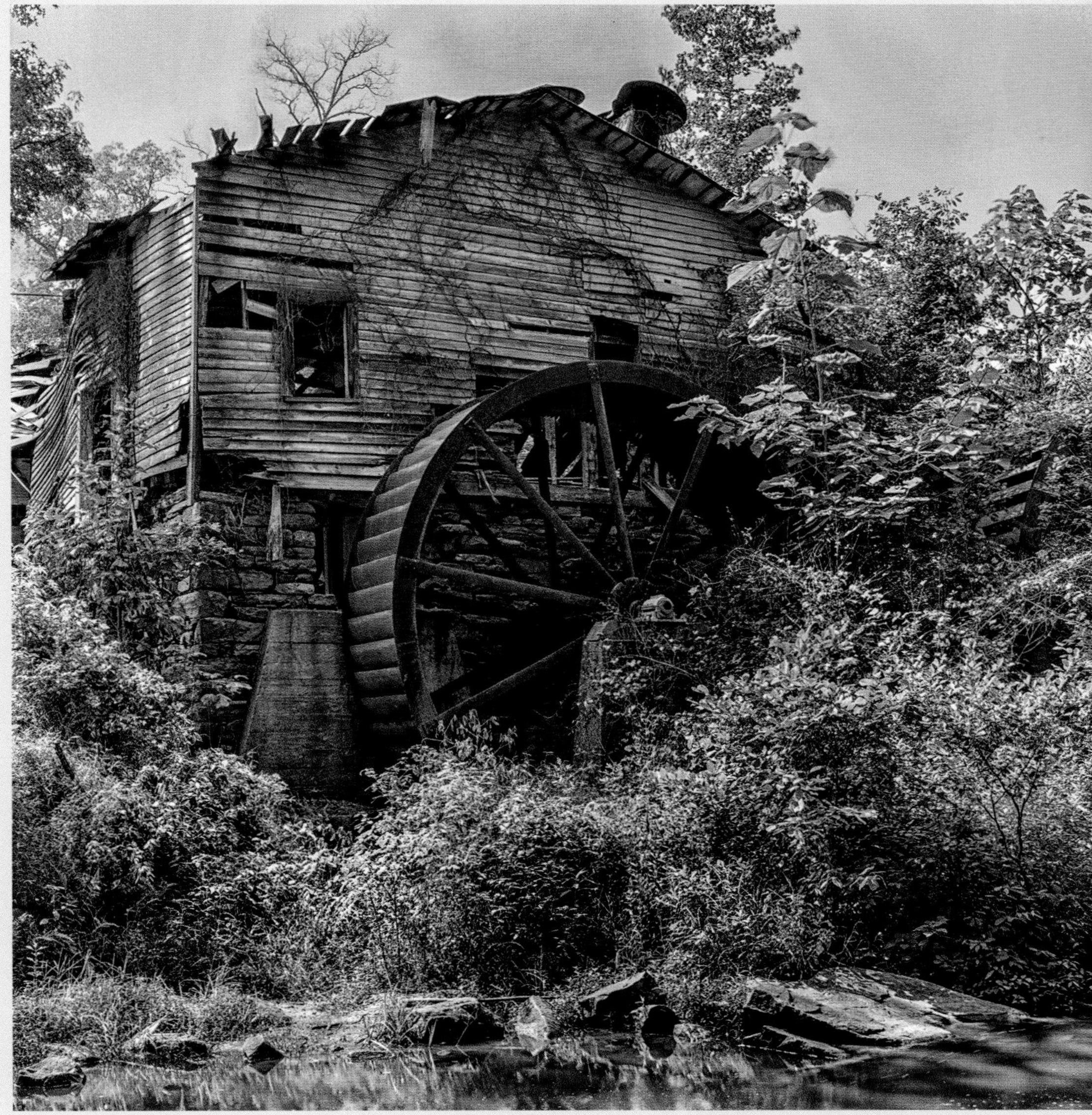

This old watermill relic appears to feature a second waterwheel, barely visible in the dense vegetation at far right. A stream can be seen in the foreground. This image was shared with the members of the Society for the Preservation of Old Mills for possible identification, but to no avail; thus it seems unlikely that anyone living today can name this old mill.

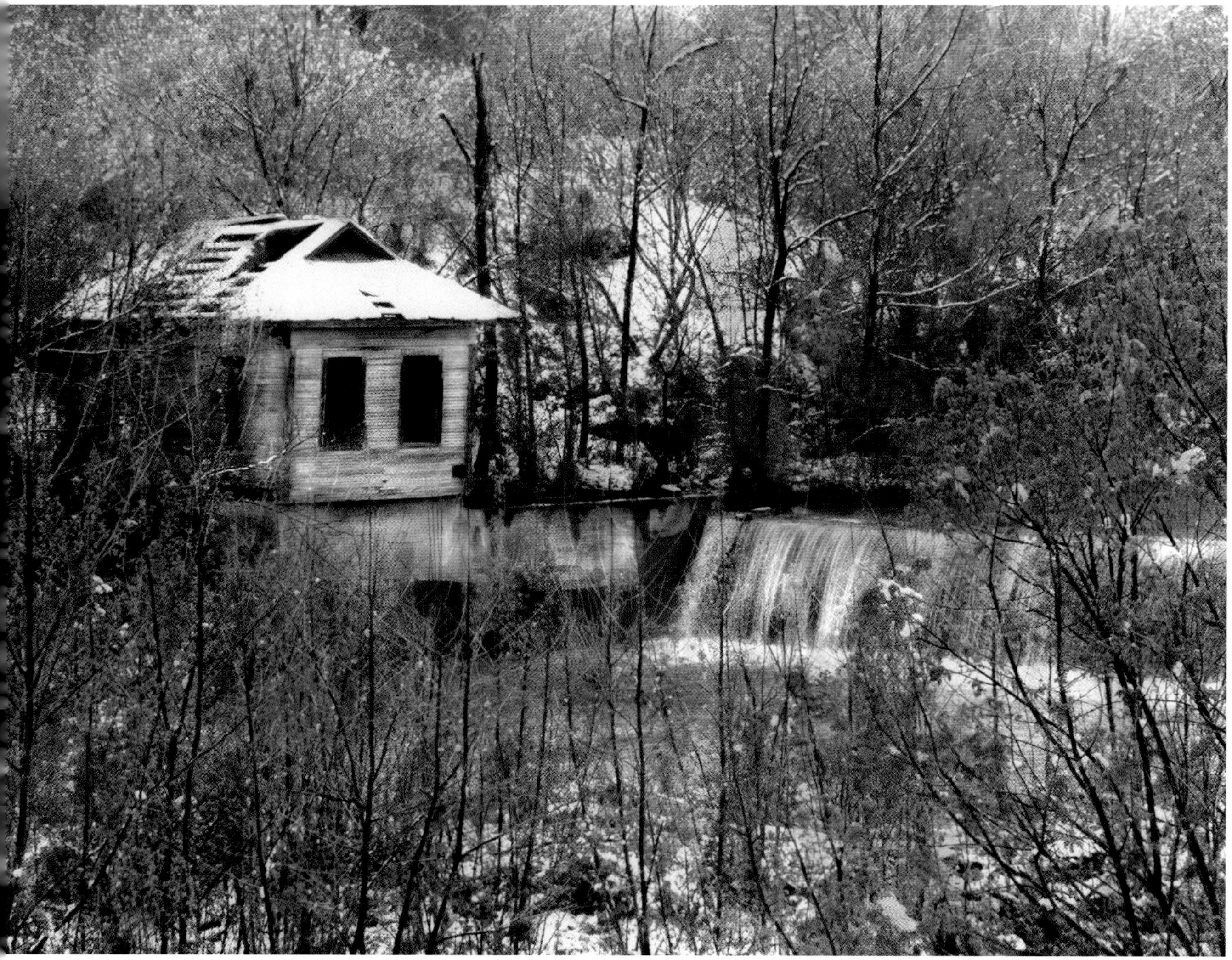

Although the dam is still in place at this site on Buck Creek in Alabama, the old Davidson Powerhouse is now gone. The area where the mill and dynamo once stood has been cleared and commercially developed with restaurants and shops. Some equipment that had been inside the structure can still be seen on the mill's concrete foundation among the modern developments.

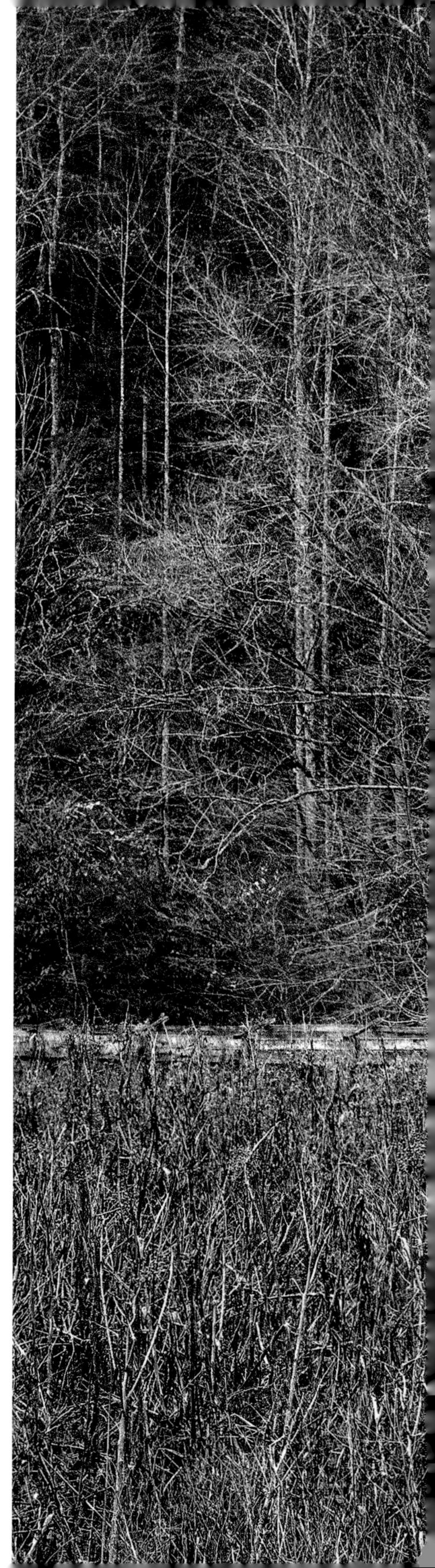

A healthy pine plantation now stands where this mill once operated in East Tennessee, with water delivered through a long and picturesque wooden flume. Years ago, the author purchased a postcard of this mill at a gift shop in nearby Gatlinburg.

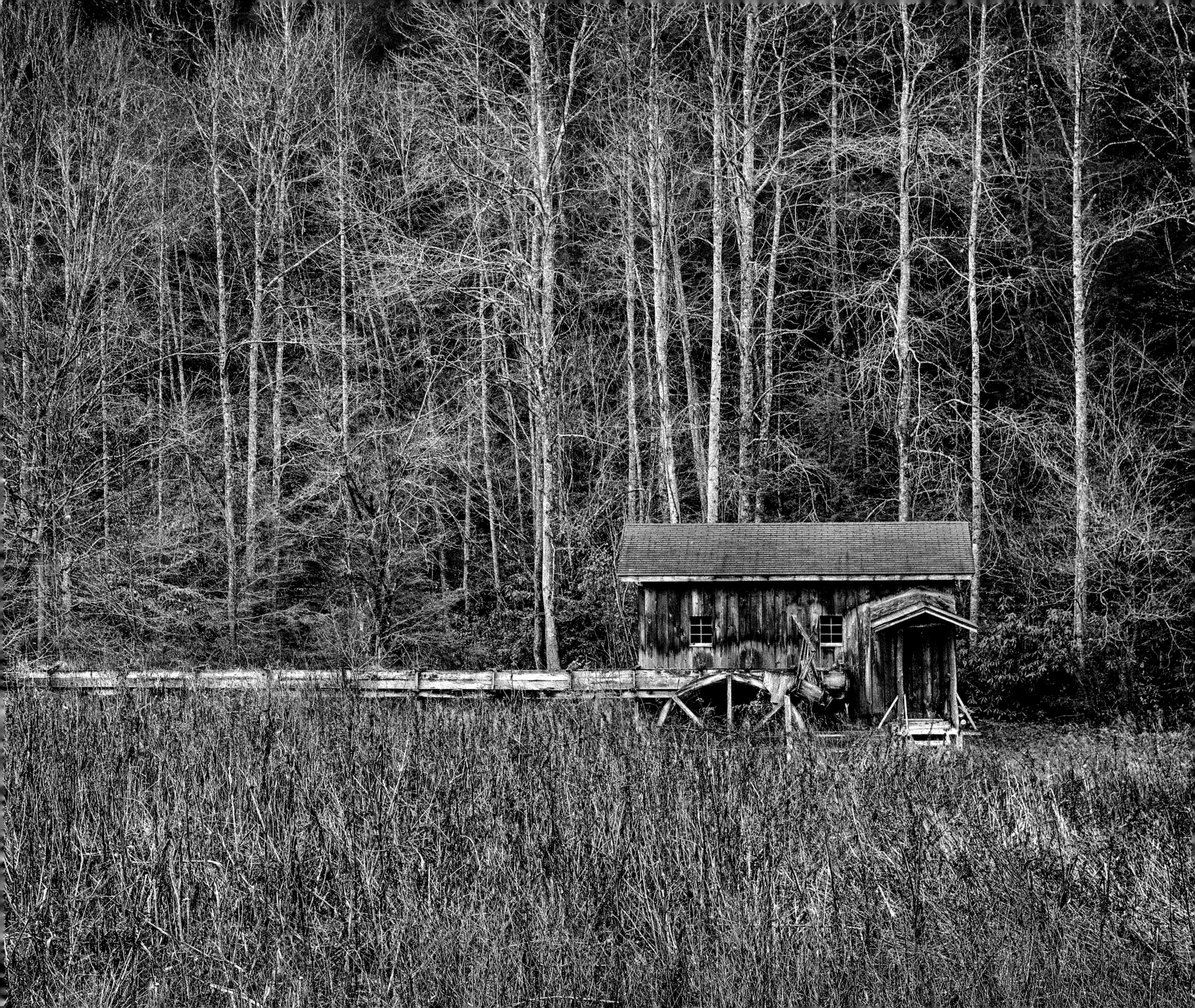

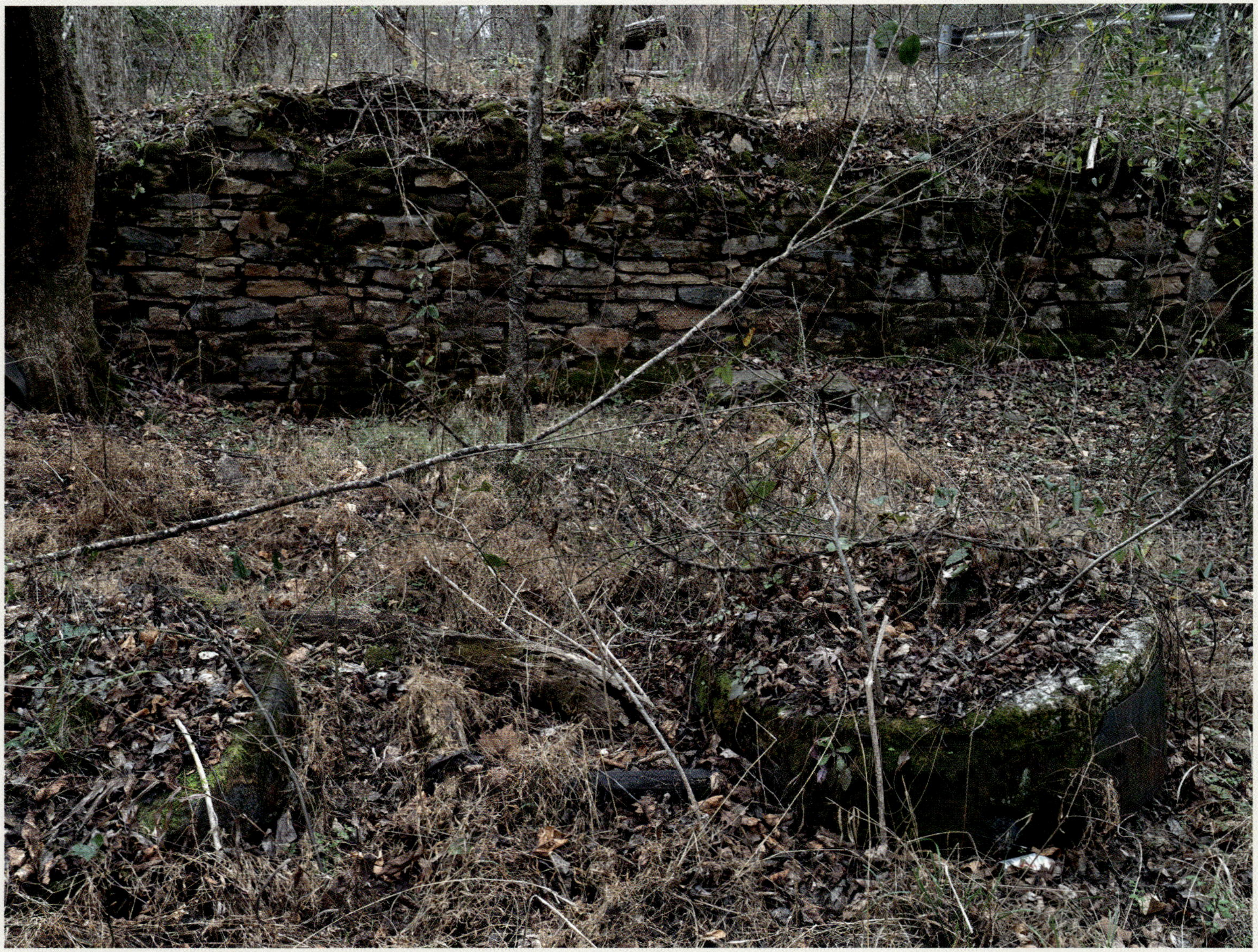

This is a somewhat typical Southern site where a gristmill and a sawmill once operated. Several millstones remain, having not yet been scavenged for display at a restaurant or in someone's front yard. Part of the structure's stone foundation is still standing in the background adjacent to the creek, and other substantial stones and building materials are spread about the area.

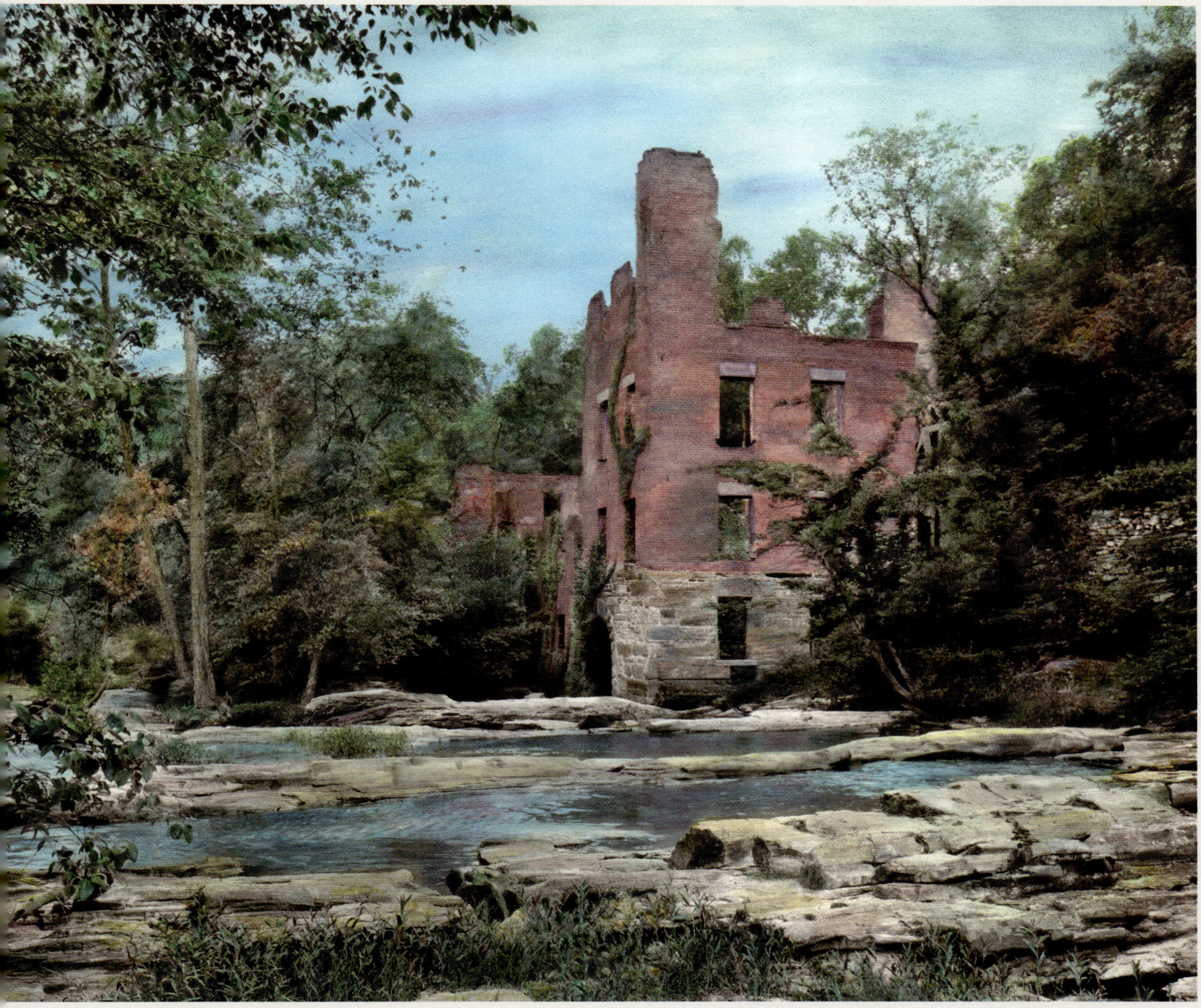

The New Manchester textile mill ruins are located at Sweetwater Creek State Park in Lithia Springs, Georgia. This massive mill opened in 1849 and continued operating well into the Civil War; in July 1864 Union forces under General William T. Sherman destroyed it, including the fifty-thousand-pound wooden waterwheel. Some ruins remain at the site, and efforts are underway to stabilize the remaining structure.

The Old Red Mill is a familiar landmark to many near Albertville on the east side of Sand Mountain in northeast Alabama. Over the years, three distinct gristmills have been located on this site, and generations visited these mills regularly to have their Sand Mountain corn ground.

Outside this immediate community, the mills were never widely known, and few have heard of the Old Red Mill or the two earlier mills. In the years since the Old Red Mill closed, the site has been most well known to area photographers, weekend hikers, and others out to enjoy the scenic beauty of the landscape.

The same is true of countless other old watermills across the South. Much of our watermill heritage is being lost year by year, as these landmarks, once essential to community life, become all but forgotten and disregarded.

The Old Red Mill, like many other old Southern watermills, is almost without any documented history. Few details of events, operations, or happenings were ever preserved and recorded—not even significant milestones. Unlike some famous and still prominent Southern watermill sites that abound with colorful and storied pasts, complete chronologies of major events and timings, and active foundations and support groups, this remote old mill never benefited from even a brief written record.

In an attempt to resurrect as much of the mill's history as possible, this author has undertaken original research for a number of years. The details provided here are derived from conversations and visits with the current owner, interviews with descendants and friends of former owners, letters, notes, brief news clippings, historic photographs, deeds, and personal observations during visits over the past forty-plus years.

As background, the Old Red Mill is set along a steep mountain slope in a deep gorge beside Martling Road. The mill is sometimes referred to as "the Old Mill at Double Bridges." This title is quite descriptive because the mill is hung on the brink of Scarham Creek just above its convergence with Whippoorwill Creek. Bridges cross each creek—hence "double bridges"—as the roadway switches back and forth to make a nearly 180-degree bend in winding down into the deep gorge and back up the steep mountainside. The scene at the mill is incredibly dramatic, with a rocky waterfall and a series of spectacularly beautiful but potentially dangerous churning rapids. The location must have seemed ideal for a watermill.

The first of the three known mills on Scarham Creek was established at the site of the current mill on Martling Road sometime after the Civil War, and it operated until the structure burned in about 1895. A single preserved and now colorized historic photograph survives to confirm the existence and appearance of the original mill, which seems to have been known simply as "the mill by the falls at Double Bridges."

A few years after the first mill burned, a second mill was built. Records show that it was operating by about 1900, and it came to be known as the W. C. Cole Mill. Subsequent deeds recorded in the Marshall County, Alabama, courthouse show that this mill changed ownership more than ten times during the first two and a half decades of the twentieth century. At one point the mill and the millpond were separately owned.

In 1912, newlyweds Alva and Annie Madilyn Dick purchased the mill from Annie's father. They operated the mill for several years, until they came to believe that it was haunted; they sold the property soon after. No one living today seems to know exactly why the Dicks thought the mill was haunted, but superstitions about ghosts in mills were common in this era. The local community also circulated stories of "haints" who lived under the two bridges adjacent to the mill. Locals believed the haints would catch the unsuspecting and carry them away!

From 1934 until well into the 1950s, the mill site was owned and operated by Irby E. Powell and his family, who built a grocery store across the road from the mill and a miller's house up the hillside. Powell initially operated the W. C. Cole Mill, but in 1951 he replaced that mill with the current Old Red Mill, which continued to grind corn until milling operations ended in the late 1950s.

Both the W. C. Cole Mill and the Old Red Mill were always turbine operated and used a single run or possibly two runs of millstones for grinding corn. Three millstones from operations at the mill have been preserved and are displayed nearby, along with a large pulley from the main turbine driveshaft. Most of a Fitz-Burnham turbine from the early 1900s has also been recovered from about thirteen feet of muck in the raceway, and the turbine parts are displayed at the miller's house.

The dam at the mill site was about six or seven feet in height at the raceway entrance but no more than about three and a half feet tall on the opposite side of the

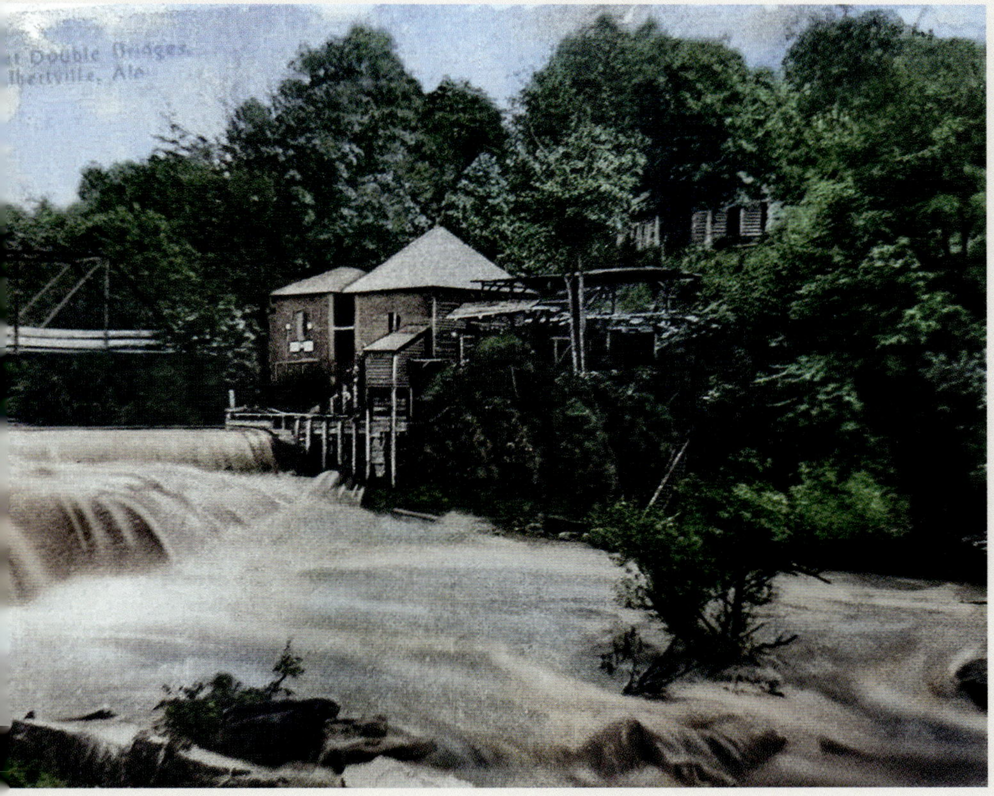

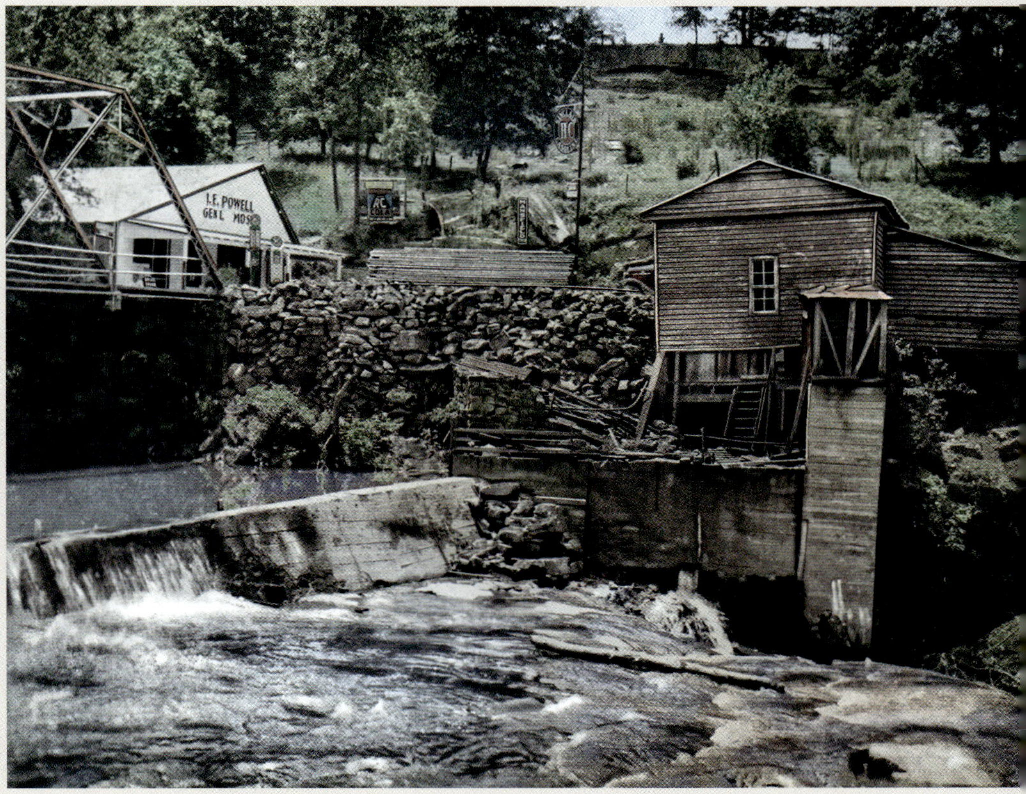

The first mill at the double bridges on Scarham Creek operated through the late nineteenth century until it burned in about 1895. The mill appears to have consisted of two structures. The structure on the right was a corn mill. The other structure is thought to have been a cotton gin, a blacksmith shop, or a general merchandise store. (This historic image, by an unknown photographer, has been digitized, restored, and colorized by the author.)

The W. C. Cole Mill was built around 1900 to replace the earlier mill on this site. It operated until it was replaced by the Old Red Mill in 1951. Note the large store with gas pumps just across Martling Road from the mill. (This historic image, by an unknown photographer, has been digitized, restored, and colorized by the author.)

creek. It was constructed of stone and rebar-reinforced concrete, using sand from the creek bed. The dam washed out in 1936 and had to be rebuilt. In recent years, a portion of the dam has washed out again. Both failures are attributed to heavy logs washing down the creek and striking the dam, much like a battering ram.

The Powell grocery store eventually burned, and after a time it was replaced with a new store by the property's next owner, L. G. Belue. This store was located across Scarham Creek and between the two bridges. It backed up to the deep gorge near the merger of the two creeks.

Belue and his family never operated the gristmill but instead repurposed the building and ran it as an outlet for new and used shoes and clothes in the 1960s and 1970s. (The author's hand-tinted photo of the mill, taken in 1982, shows signage painted on the mill building advertising "shoes and clothes" for sale.) By 1980 the Belue store had also burned, and all businesses at the mill site had closed. Since then, the mill and its site (with the exception of the miller's house) have stood vacant.

In about 2001, Lloyd Johnson purchased the twelve-acre site, including the mill and the miller's house, and he still owns the property today. By the time Johnson bought the mill, most of the equipment had been removed; only some shafts and pulleys remained inside. He cleaned out the mill, spruced up the exterior, recovered the turbine, and moved into the miller's house. Today, he keeps the grounds clean and the grass cut. Johnson can be credited with saving the mill.

In the early 2000s, the state constructed a highway bypass across the gorge to reroute traffic past the deep, narrow, and somewhat perilous bend in the road at the bridges. The double bridges are now closed to through traffic, so fewer people are aware of the mill's imperiled condition or that the mill even exists.

There were newspaper reports back in 2017 that the mill might be torn down, but that information appears to have been incorrect. As of this writing, Johnson has no plans to demolish the mill, but restoring the structure seems beyond his personal and financial capabilities. Restoration and preservation would require work by and funding from a foundation, corporate sponsor, historical group, or local, state, or federal program. In reality, preservation of the Old Red Mill would be costly and would seem of questionable value to many in the community; thus it is unlikely to occur. At this point, the building continues to deteriorate year by year.

The Old Red Mill is the last of at least twenty watermills that once served the Sand Mountain farming community, and in an ideal world, it would be preserved. The mill and the site remain charming, emotionally captivating, and inspiring to all who visit, but this mill is not likely to be here for future generations to appreciate.

OPPOSITE: This 1982 image of the Old Red Mill is a hand-tinted black-and-white photograph. Notice the words "Shoes and Clothes" painted in white on the side of the mill.

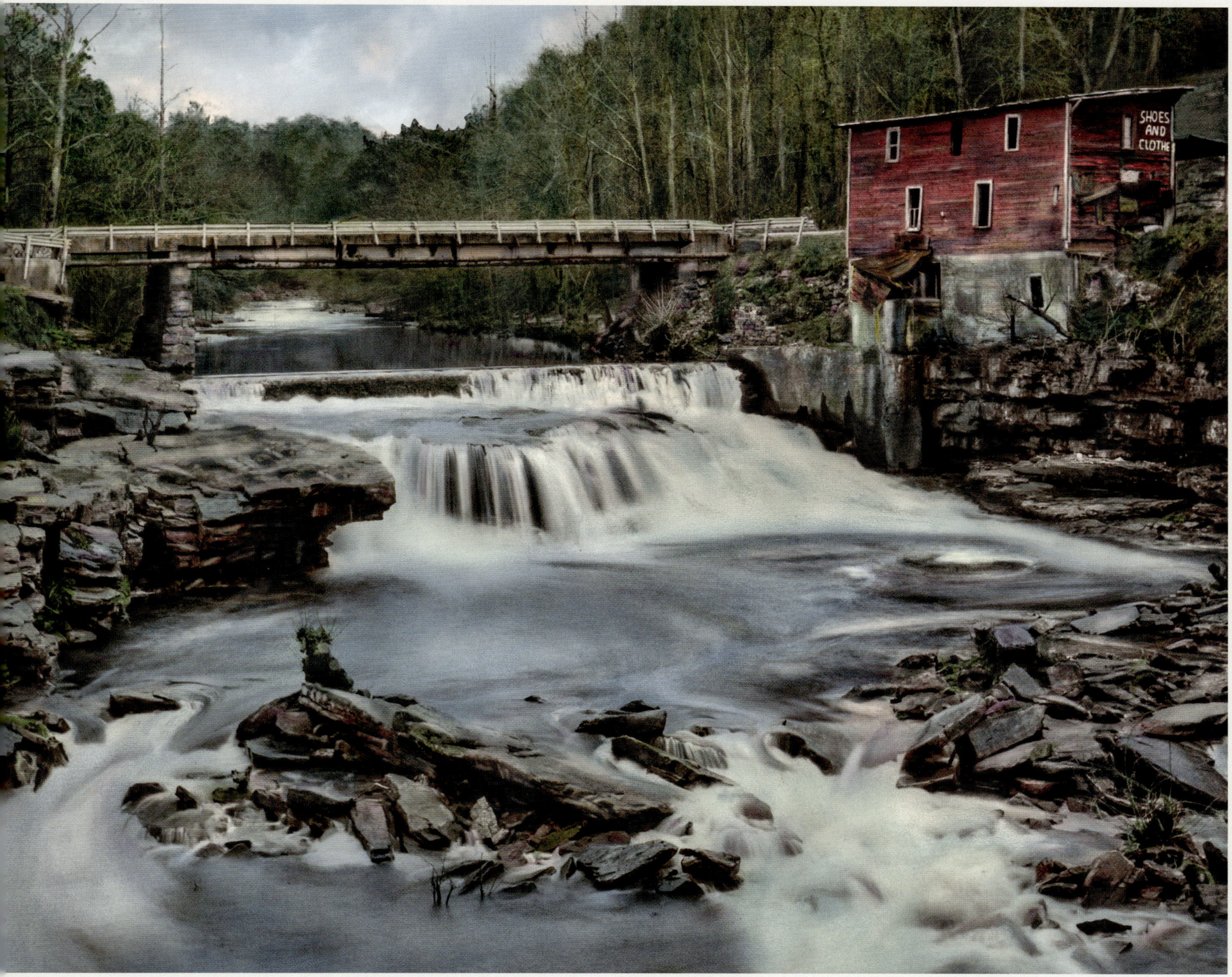
SHOES
AND
CLOTHE

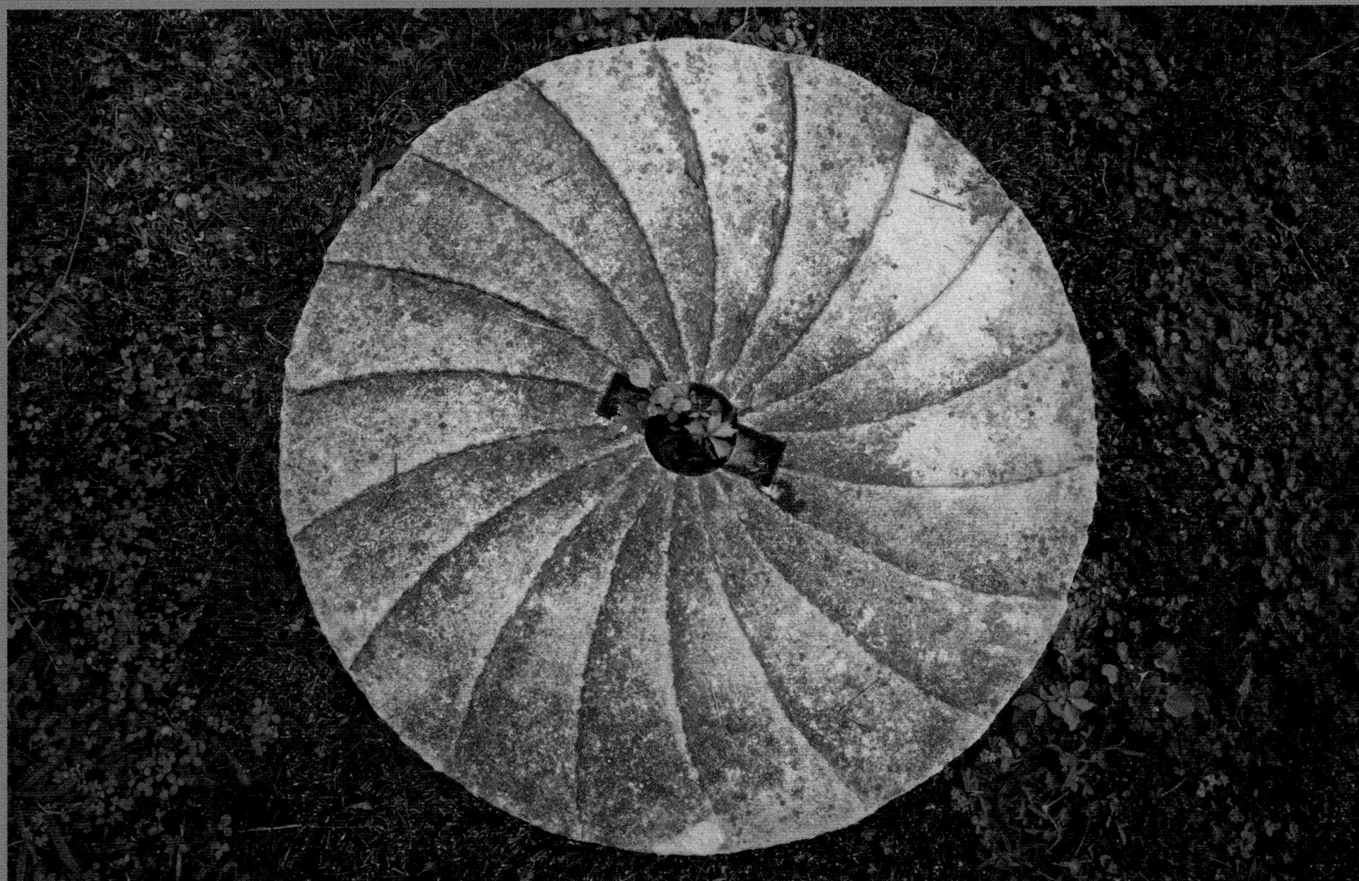

THE WATERMILLS

Aderholdt's Mill

Jacksonville, Alabama, Late 1830s / 1990s

ADERHOLDT'S MILL is unquestionably a beautiful and picturesque Southern watermill. Set along a reflective basin, with a vibrant red overshot waterwheel, the mill exudes a classic colonial charm that is enchanting. Constructed in an era when most Southern watermills were fairly small and simple structures built with logs and hewn wood, Aderholdt was designed as a two-and-a-half-story structure of dark red brick on a granite foundation. Today it looks more like something from the mid-Atlantic region than a typical mill in the Deep South. (This mill replaced an 1820s mill that had burned. The original mill dated to an era when the surrounding area was still part of the Creek Nation.)

The mill features double second-floor doors for unloading sacks of corn and wheat and a large access door on the first level. From 1836 until 1860, the mill operated on a wooden tub wheel that was powerful for its time. Then from 1860 to 1935, a large wooden waterwheel provided the power for two sets of millstones. In about 1936, an iron turbine replaced the old waterwheel, and the mill continued operations until it was retired in 1976.

The Little Tallasseehatchee Creek provided plenty of waterpower from a large millpond above the mill. The same creek also powered Broadwell Mill, located just a few miles downstream; only the foundation and millstones remain at that site.

As with most watermills in the mid- to late twentieth century, Aderholdt's Mill was falling into disrepair in the years leading up to its retirement. After the mill closed, its condition deteriorated quickly. In the early 1990s, a retired couple purchased the mill site and, with creative and energetic vision, began an extensive restoration. An eighteen-foot Fitz-brand red steel waterwheel was installed at the location of the original wooden wheel to replace the turbine. The building was reworked inside and out to provide a spectacular home. The site was recently sold due to the death of one of the owners. Fortunately, some original milling equipment was retained inside, and the new owners can still grind corn!

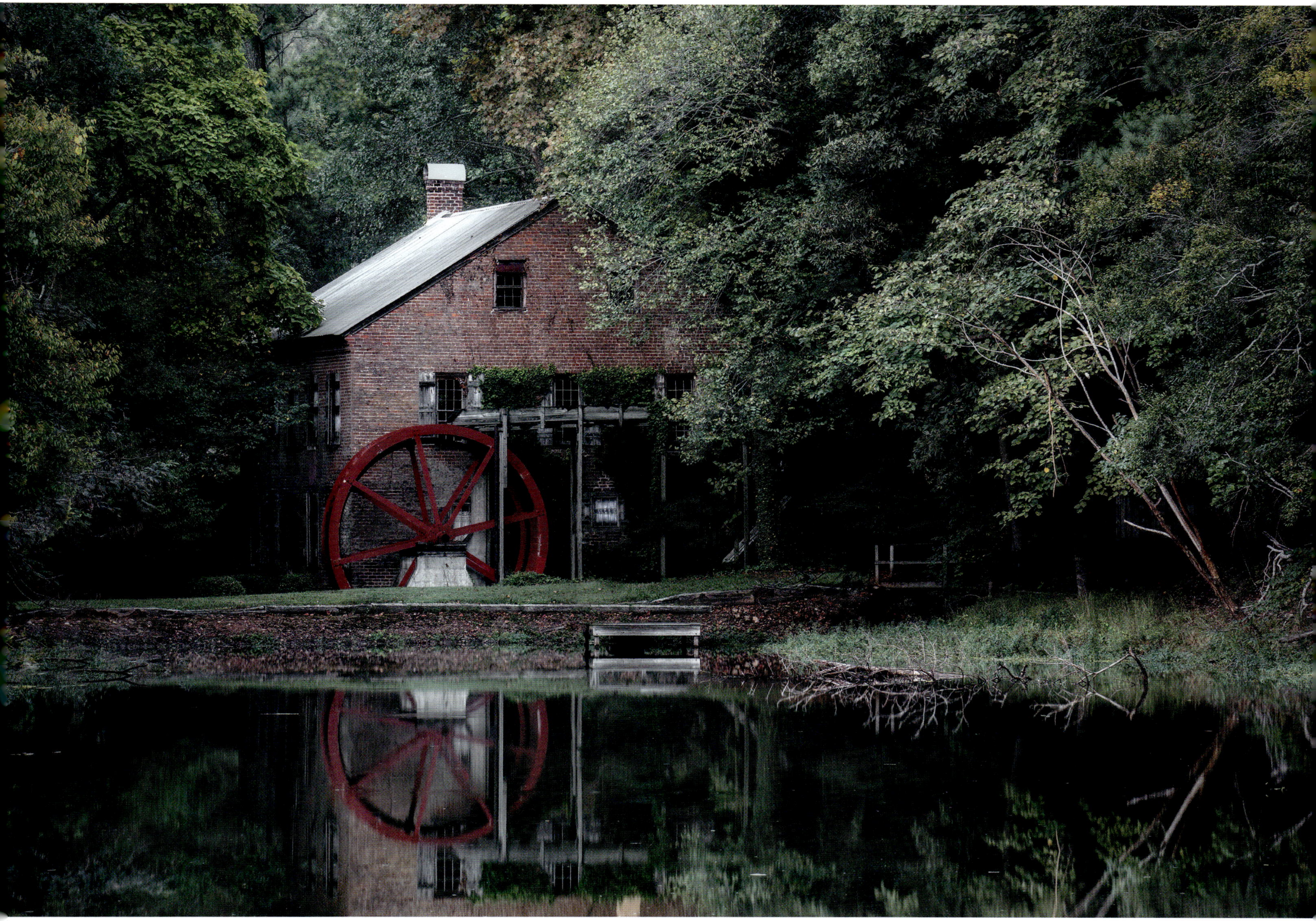

Handsomely restored, Aderholdt's Mill is nestled in the lush forests of northeast Alabama.

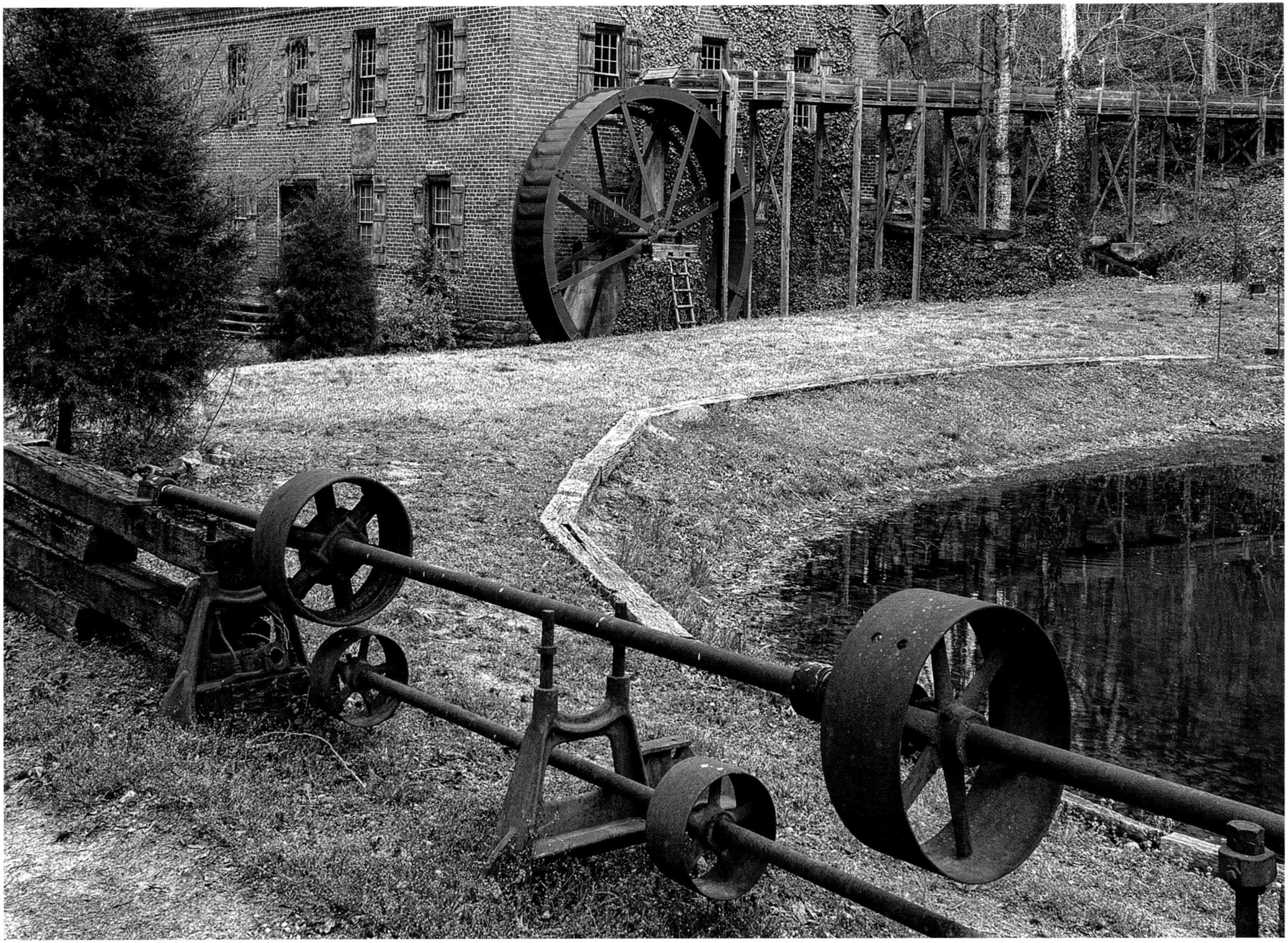

These intriguing old pulleys and shafts that once delivered power to machinery in the mill have taken on a new life and purpose, as they line a pathway along the pond below the mill.

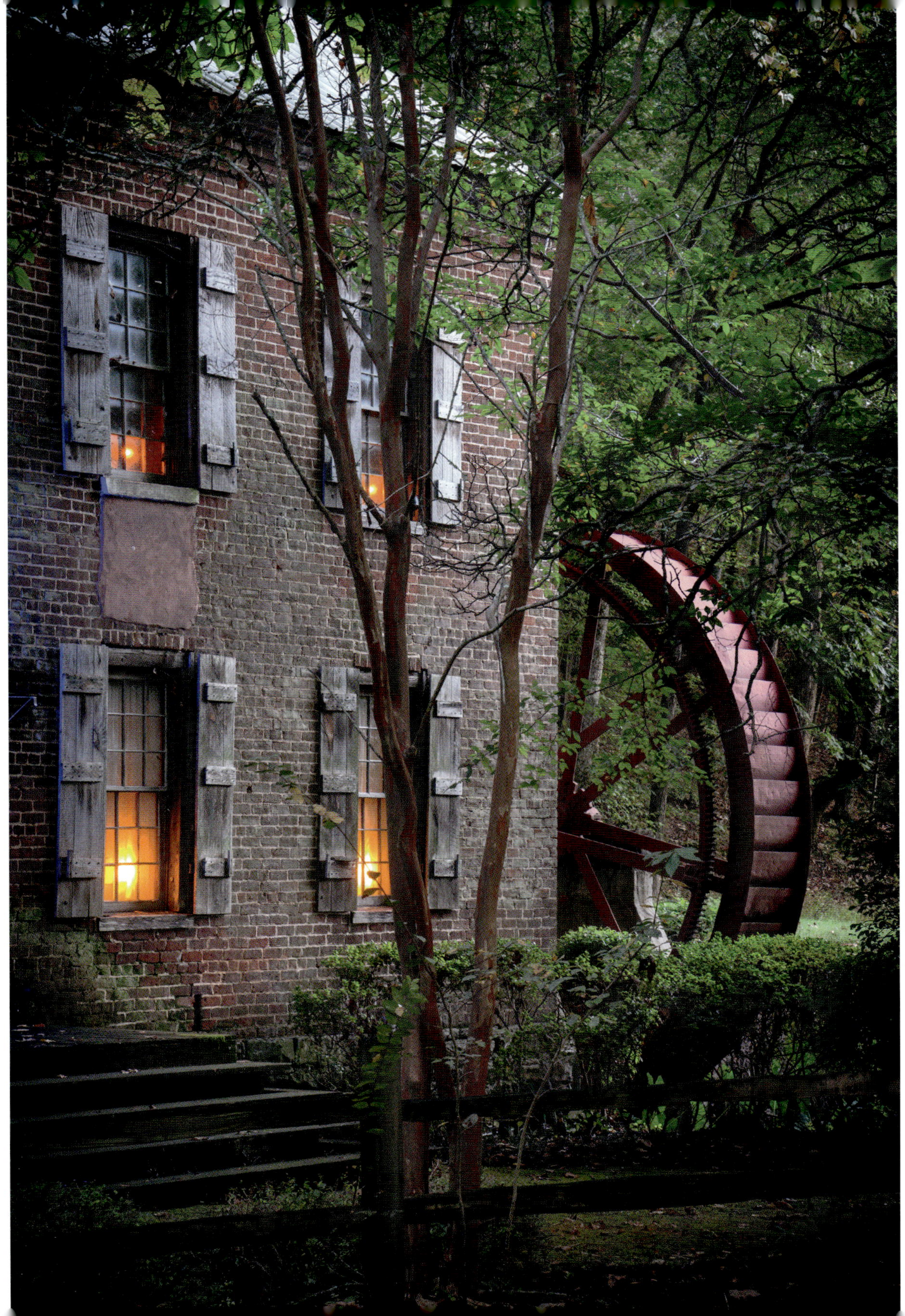

Aderholdt's Mill lives on as a warm, inviting, and peaceful residence, with its large red waterwheel in the background serving as evidence of its historic past.

Watermills at Tannehill Ironworks Historical State Park

McCalla, Alabama, 1860s / with Bicentennial Reconstructions

THIS LARGE PARK is the home of two restored and preserved waterwheel installations. The John Wesley Hall Grist Mill and Cotton Gin is located on the east side of the park along Mill Creek. The Tannehill Ironworks and the Roupes Valley blower house are farther to the west above Mud Creek.

The present gristmill is actually the third gristmill at Tannehill. The first mill was sited about a mile to the west on Mud Creek but was burned by Union forces in 1865 as they made their way to the nearby ironworks. (The Union army destroyed everything of value at Tannehill, including the cabins used by enslaved families.) A second mill was constructed in 1867, and it operated for 64 years, until 1931.

In 1976–77, as part of the state's celebration of the American Bicentennial, the present mill was built from the original 1867 plans on the site of the second mill. The reconstructed mill is fully operational and demonstrates the grinding of corn seasonally on weekends and during special events. Mill Creek supplies water from a millpond located a short distance upstream. A stone gateway and large wooden flume deliver water to a ten-by-three-foot metal overshot wheel. Inside the mill, most of the linkage is accomplished with substantial belts and pulleys, rather than through gears and shafts as at many gristmills.

The nearby ironworks is one of the finest examples of an early iron furnace in the United States. Nearly half a million visitors tour the site each year. Iron making first began at Tannehill in about 1830. Work on three large charcoal blast furnaces began just before the Civil War. At their peak of production, the furnaces could provide twenty-two tons of iron per day to the Naval Gun Works and Arsenal at Selma. The furnaces were destroyed near the end of the war, but by that time the operation had fully demonstrated that Red Mountain iron ore, which was abundant in the Birmingham area, would have a valuable future in the industrial development of the United States.

Like the gristmill, the ironworks were rebuilt to help commemorate the Bicentennial. Work began in the early 1970s and continued into the 1990s. Iron making was first demonstrated at the reconstructed site in the late 1970s. The waterwheel was added later but is not operational due to cost limitations. The blower house actually has space for a second waterwheel and is set up inside for bellows to be added.

To offer a brief historical and semitechnical review: Until the twentieth century, iron furnaces used waterwheels to power huge bellows that blew air into molten iron mixtures within stone furnaces. These "blasts" of air raised the temperature of the furnace mixture, producing a much higher grade of iron. At Tannehill, a canal to deliver water to the wheel does not currently exist; the restored furnaces are equipped with electric blowers for iron-making demonstrations. Along with waterwheels, early steam engines were sometimes used to blow furnace air during the Civil War. Steam power was part of the Tannehill operation.

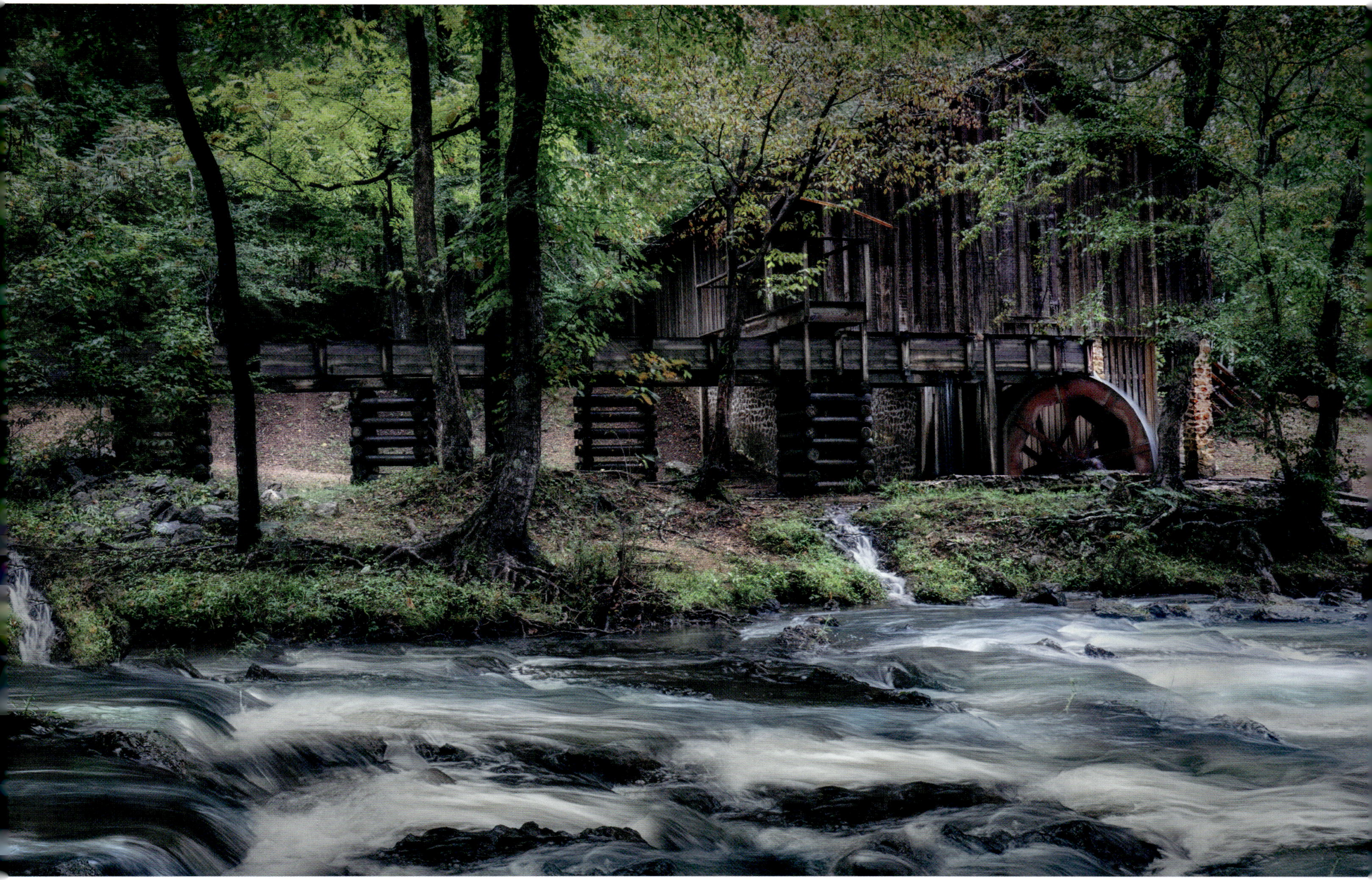

John Wesley Hall Grist Mill and Cotton Gin is sited along Mill Creek and is actually only a gristmill; a cotton gin is located in another area of the park.

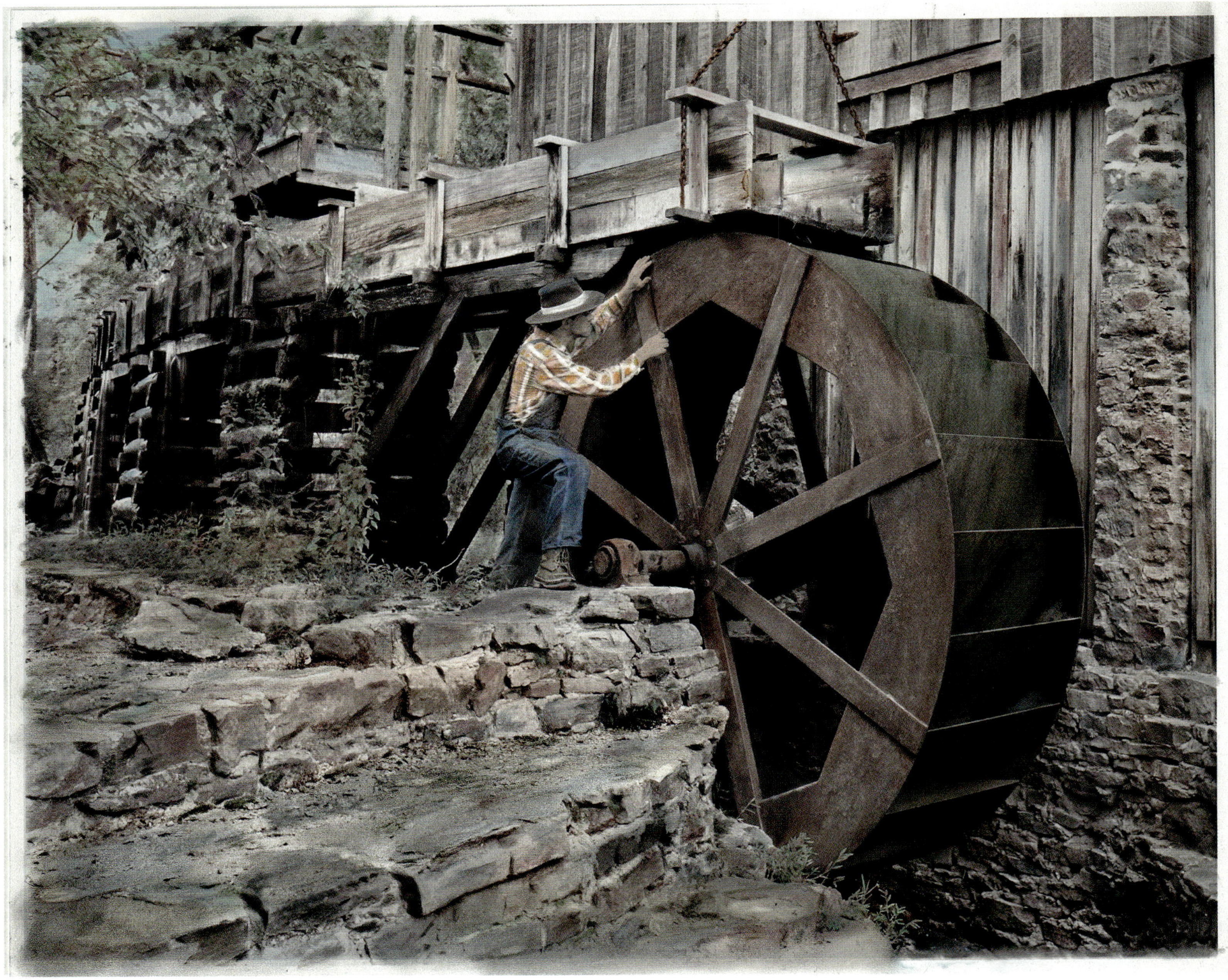

The waterwheel at the John Wesley Hall Grist Mill is ten feet in diameter and three feet wide. The wheel turns at a rate of about ten to twelve revolutions per minute during operation. In this hand-tinted self-portrait, the author can be seen inspecting the wheel.

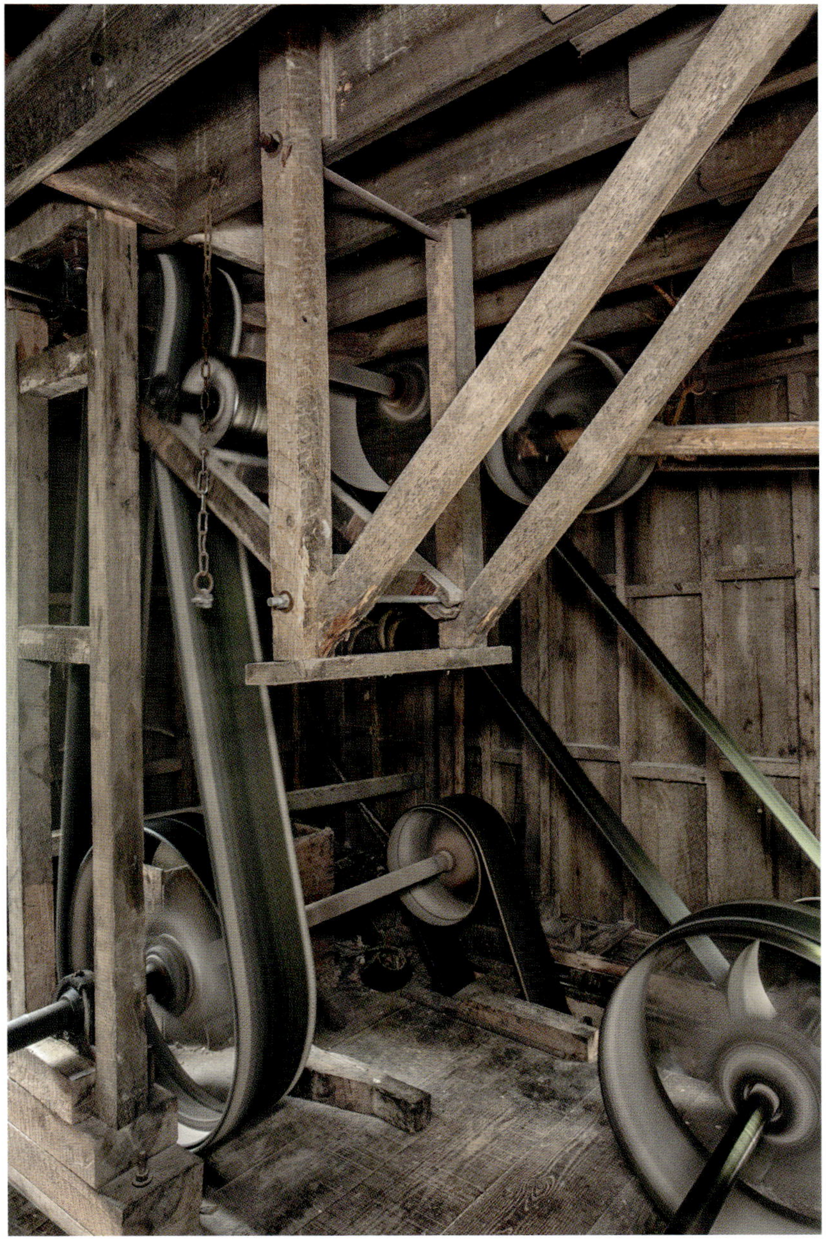

The mill is operated by a series of whirling belts and pulleys that are connected to the waterwheel just outside the mill. The diameter and sequence of pulleys increases the rate of rotation from about 10 to 12 revolutions per minute at the waterwheel to about 100 to 120 revolutions per minute at the millstones.

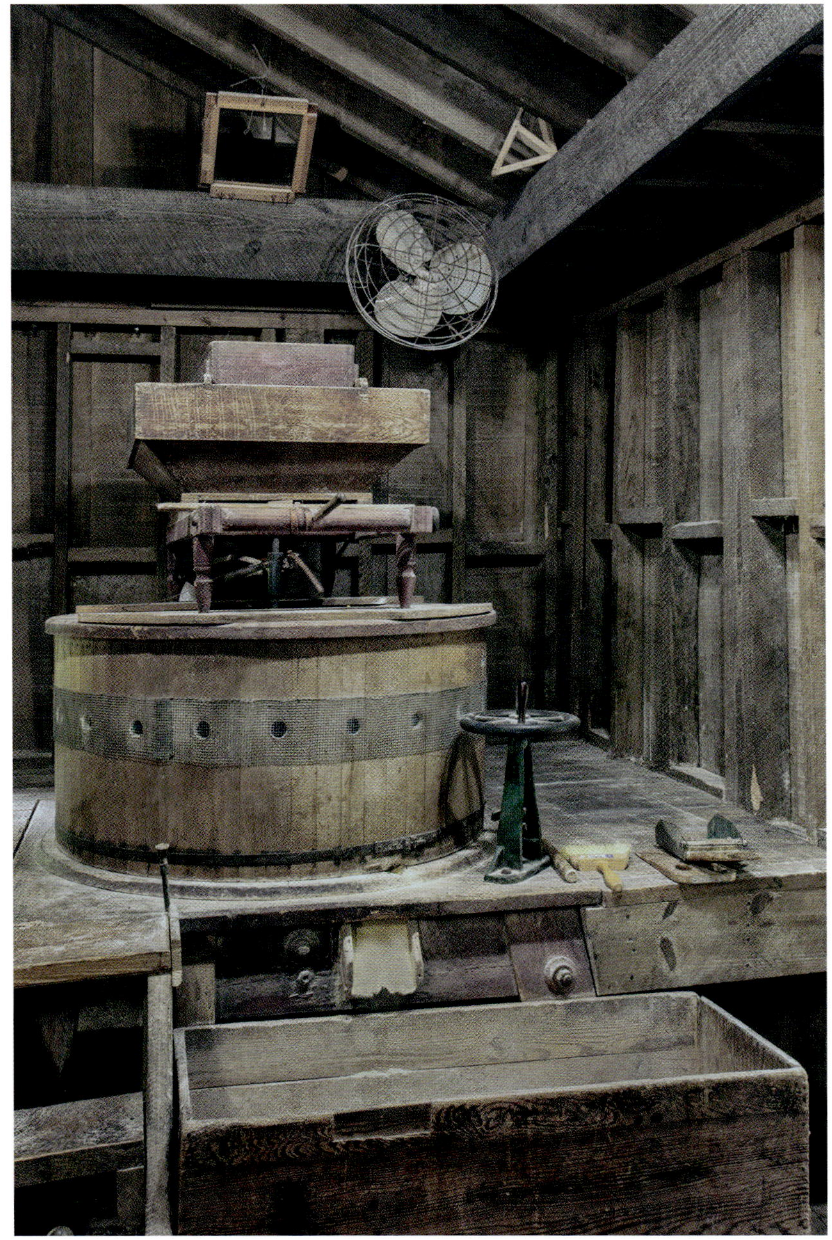

The mill grinds corn into cornmeal and grits for demonstrations. The meal and grits are collected in the bin at bottom in the photo. The small green metal wheel at right is used to adjust the spacing between the stones.

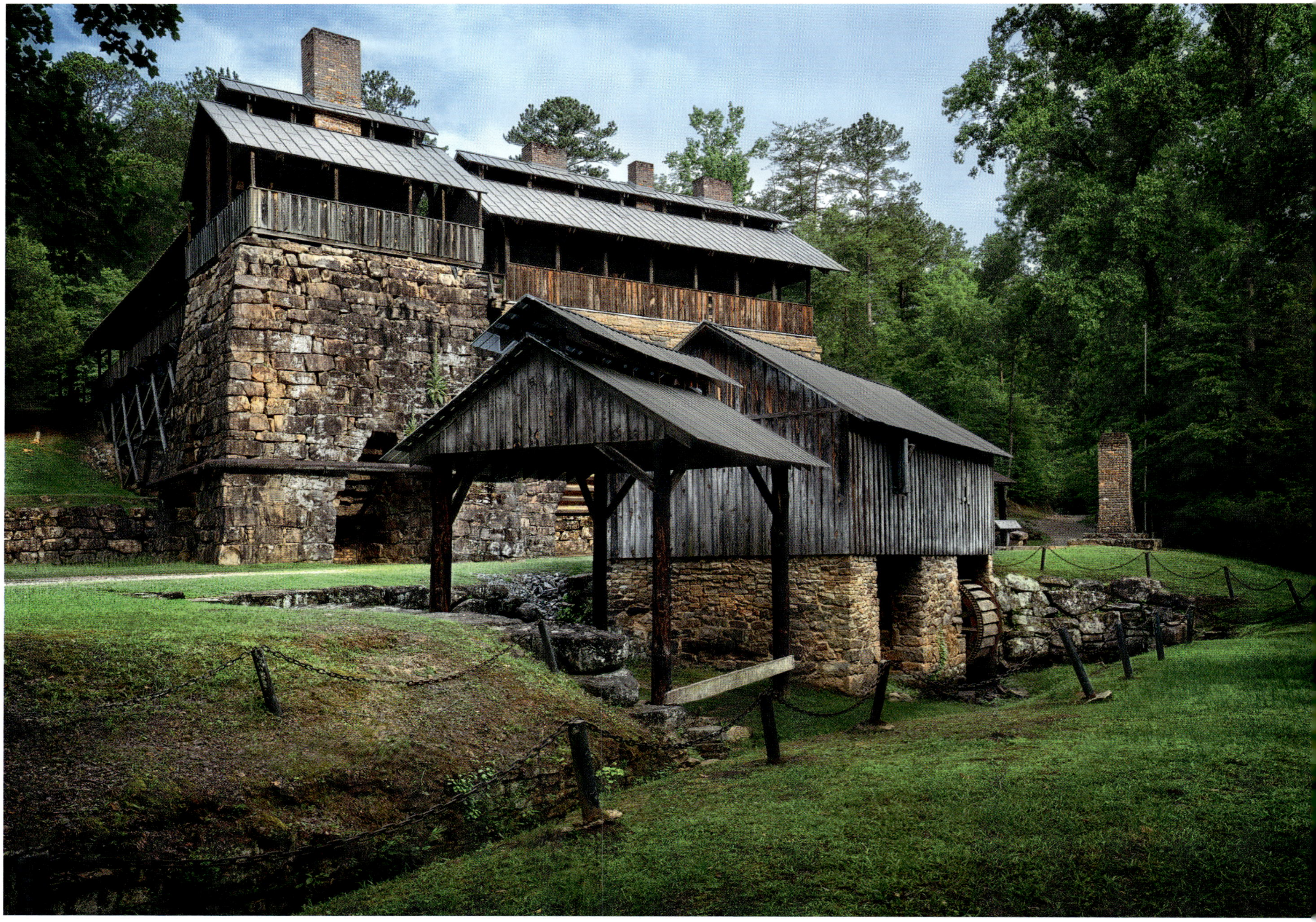

The Roupes Valley blower house, featuring a waterwheel, is shown here
with the restored and operational iron furnaces in the background.

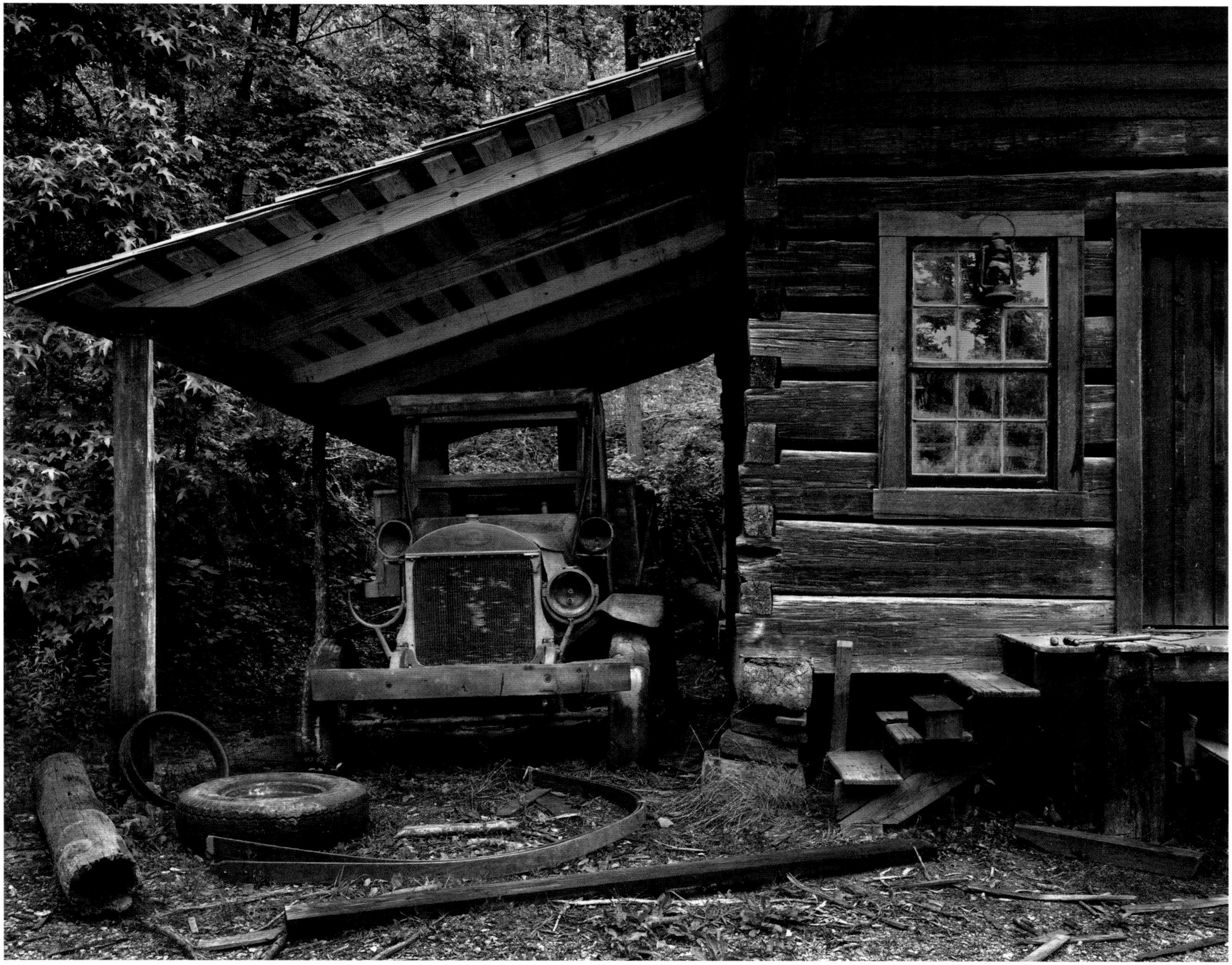

This old chain-driven Mack truck at Tannehill is typical of the type of vehicles used by millers and farmers from the early twentieth century until the World War II era.

Pratt Cotton Gin Mill

Prattville, Alabama, 1830s / with Several Later Additions

EQUAL IN SCALE to the massive water-powered industrial complexes at Augusta and Columbus in Georgia, the Pratt Cotton Gin Mill was the largest cotton gin factory in the world for many years. In the mid-1830s, Daniel Pratt purchased land along Autauga Creek, which had previously powered a sawmill and several other small factories. Pratt began his gin production there as a small operation, but he was from New England and was familiar with much larger industrial watermills; his goal was to bring that scale of operation to the South. Most of the structures shown in the image here date to 1848, 1852, 1854, or 1912. Some records show that the section of the structure nearest the dam was actually a gristmill in the early 1840s and may have been converted. Along with cotton gins, the mill also produced wooden construction items, including doors, windows, and sashes for the state capitol building in nearby Montgomery.

As a major social legacy, history has shown that Pratt's watermill contributed to an increase in demand for enslaved workers on cotton plantations because the cotton gins made it possible to process so much more cotton. In the period leading up to the Civil War, the mill was operated by both enslaved and paid workers. Interestingly, the mill was spared during the Civil War even though Union troops operated in the area.

After the war, the mill continued to prosper and, in addition to cotton gins, produced other machinery for processing cotton, pecans, and fruits. In the years before the mill ceased operations in 2012, machine gun mounts, cast-iron fences, electronic control systems, electric motor controls, aerators, video scanning controls, and a host of other items were manufactured there.

The site is presently being redeveloped as a commercial and residential property. The developer agreed to retain the historic appearance of the old mill from the creekside view.

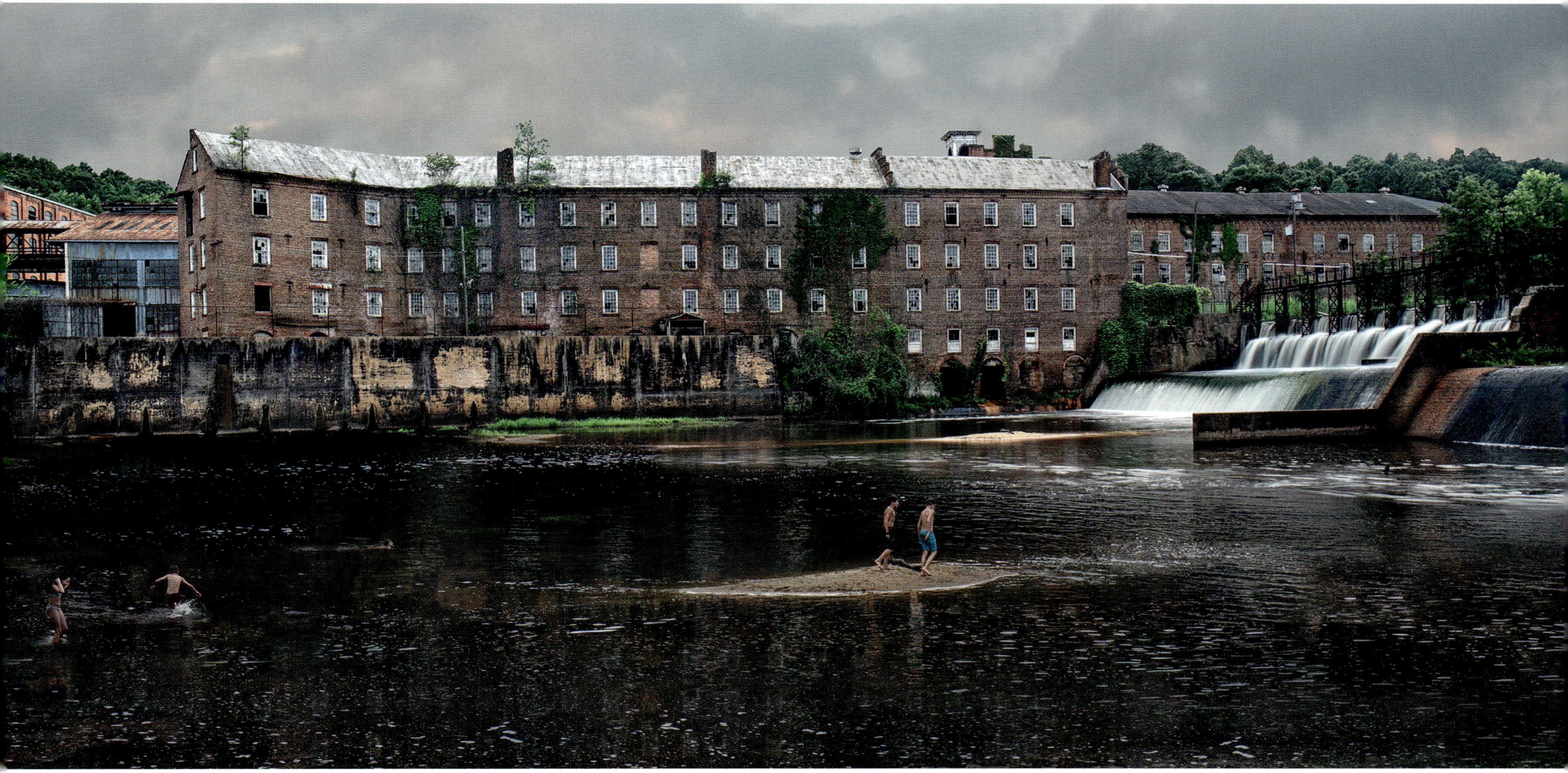

The wide, shallow pool of water below the Pratt Cotton Gin Mill on Autauga Creek has long been a popular recreational area for children during the summer months. The bottom of the shallow creek is concrete, which was thought to have helped to control erosion when the mill was operating.

Estate Mills

Swann Millhouse, 1900; Robert Jemison Millhouse, 1926; Sweetwater Falls Millhouse, 2003 / Birmingham Area, Alabama

THROUGH THE LATE nineteenth century and into the early twentieth century, larger cities across the South became more congested, noisier, and more polluted. At the same time, transportation for traveling to and from these cities continued to improve. As a result, wealthier families began to move farther from the city centers and onto larger, estate-size properties out of town. These properties were to become the region's early suburban areas, and though generally not as extravagant as Gilded Age estates in the Northeast, Southern estate homes might include elaborate landscaping, gardens, pools, statues, and other fine appointments.

Birmingham offers a perfect example of urban flight by families that had made fortunes in the city's mining, ironworks, piping, chemical, and manufacturing operations. As the iron and steel center of the South, this city hummed with the activity of foundries and steel mills as well as hundreds of steam-driven locomotives every day. The skies of Birmingham had become dirty and smelly, and what had emerged from farmland after the Civil War to become the "Magic City" was now jokingly referred to as "Smoky City."

Geographically, Red Mountain, famous internationally as a source of iron ore, separated the city to the north from the pristine forested lands and valleys to the south. With roads having been completed over the mountain and with automobiles now available, Shades Valley beyond Red Mountain became a choice retreat for families who could afford the move. These families relocated along Shades Creek, the Little Cahaba River, and other picturesque waterways that flowed through the valley.

By the turn of the twentieth century, the Theodore Swann family, prominent in business and chemical manufacturing, had begun to develop property "over the mountain." In 1900 and the years that followed, they built what was at the time the largest mansion in the region, as well as the beautiful Swann Millhouse. Even today, situated in a wooded bend along the Little Cahaba River, the Swann Millhouse is reminiscent of a scene from a classic fairy tale. The mill was restored in 2015 and features a wooden undershot waterwheel positioned on a stone raceway adjacent to a spectacular dam and spillway.

During this same era, Robert Jemison Jr., noted Birmingham developer and businessman, purchased property along Shades Creek and began planning the development of an exclusive area that would become known as Mountain Brook. To further promote this development, he also constructed a beautiful watermill, the Robert Jemison millhouse. (An interesting sidenote: Jemison was related to famed physicist Robert Jemison Van de Graaff, who developed the Van de Graaff generator.)

Today the Robert Jemison millhouse is considered to be one of the most beautiful watermills in the United States. Jemison commissioned landscape architect William H. Kessler to create the plans for the mill. The Jemison mill replaced and is built on the foundation of an earlier 1879 gristmill known as the Perryman Mill; that mill ceased operations in 1887. Kessler designed his mill with oak clapboard siding and a tall sloping roof of hand-split shakes. The hardwood floors were pegged construction, and the fireplace incorporated an original millstone from the Perryman Mill. The wooden undershot waterwheel was and still is operational, but this structure was never intended to be an authentic, functional mill. The only access to the mill has always been across a footbridge over the creek.

The mill initially served as a tearoom and eatery, but it was sold in 1952 and repurposed as a private residence, with careful ongoing preservation. In recent years it has served as a guesthouse, an event venue, and a tearoom.

To the delight of watermill fans in the North Alabama area, a new millhouse joined the Swann and Robert Jemison millhouses in 2003. Sweetwater Falls Millhouse, a massive Renaissance-themed, nonoperational replica, is located in an exclusive residential community along the Robert Trent Jones Golf Trail near historic Ross Bridge, only a few miles west of the other estate watermills highlighted here. This is a beautiful and costly structure; such watermill projects are rare today.

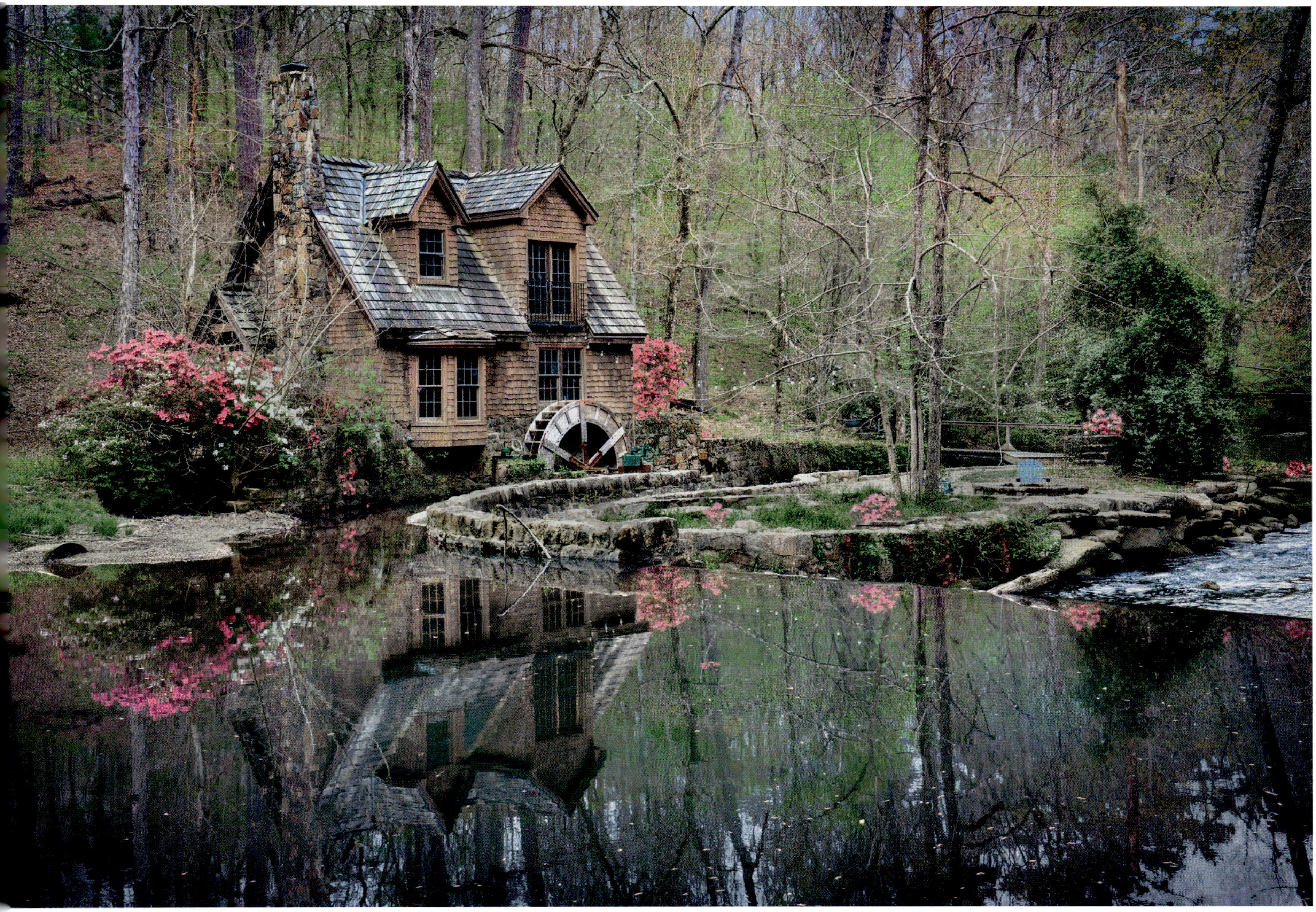

Framed by azaleas and the light green foliage along the millpond, Swann Millhouse emerges from the landscape in the spring like a fairy-tale scene.

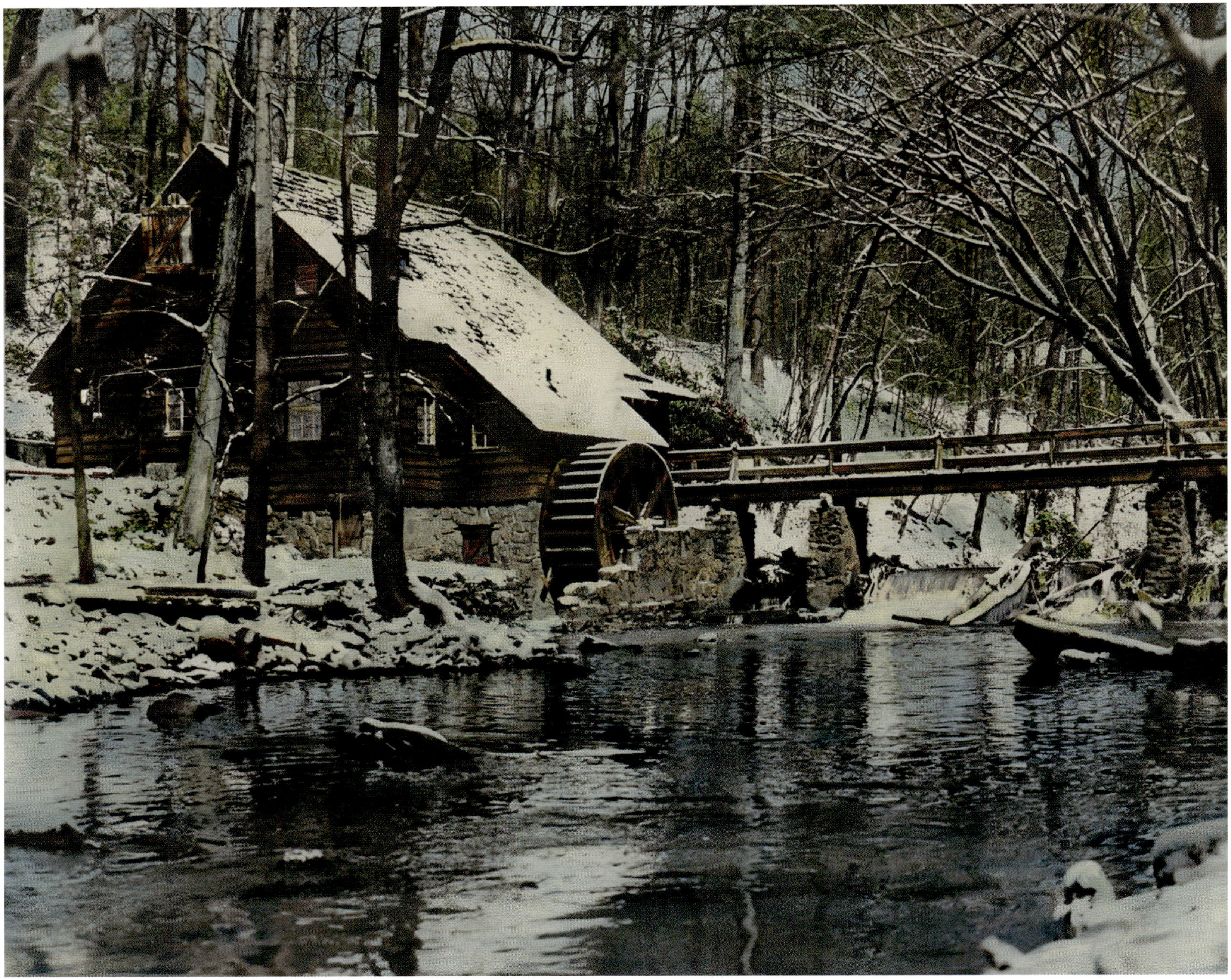

The classic charm of a watermill is undeniable in this hand-tinted image of the Robert Jemison millhouse captured on a snowy morning in the 1970s.

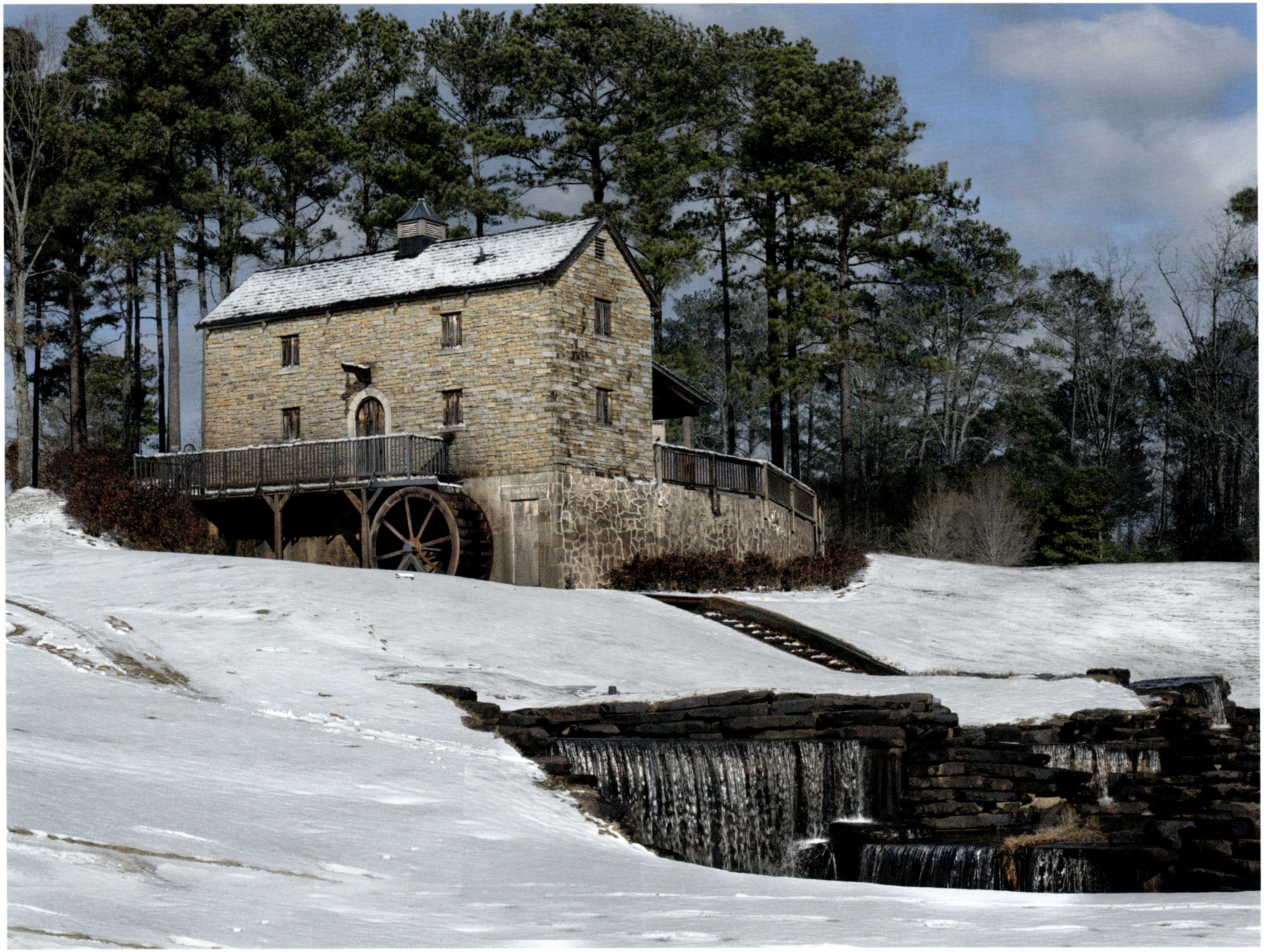

Water from a lake above Sweetwater Falls Millhouse flows across the mill's impressive nine-foot steel overshot waterwheel and cascades a total of some eighty feet on its way down to a pond below the mill.

Kymulga Grist Mill

Childersburg, Alabama, 1864 / Restorations in Recent Decades

KYMULGA GRIST MILL is a large mid-nineteenth-century commercial-scale gristmill in east-central Alabama. Still operational, the mill could process a million bushels of corn annually as recently as World War II, with only waterpower turning five sets of millstones that date to the 1860s. The millstones, which weigh as much as 3,200 pounds each, were shipped from France during the American Civil War, bypassing the Union blockade around Mobile Bay.

The mill was never powered by a vertical waterwheel, the most common technology in the 1860s, but instead relied on three high-horsepower iron water turbines installed beneath the mill. Grain arrived from local farms in baskets and sacks, as at smaller mills, but was also shipped in by railcar from the Midwest. One of the oldest rail lines in Alabama was built to service the mill.

The team that built the mill consisted of G. E. Morris, an ironworker from South Carolina, and twenty leased enslaved craftsmen. After the Civil War, the freed craftsmen and their White former overseer formed a company/partnership and went into business together; they traveled the country building watermills.

This mill was designed as an automatic mill with a complex system of gears, pulleys, and belts to handle the heavy lifting. Most of the grain was transferred from railcars into a three-thousand-bushel bunker below the mill. After the grain was unloaded into the bunker, the millers did not see the product again until it was ready for packing into bags and barrels for loading onto wagons for distribution and sale.

In 1941, the U.S. Army purchased the surrounding town of Kymulga, along with thirteen thousand to sixteen thousand acres across the area, for construction of a massive ammunition plant known as the Alabama Ordnance Works. What few knew at the time was that although the site was an ammunition plant, it also housed a top-secret heavy-water production facility for the Manhattan Project!

Following the war, the ordnance works closed. The Kymulga mill continued to operate under various owners and at reduced scale until 1977. With almost no maintenance or upgrades for several decades, the mill had, like many other watermills of the time, fallen into disrepair and was subsequently abandoned.

Both the mill and the adjacent covered bridge were listed on the National Register of Historic Places in 1976, but this did not assure their preservation, and the condition of the mill continued to worsen.

In 1988, the Childersburg Heritage Committee purchased the site and, with help from the Alabama Historical Commission and the Talladega County Commission, established the current historical park. In 2011, the City of Childersburg acquired the park, and it is now managed by the Childersburg Historic Preservation Commission. Under the direction of Ron Smith, senior miller and park manager, restoration work on the mill has been extensive and is still ongoing.

During a major flood in 2012, the mill suffered a catastrophic foundation collapse, even though much of the mill had been restored by that time. The huge sixteen-inch heart pine beams that had supported the mill since the mid-1800s appeared substantial but had hollowed out on the inside over the years, and they washed from their supports during the raging flood. At a cost of $265,000, with construction and labor services provided at no charge, a steel and concrete foundation was installed to stabilize and secure the mill.

Today the mill is a popular tourist attraction and hosts festivals, business retreats, weddings, and events throughout the year. The mill is open to visitors Wednesday through Sunday and serves as a window into life in the South during the late nineteenth century.

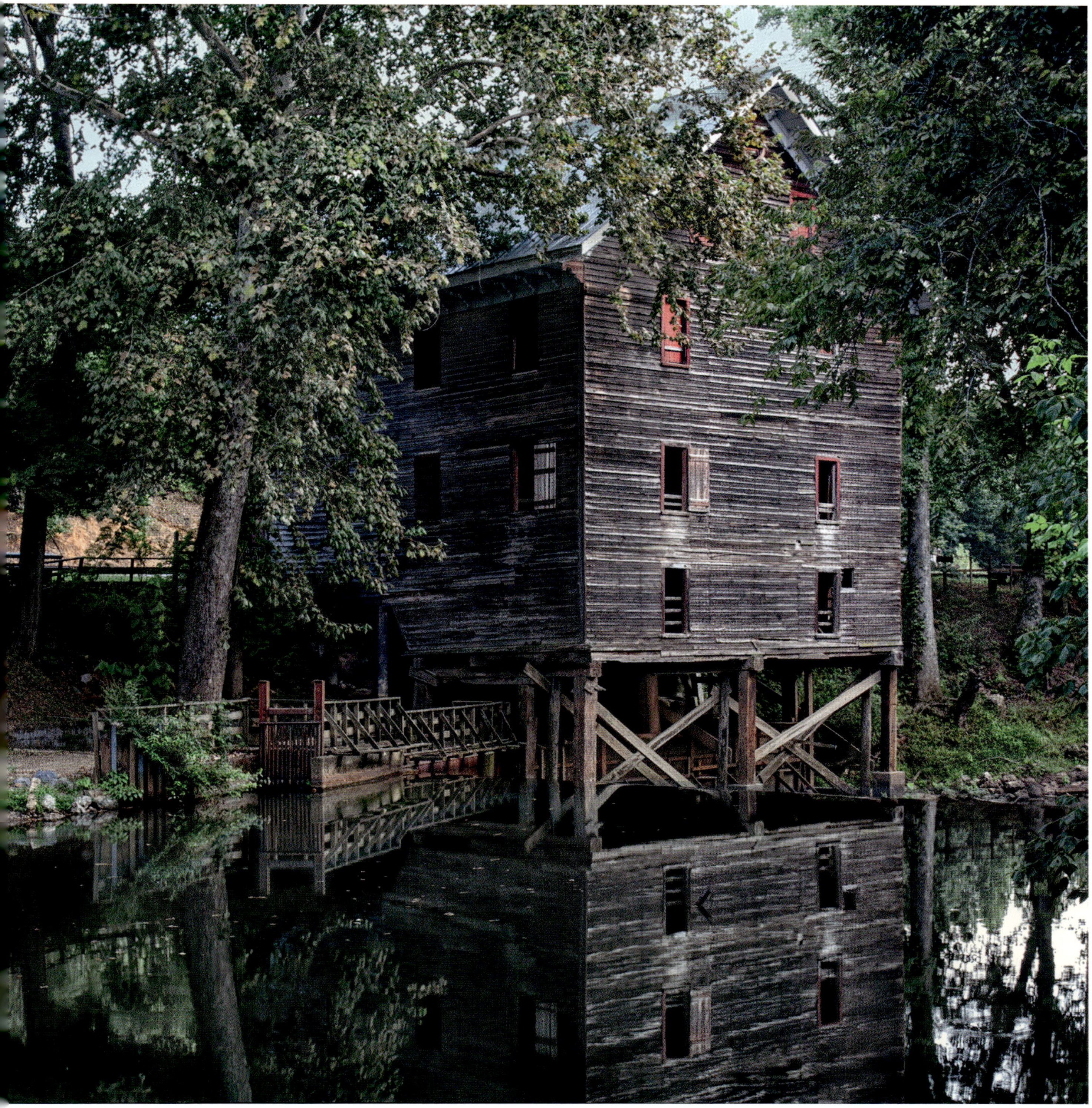

Kymulga Grist Mill is quite picturesque as viewed from the adjacent covered bridge over Talladega Creek.

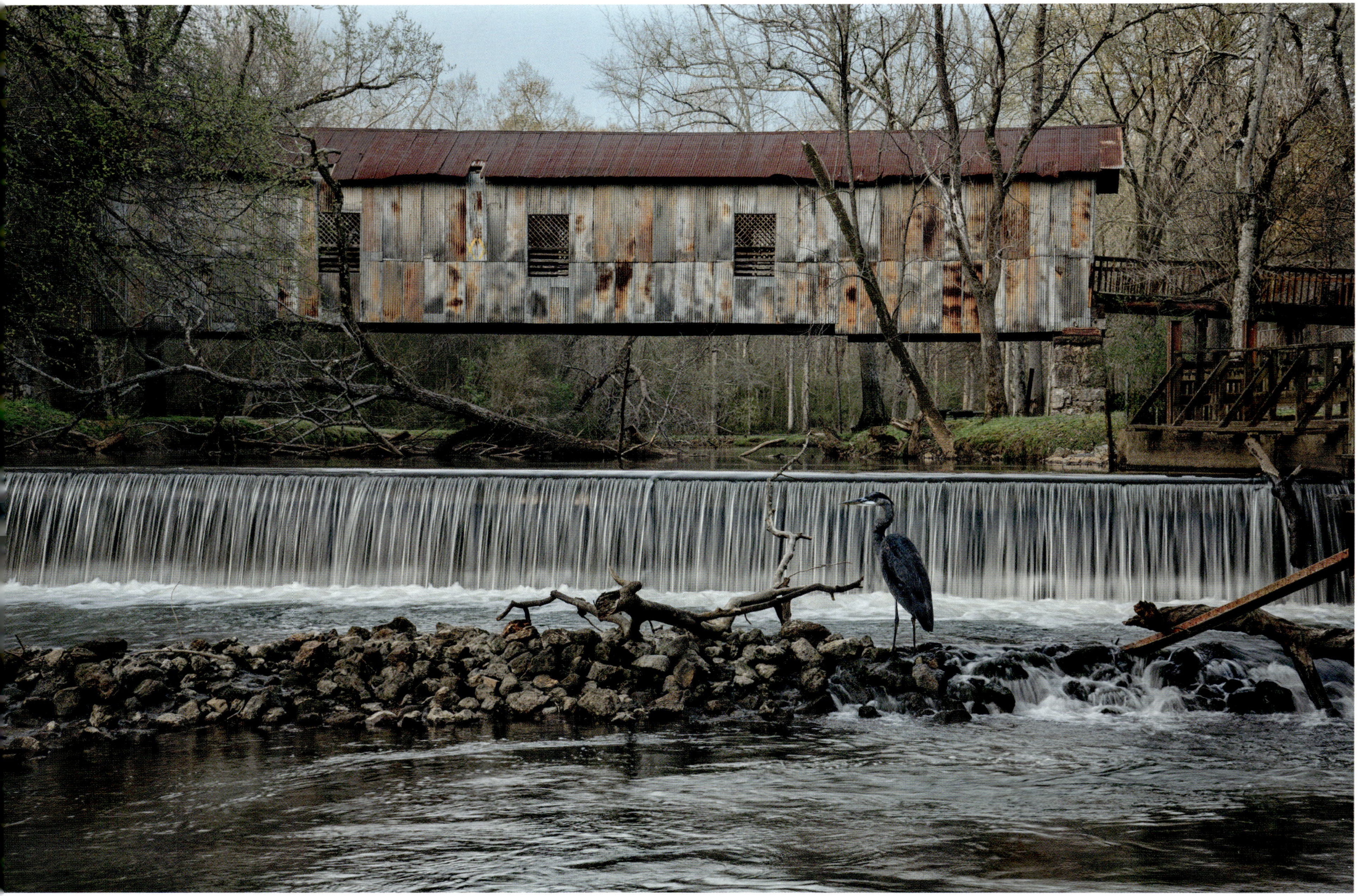

Before the mill was constructed, a covered bridge was built across Talladega Creek in 1861. The bridge was necessary to provide access to the ancient timber forests beyond the creek as a source of construction materials for the mill. The bridge is one of only two nineteenth-century covered bridges in Alabama that have not been rebuilt or relocated.

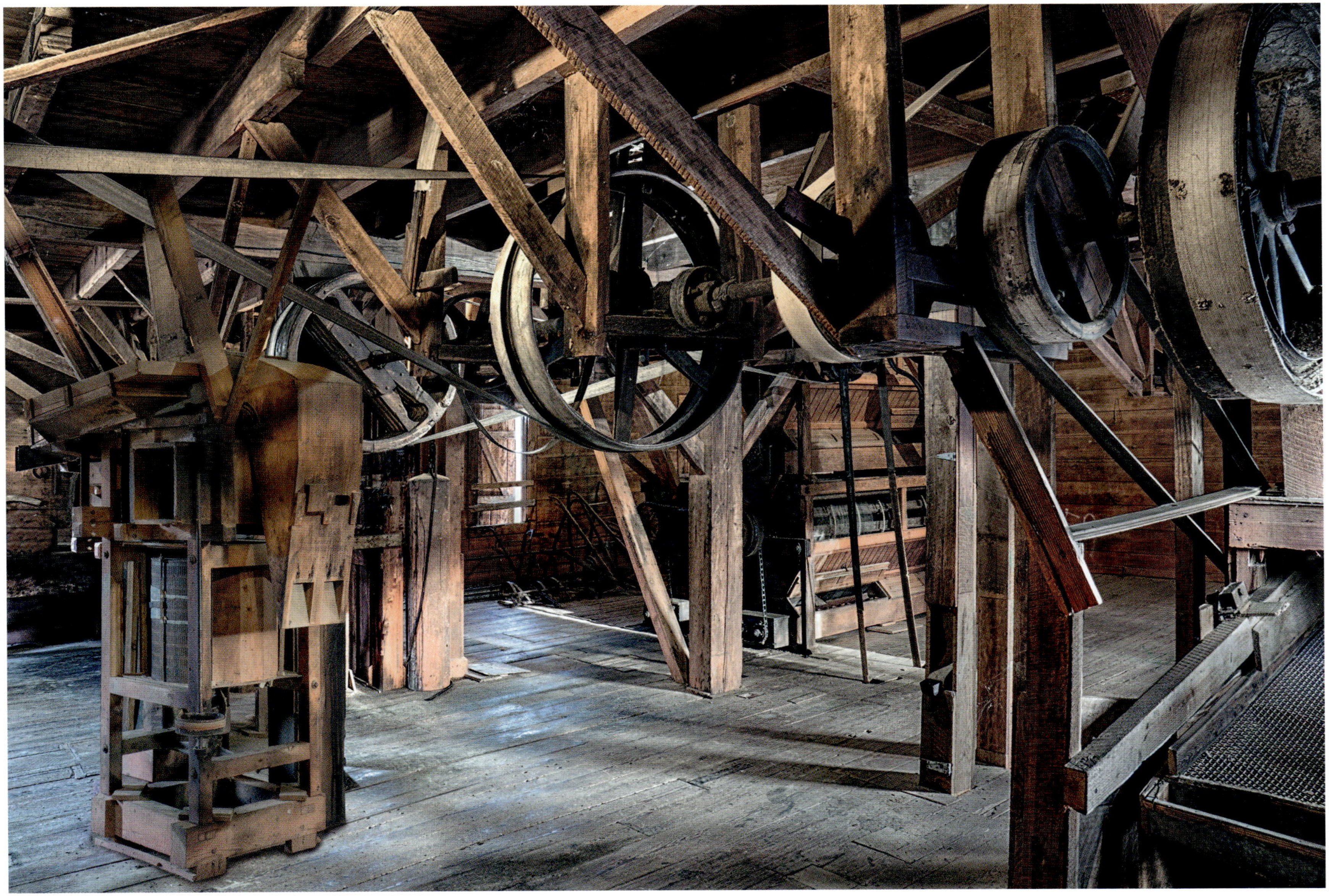

The complexity of the automated mechanical milling equipment is evident in this image.

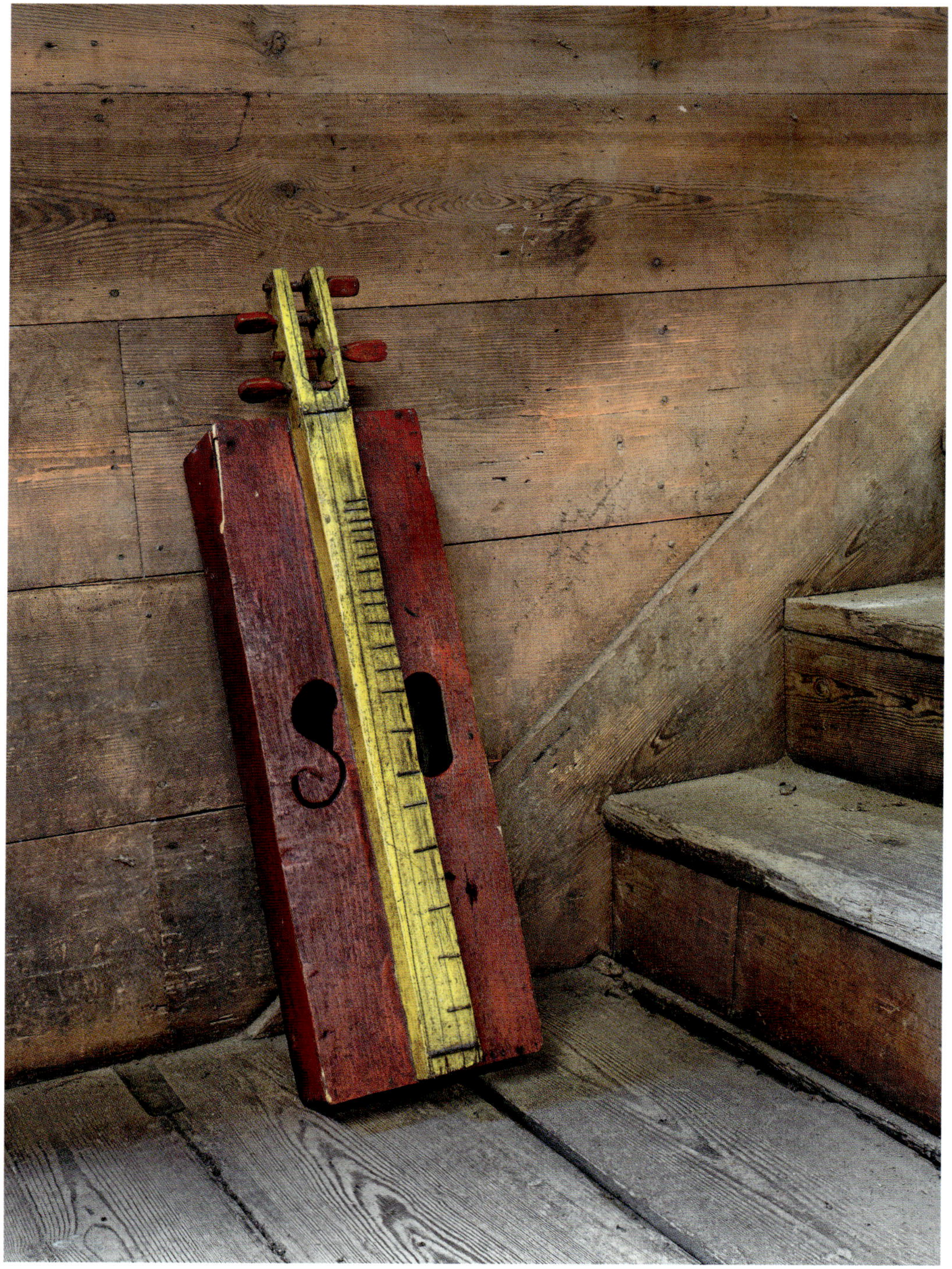

This dulcimer was made at Kymulga in 1897. The back side is engraved with the date, artist's name, and location. The instrument is now displayed at the mill.

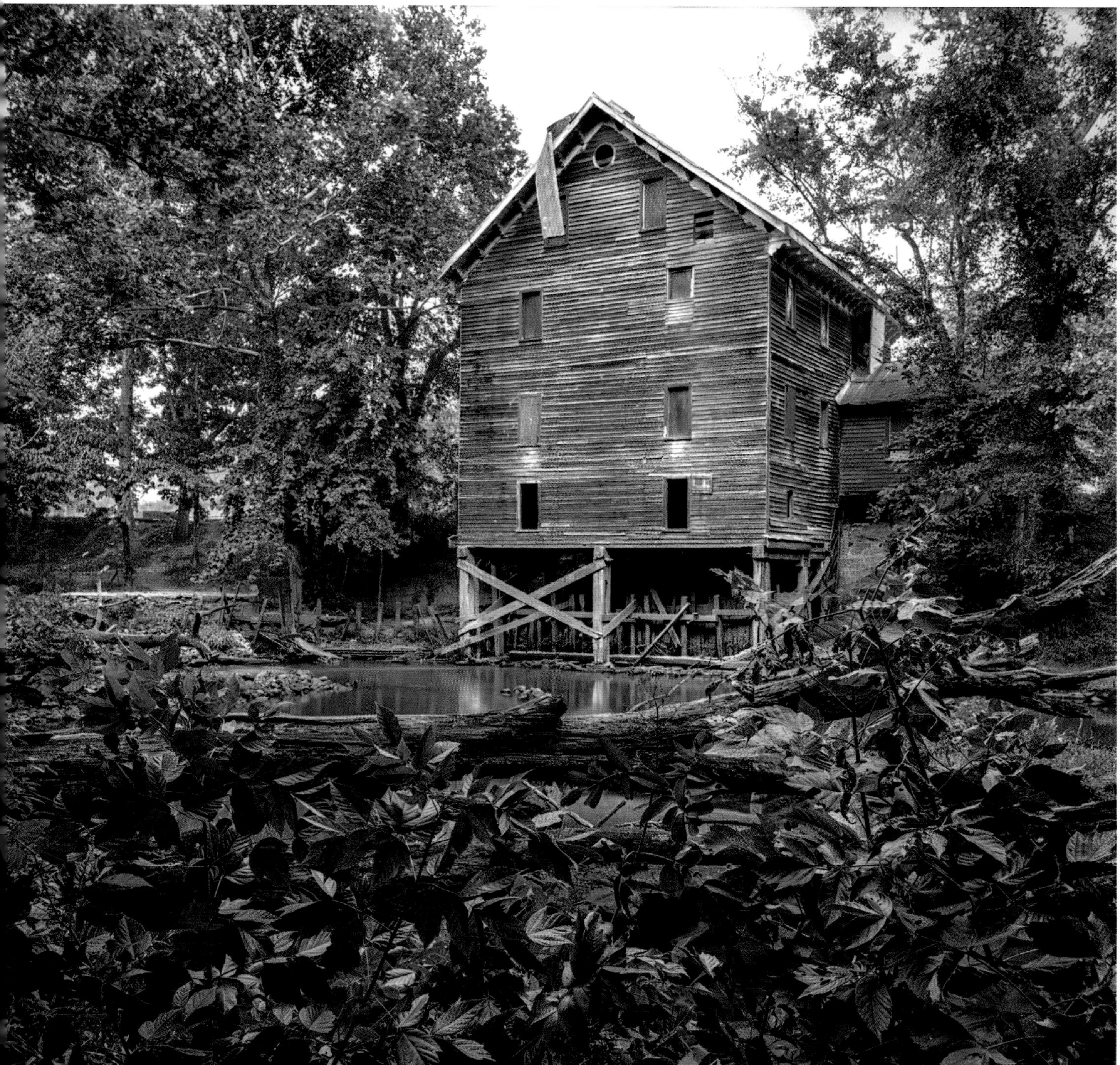

In this photograph by the author, Kymulga Grist Mill is shown in peril during the early 1980s. As is evident in this photograph, the mill had become structurally unsound; tin had blown from the roof, and interior floors had rotted.

Alabama's Historic Textile Watermills, 1800s

Tallassee Mills, Tallassee, Alabama, 1844; Wehadkee Yarn Mills, Rock Mills, Alabama, 1881

TEXTILE MILLS began to appear in the South in the early 1800s, but at first these mills were small and produced materials that required additional finishing, often in New England. However, Southern investors and industrialists quickly saw the advantages of establishing end-to-end textile production in the South, where cotton was locally available and already king of the agricultural economy.

In the years leading up to the Civil War, Southern textile mills grew rapidly, as thousands of poor sharecropping families saw life in a mill town as a better option than hard, impoverished farm life. Enslaved laborers were also a part of the growing watermill textile industry.

In terms of technology, the textile mill introduced an entirely new scale to the size and power requirements of a watermill. Unlike a waterwheel at a gristmill or sawmill that turned simple stones or actuated a saw blade, the much more powerful waterwheel at a textile mill might operate dozens of machines on several floors of a building to manage the various steps in textile production, from ginning or carding to spinning, weaving, and sewing.

Because massive waterwheels—and later powerful water turbines—were needed to drive the textile mill machinery, larger streams and rivers were necessary for their operation. Many textile mills in the Carolinas, Georgia, and Alabama were located along the fall line where the Piedmont Plateau geographic region meets the Coastal Plain. Waterfalls and rapids were common along this abrupt transition in terrain elevation, where the fall and flow of the water provided the tremendous power needed for running large mills.

As mills were built, mill towns grew up along the waterways, often under the sponsorship of the wealthy mill owners. The town of Tallassee, Alabama, is a classic example. A little further east toward Georgia, on the rushing streams feeding the Chattahoochee River, were additional textile operations, such as Wehadkee Yarn Mills. The town there was known as Rock Mills.

As the mills thrived, the demand for labor grew. Some of the jobs were in tight spaces that were well suited for small workers, especially children. In 1909, the State of Alabama passed a major child labor law in an attempt to eliminate or reduce the use of child laborers in the Tallassee Mills and in similar mills and factories across the state. The law prohibited children under the age of twelve from working in mills and factories, ensuring that they could attend school. Children twelve and over could still be mill workers, however.

Waterpower continued as a driving economic force in Alabama until the transition to steam and internal combustion engines, and ultimately to electric motors. Into the 1980s, textile mills such as the Tallassee Mills and Wehadkee Yarn Mills were large employers and important industries in the South. However, beginning in the 1970s, Southern textile mills fell victim to more productive and automated competition in Asia, Mexico, and Canada.

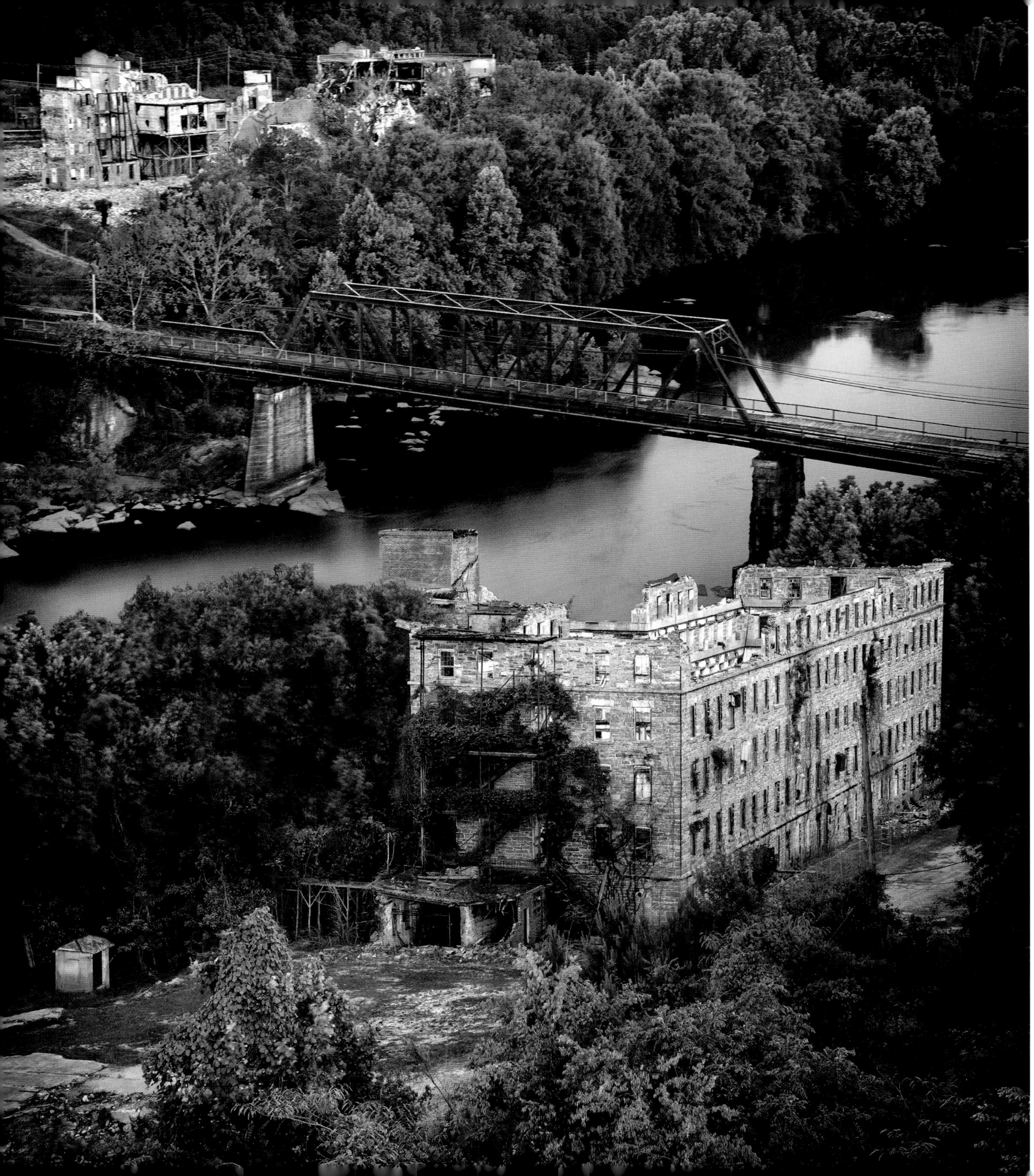

The Tallassee Mills in Tallassee, Alabama, began operation along the Great Falls of the Tallapoosa River in 1844 and expanded in the 1890s and the 1920s with the mills seen in the background across the river, all now in ruins. These mills would become the oldest continuously operating textile milling facility in the United States before finally closing in 2005. The mills served the armed forces in three major wars—the Civil War, World War I, and World War II. In 1943, 1944, and 1945, the Tallassee Mills received the Army-Navy "E" Award from the U.S. government for excellence and efficiency in the production of war materials.

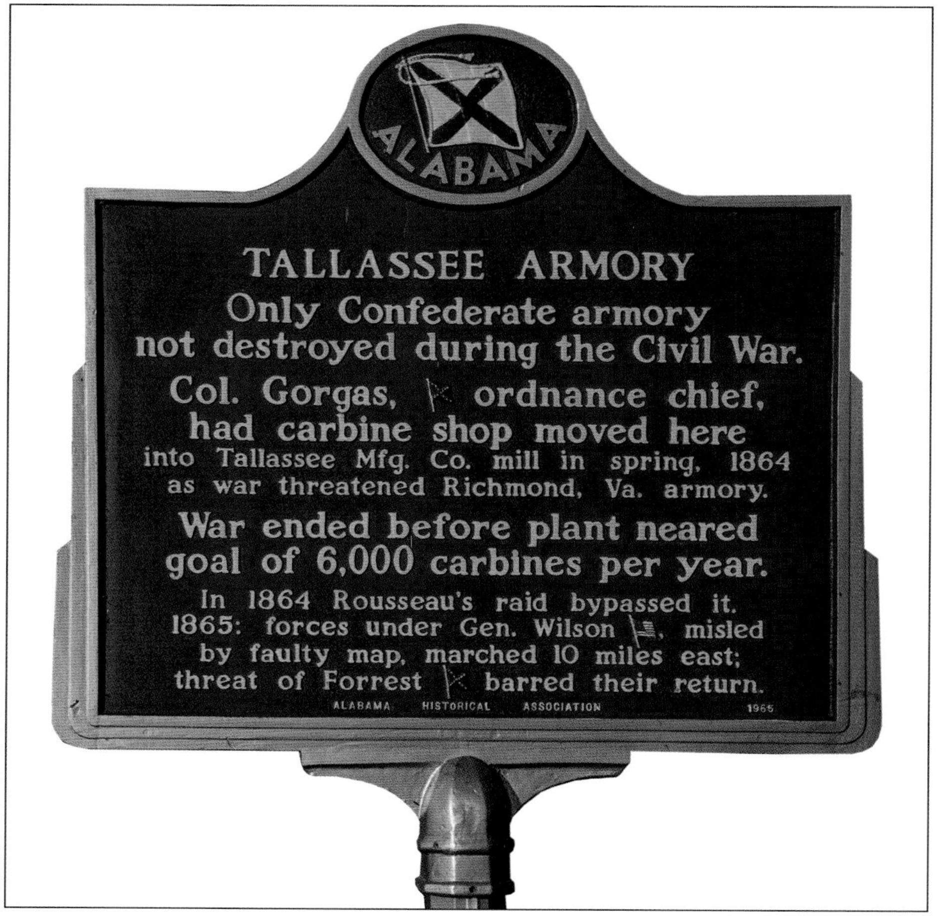

The Tallassee Mills were an invaluable source of clothing, blankets, tents, and other materials for the Confederate army throughout the Civil War. This historic marker adjacent to the Tallapoosa River at the site of the mills commemorates the armory operation at the site. The structure that housed the armory has been preserved and may serve as a future museum.

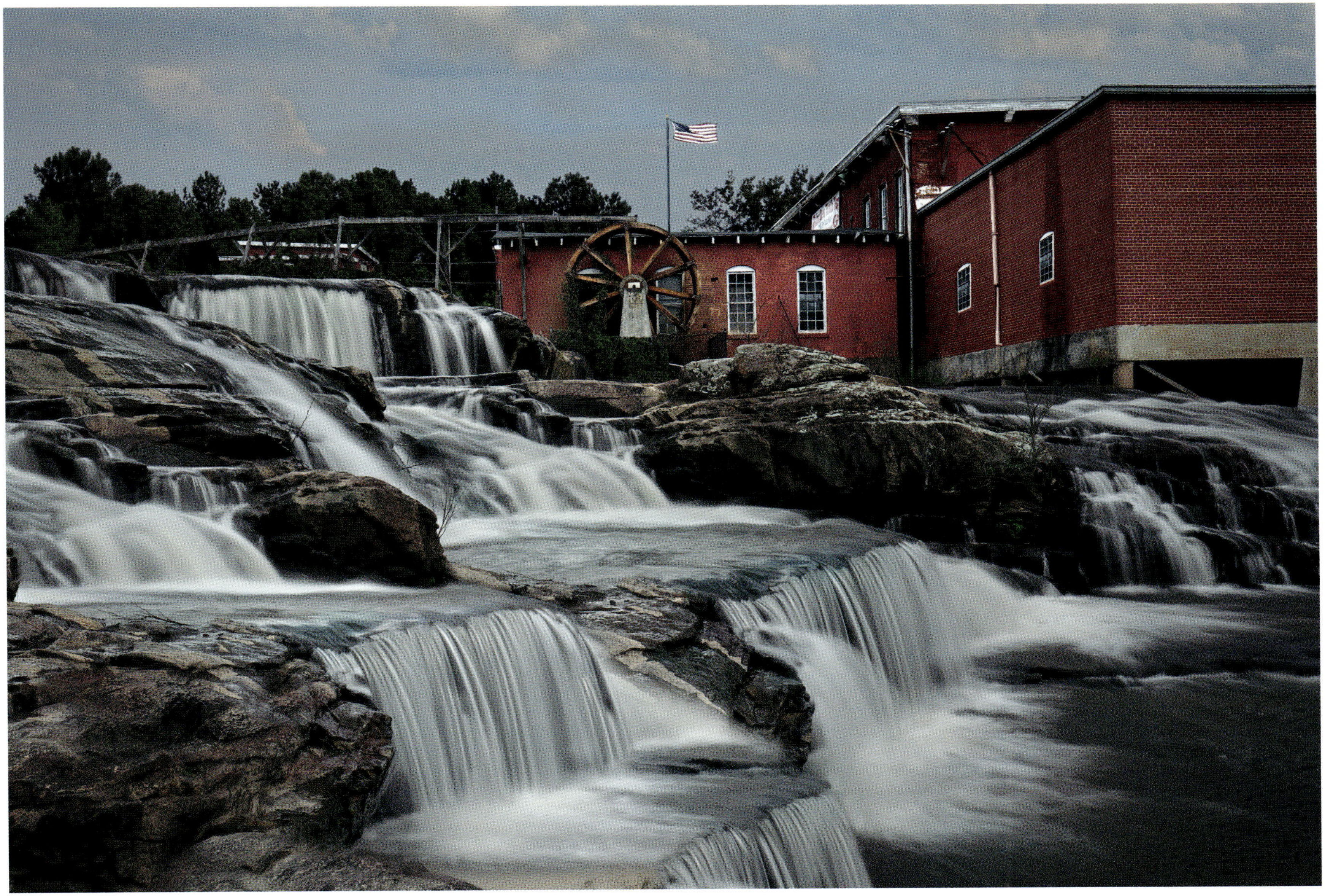

Wehadkee Yarn Mills, on the banks of Wehadkee Creek in Rock Mills, Alabama, dates to 1881; the mill was founded on a site that had been home to Indigenous people for centuries before the 1832 Treaty of Cusseta. A sawmill and blacksmith shop first began operations here in 1837. This image shows the power of the waterfalls at the transition from the Piedmont Plateau to the Coastal Plain. The mill, with offices in nearby West Point, Georgia, produced 100 percent cotton yarn. As happened for many other Southern textile mills, work dwindled in the late twentieth century, until the mill closed in the face of more modern, automated, and productive global competition.

Kenan's Mill

Selma, Alabama, 1861 / Restoration in Early 2000s

KENAN'S MILL is located on Valley Creek in a beautifully oak-wooded, Spanish-moss-draped forest just north of Selma. The turbine-powered mill is a rare fully operational treasure for the Alabama Black Belt region.

The mill served the community by grinding corn and other grains for more than one hundred years until it closed in 1968. A miller lived on the property until that year. As with many Southern watermills, the closure of Kenan's Mill was primarily a result of regional lifestyle changes that came with the rise of modern supermarkets.

Records indicate that a mill was probably located on the site as early as the 1830s. The Kenan family arrived in the area around the time the Civil War broke out and constructed the current mill. The family owned the property until 1997, when Elizabeth Kenan Buchanan donated the site to the Selma–Dallas County Historic Preservation Society with an understanding that it would be used for educational purposes and community events.

The Kenan's Mill site has a colorful past. Southern mills were often the hub of local activities, and this mill was no exception. The property was always bustling and was even home to an early water park. Back in 1904, Owen Kenan not only made ongoing improvements to the mill but also added a swimming pool, a dance floor, and a picnic island. As is too often the story, a raging flood in 1939 along rapidly flowing Valley Creek destroyed most of the additional buildings and even washed the island away. Fortunately, the mill survived.

The mill structure is a substantial wood-framed construction on a concrete and stone arch foundation. The exterior has never been painted. The iron turbine is installed just below the millstone floor, and an underground tunnel delivers water from the creek to the turbine. Two sets of forty-eight-inch millstones are housed in and above an elevated platform on the main level of the mill.

The mill welcomed both Black and White farmers and their families. Inside the mill, Black and White millers worked at the millstones and maintained the structure, dam, and site.

Turbines usually require less maintenance than waterwheels, but when major repair and rework are required, it is often a significant project. In 2000, Kenan's heavy turbine was removed from beneath the mill and trucked to Covington, Georgia, for restoration. The turbine was gone for eighteen months before it was returned for reinstallation.

By March 2002, the mill was back in full operating order, and the first corn was ground. Corn grinding demonstrations have since been a feature at special events on the site, as stipulated by Elizabeth Buchanan when she donated the mill. Since the onset of the COVID pandemic, and after several years of low water in the autumn season, scheduled grinding has been a challenge, but the mill remains in good working order.

The mill grounds include the original gristmill, a large nineteenth-century brick charcoal kiln, an early 1900s miller's cottage, an iron bridge, and a swinging bridge over Valley Creek. Restoration and maintenance work at the site are continuing, and the mill is currently open on request and for special events.

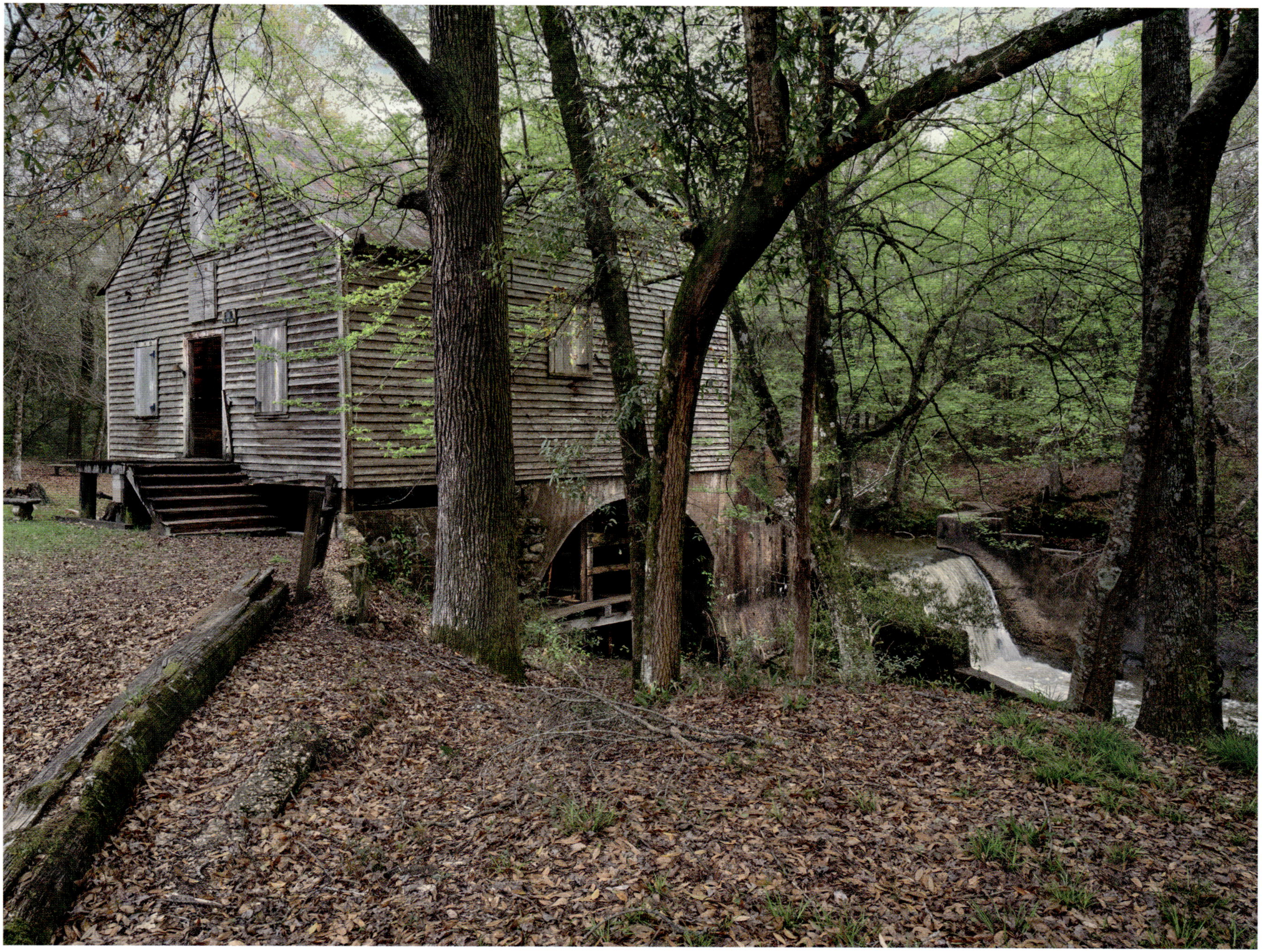

This image of Kenan's Mill shows the front entrance, the archway beneath the mill where the turbine is housed, and the concrete dam across Valley Creek.

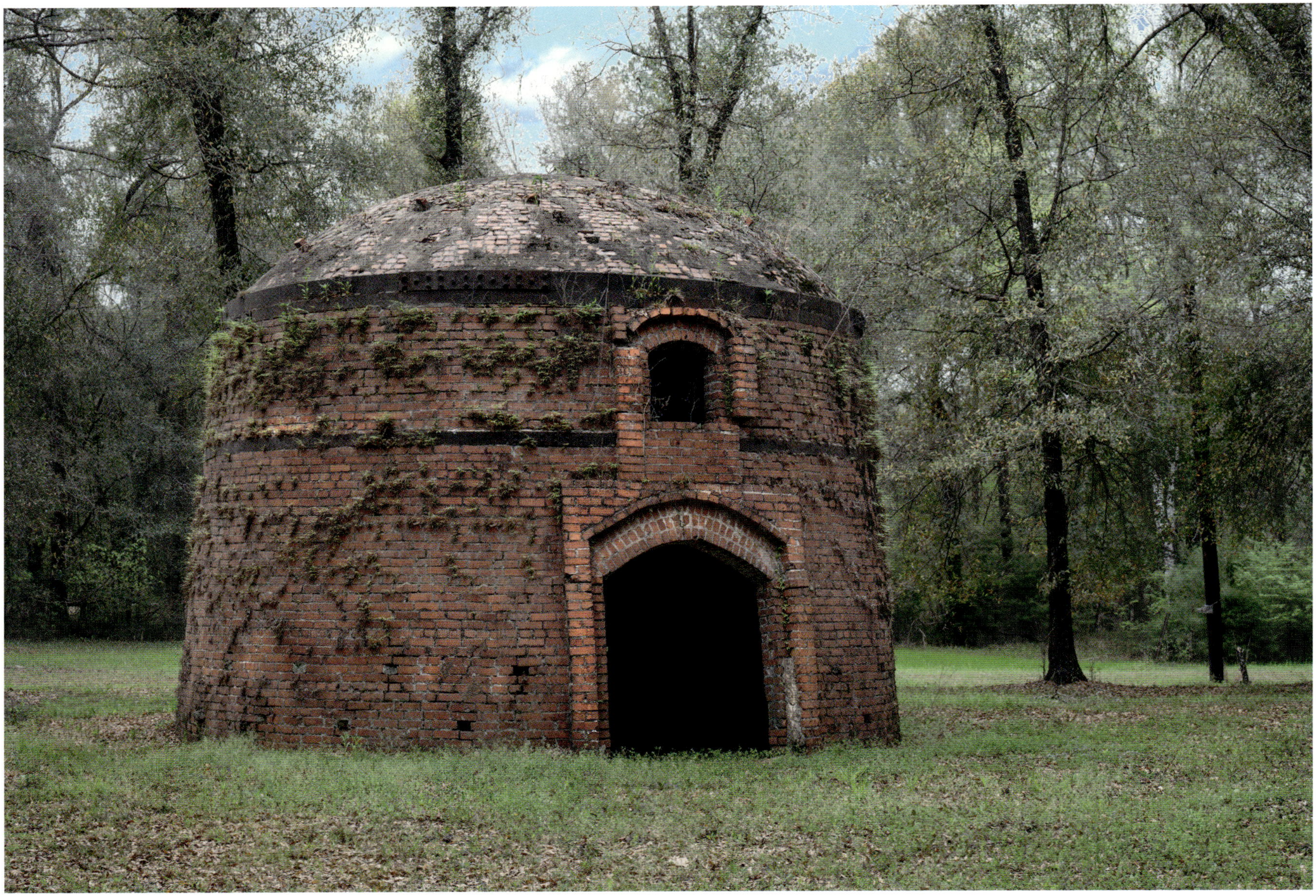

This large brick "beehive" charcoal kiln is a rare example of historic preservation on the Kenan's Mill property. Charcoal was a crucial fuel for iron furnaces during the 1800s.

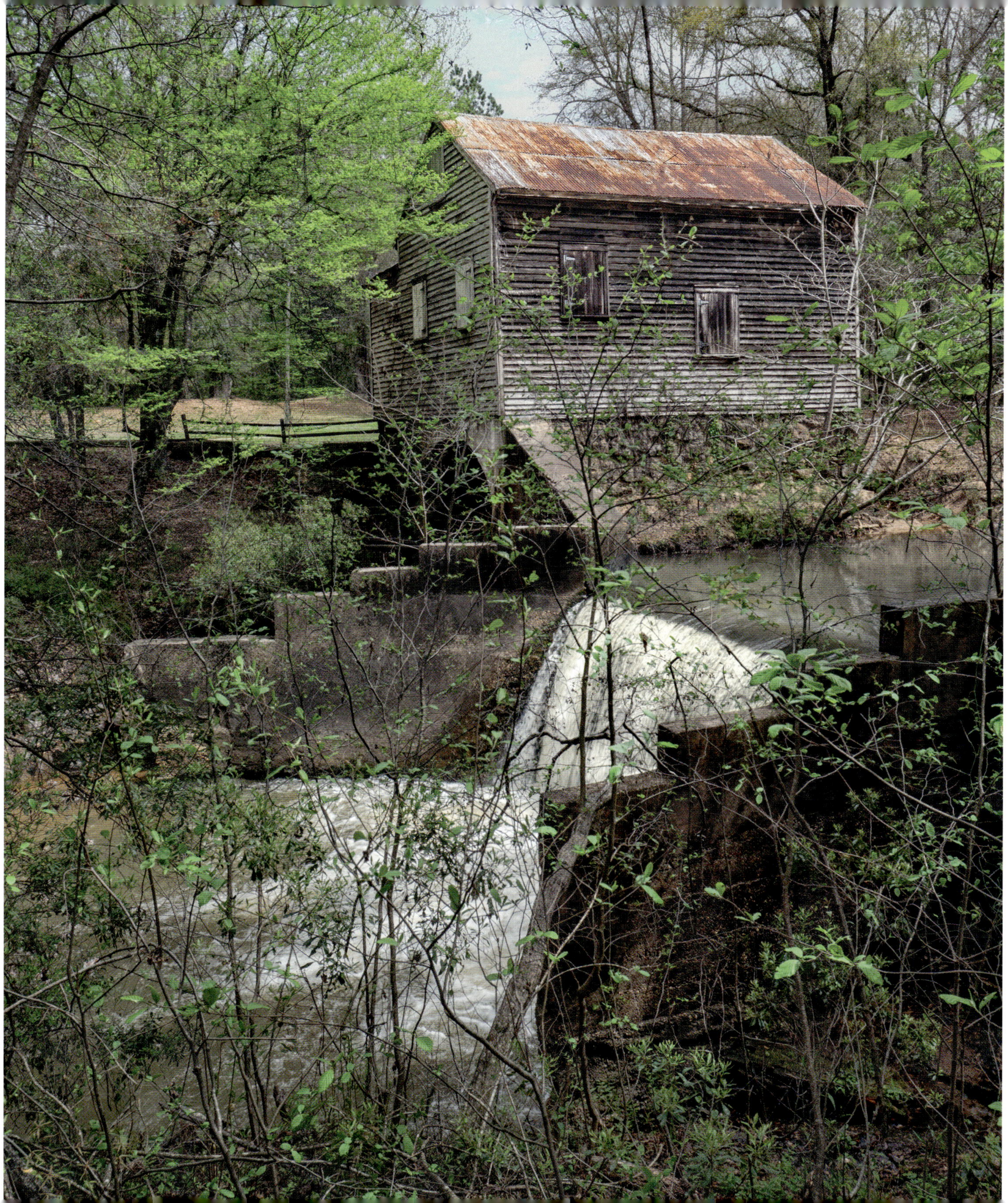

Valley Creek is a rushing, notably high-energy stream in the relatively flat Black Belt region of Alabama. The creek provides an ideal site for a mill this far south.

Johnson Mill

Johnson, Arkansas, 1867 / 1990s

THOUGHT TO BE the oldest business in Arkansas, Johnson Mill was originally located in a rural setting, long before anyone could have imagined that the largest retailer in the United States would establish its international headquarters in nearby Bentonville. As is the case for many remaining old Southern watermills, commercial development now surrounds this property in northwest Arkansas.

The mill is a large wood-framed structure and is two and a half stories tall. It was always powered by a waterwheel. Over the years, the mill was known variously as Truesdale Mill, Sutton Mill, and Spring Mill. The business was a cornmeal-only operation; it never produced flour from wheat. Some watermills located as far south as Johnson Mill did grind wheat-like grass seed, but corn was and still is the most important cereal crop in northwest Arkansas.

The current mill was erected in 1867. An even earlier mill was located at the site in 1835, and some records suggest a mill may have stood here as early as 1805. The 1835 mill was partially burned during the Civil War following the nearby Battle of Pea Ridge. In 1887, a large brick house was constructed adjacent to the mill, and it remains a part of the mill property. The bricks were locally fired.

The original thirty-foot waterwheel was overshot, but a massive and relatively rare thirty-foot wooden pitchback wheel was installed further toward the front of the mill building during restoration in the 1990s. That wheel is no longer operational, but it creates a striking visual image for the structure.

Today the mill displays historic milling equipment inside and outside the structure and has been converted into an award-winning hotel. Guest rooms at the Inn at the Mill are fashioned in the styles of various artists and architects, such as Claude Monet and Frank Lloyd Wright.

Like many historic Southern watermills, this complex is listed on the National Register of Historic Places.

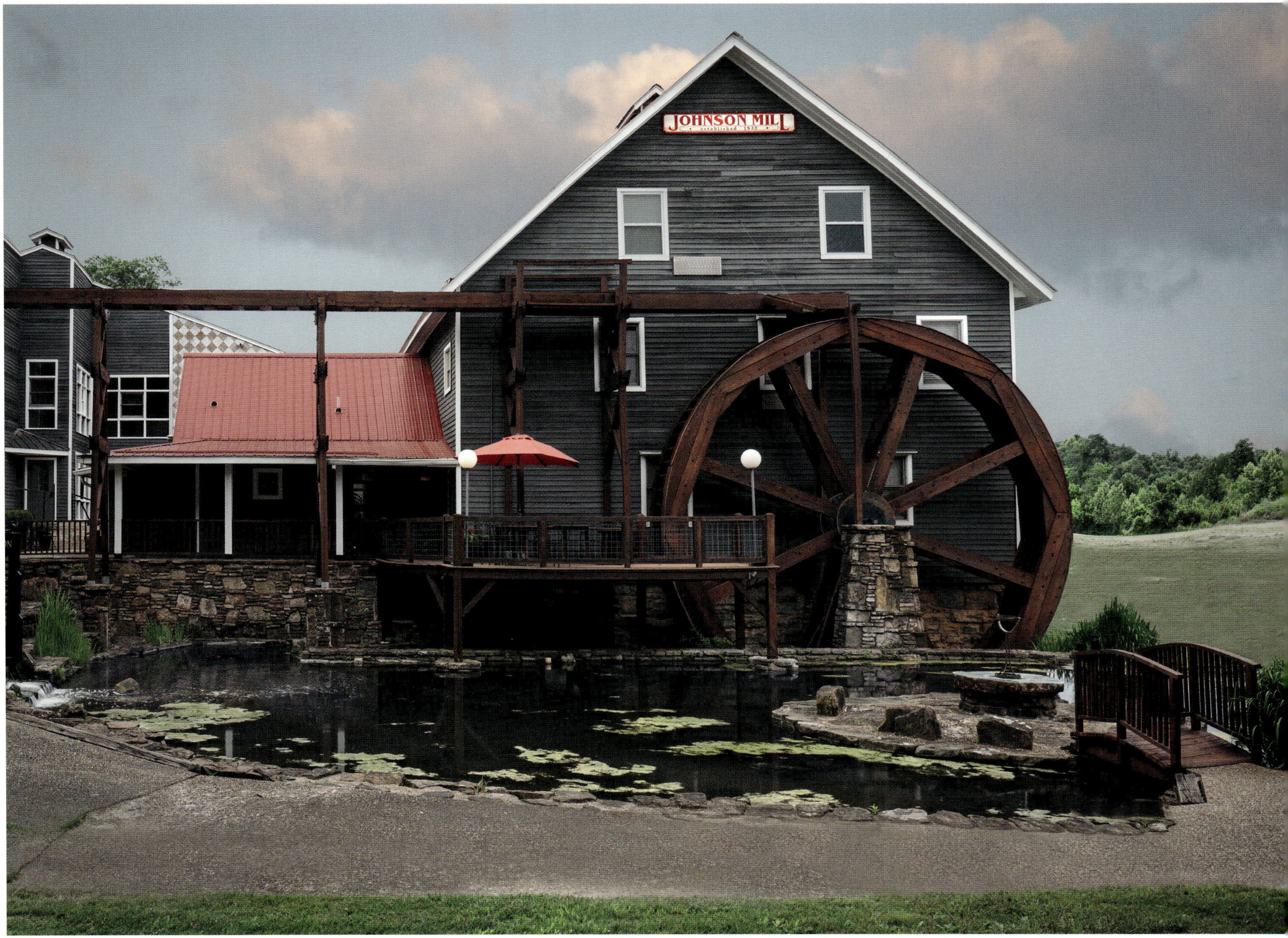

Johnson Mill, with its thirty-foot wooden waterwheel, is beautifully restored and now serves as an award-winning hotel.

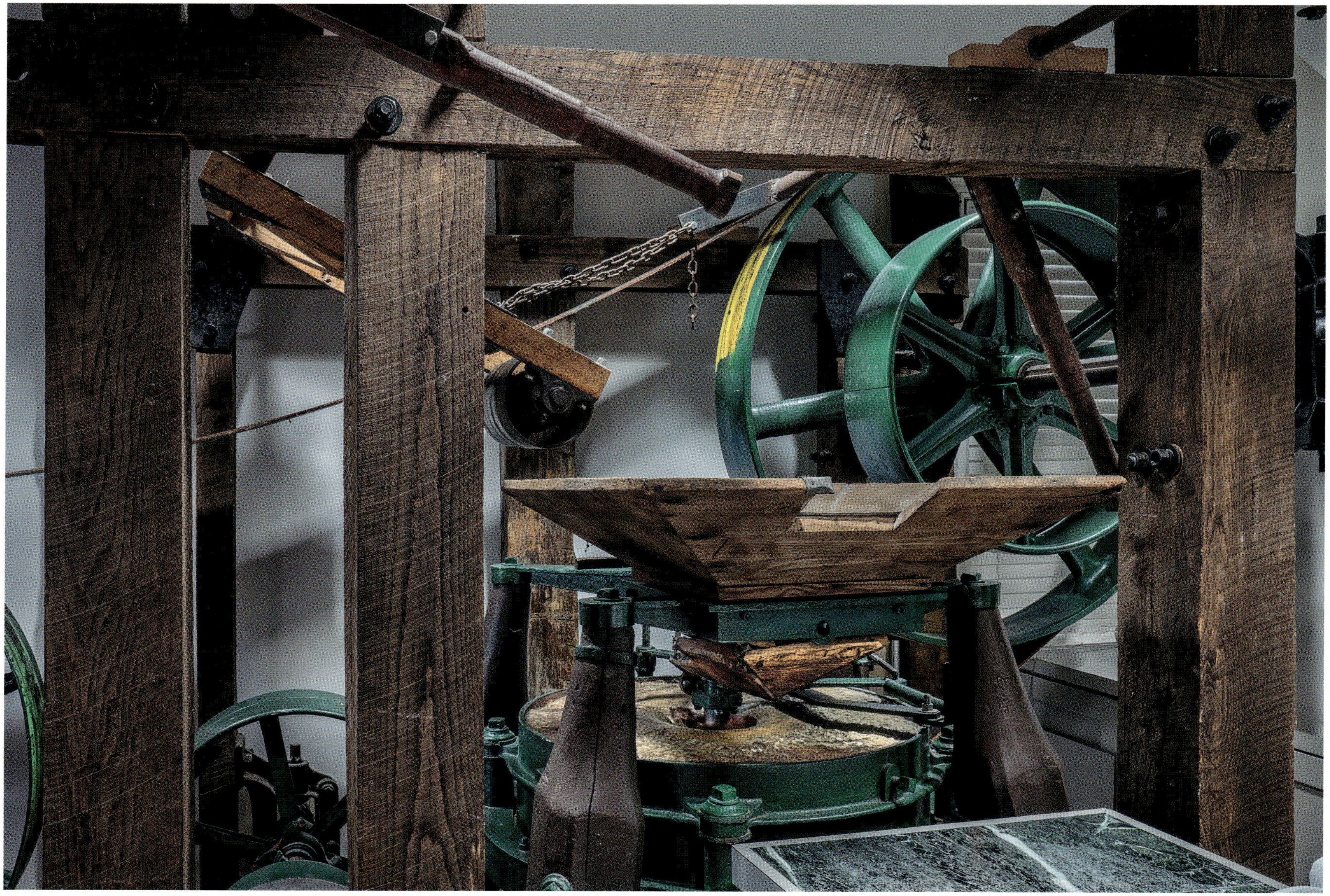

Inside Johnson Mill, vintage millstones, a wooden hopper, belts, pulleys, and other linkages and equipment have been restored to museum display condition.

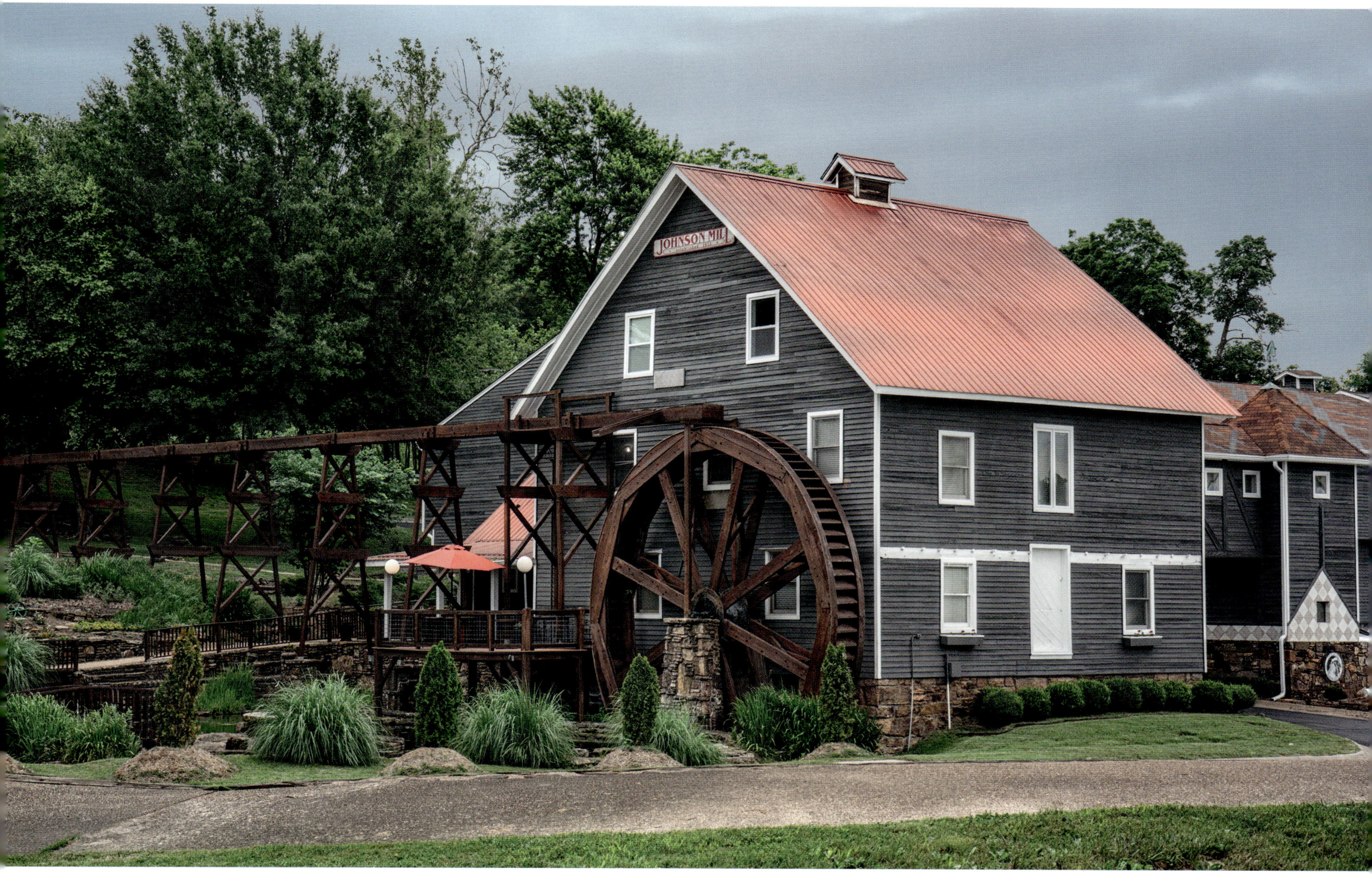

This view of Johnson Mill from across the nearby road shows some of the matching guest rooms that have been added to the inn.

Pugh's Mill (the *Gone with the Wind* Mill)

North Little Rock, Arkansas, 1933

THE OLD PUGH'S MILL has an especially interesting backstory. This is the mill that appears in the opening scenes of one of the most famous movies in history, *Gone with the Wind*. Millions of people have seen the movie and the image of the mill.

The mill dates to 1933 and was constructed as a replica of an earlier 1820s mill on the site. It is primarily a stone and concrete structure and was never intended to be an operational watermill. The replica mill and the landscaping were designed by architect Frank Carmean.

Artist and architect Dionicio Rodríguez sculpted the concrete work. The waterwheel, handrails, and other details are made of concrete that has been formed to look like wood. Throughout the park, trees, bushes, statues, walking bridges, and other displays are all made of concrete! As a relief from all the concrete and stone, wood is used inside for the millstone frame.

The concept was for the mill to look like a ruin from the pioneer era in Arkansas. This, it was thought, would visually tell the story of the loss of such an important period and technology in Southern history. The inclusion of the mill in *Gone with the Wind* certainly reinforced the concept.

The mill also serves as a memorial to Thomas R. Pugh, and when the wheel turns, it is said to symbolize Pugh's eternal energy. The structure was renovated and refreshed in 1991 by Carlos Cortés, a descendant of Dionicio Rodríguez.

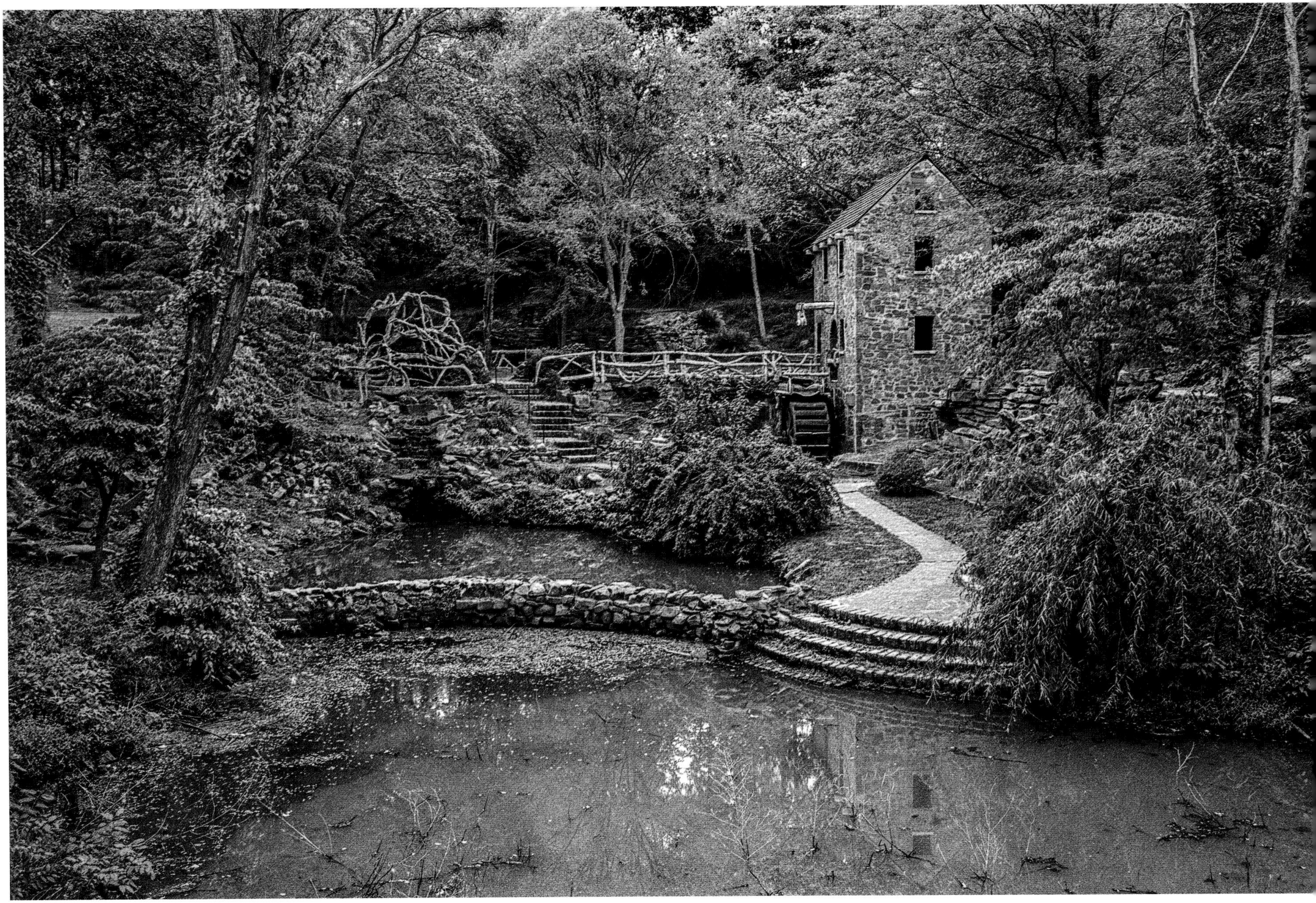

Pugh's Mill is located in a popular park and is adjacent to a spring just above sixty-acre Lakewood Lake Number Two.

War Eagle Mill

Rogers, Arkansas, 1832 / 1973

WAR EAGLE MILL is said to be the only commercially operating mill in the United States that is powered by an undershot waterwheel. As clarification, a number of other mills are still fitted with undershot wheels, but if they operate, it is only for the occasional demonstration.

The current mill was built in 1973 and is the fourth mill on the site. The first mill was erected in 1832 by a young couple from Tennessee, Sylvanus and Catherine Blackburn. That mill washed out in an 1848 flood that destroyed almost everything along the War Eagle Creek Valley.

The couple rebuilt the mill, this time expanding the structure so that it could operate not only as a gristmill but also as a sawmill. This second mill was burned early in the Civil War.

After the war, the family's son constructed the third mill at the site, finishing the project in 1873. By this time, technology had advanced to the point that a turbine was used instead of a waterwheel. The turbine powered both a gristmill and a large sawmill. Thanks to hard work by the family and others, the sawmill became the largest in northwest Arkansas. Sadly, the mill burned in 1924.

In 1973, one hundred years after the third mill was completed, the current mill rose from the ashes. This time around, an eighteen-foot cypress undershot waterwheel was chosen as the source of waterpower. With the exception of the waterwheel, the fourth mill is modeled after the third mill. Some of the milling equipment is 150 years old.

The mill shares the site with a historic one-lane steel truss bridge built in 1907. The bridge span across War Eagle Creek is 187 feet long, and the structure is floored with wooden planks that go clackety-clack as vehicles pass by the mill.

Today War Eagle Mill conducts daily milling demonstrations and hosts educational tours. The mill includes a large gift shop and a restaurant. It is also home to an annual craft fair and numerous special events throughout the year.

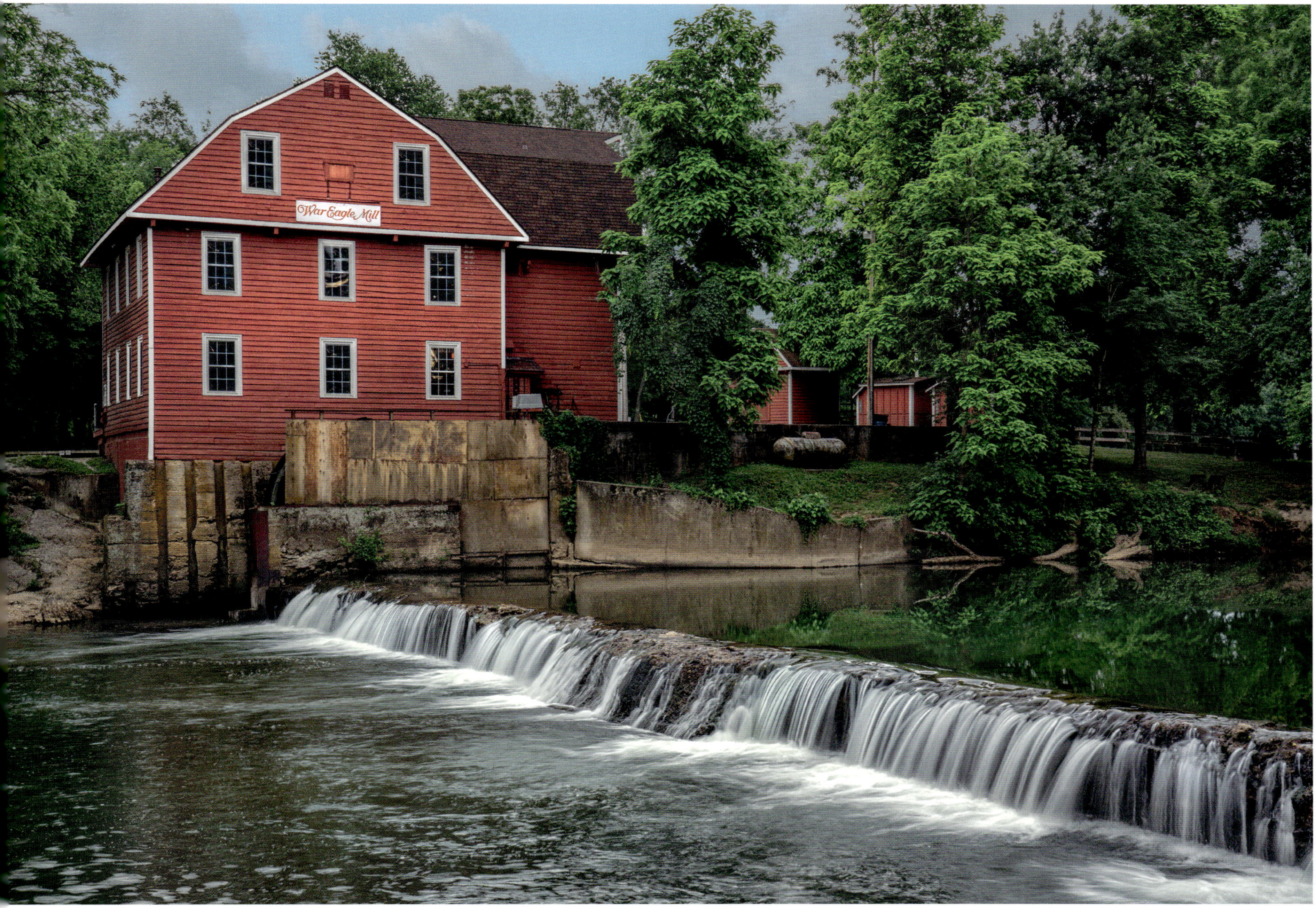

Drawing on an abundant supply of waterpower from War Eagle Creek, this large mill can operate successfully with a low dam and an undershot waterwheel.

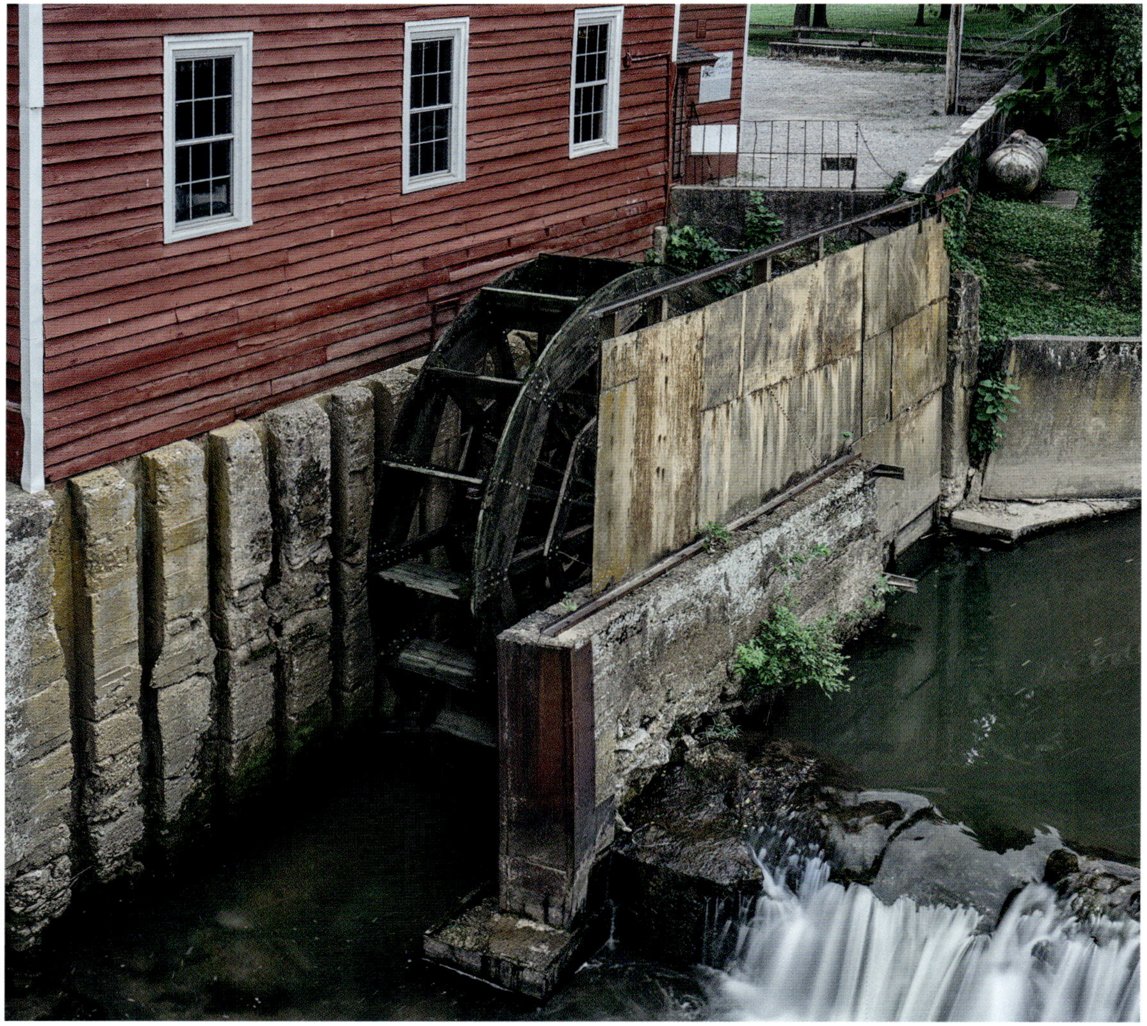

The undershot cypress-wood waterwheel at War Eagle Mill is eighteen feet in diameter.

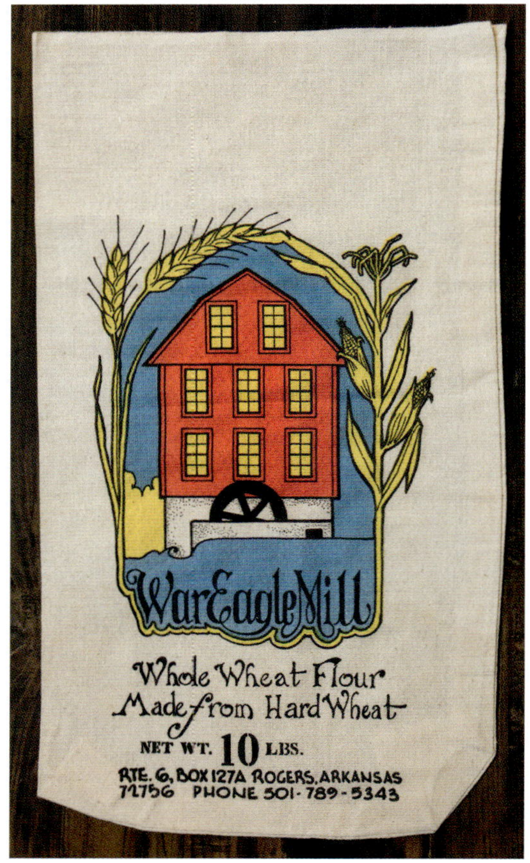

Millers often demonstrated pride in their work by packaging their flour and meal products in custom-printed cloth bags and paper sacks. The printing was sometimes done at the mill with a printing stamp machine.

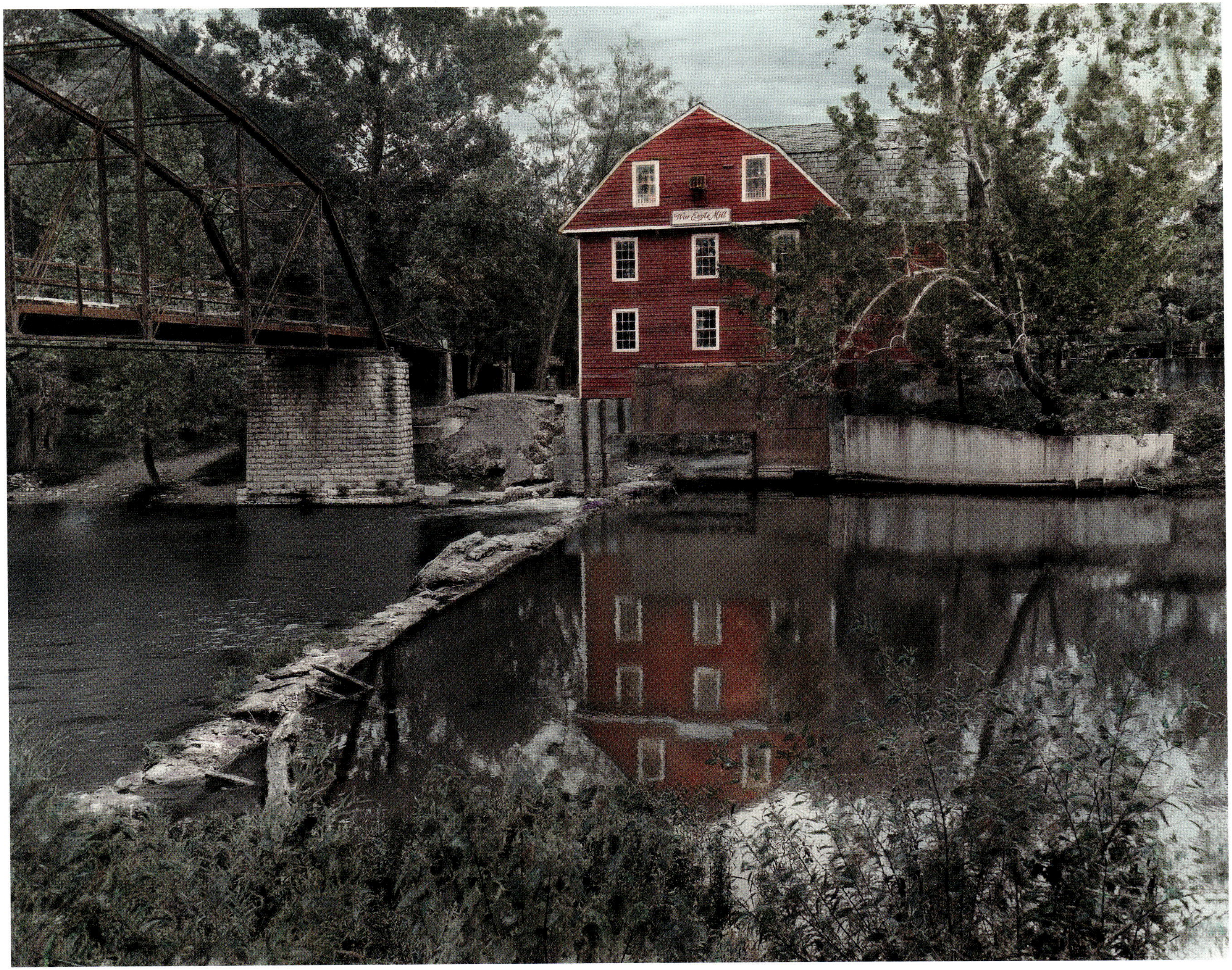

This vintage hand-tinted image of War Eagle Mill dates to just a few years after the 1973 mill was completed. The historic one-lane truss bridge can be seen to the left of the mill and dam.

Peirce Mill

District of Columbia, 1829 / with Several Restorations and Renewals over the Years

WHEN GEORGE WASHINGTON surveyed the area that would become the District of Columbia, he noted that the area had "many fast-flowing streams that could power watermills." Since Washington was a miller himself, mill siting was an important consideration for him.

In the mid-1780s, Isaac and Elizabeth Peirce moved from Pennsylvania to the Rock Creek area of Maryland, north of what would become the National Mall in Washington, D.C. Like George Washington, they were experienced millers, and they purchased 150 acres of land that included a mill and other structures.

In 1829, they replaced the earlier mill along Rock Creek with the current beautiful stone mill (some records date the mill construction to 1820) and began operating it as a gristmill for corn, wheat, and rye on what had by that time become a two-thousand-acre estate. The new mill was based on the Oliver Evans automated mill technology. A carriage barn, distillery, sawmill, springhouse, nursery, and residence were also located nearby on the estate. All were painstakingly constructed by enslaved workers with stones and other materials from nearby quarry operations. The mill continued to operate throughout the nineteenth century, including the Civil War era.

In 1890, Congress created Rock Creek Park in the District of Columbia, and the mill was included as part of the park property. Milling operation ceased in 1897 when a shaft broke and was not repaired. In the 1920s, most of the milling equipment was removed, and the building became a tearoom.

The National Park Service took over the site in 1936 and restored the mill as a Public Works Administration project. A breastshot waterwheel was installed, and machinery from other old mills in Pennsylvania and Maryland was brought in to make the mill operational. For the next twenty years or so, the mill was used to produce flour for area government food services and cafeterias. It continued to run, at least intermittently, until 1958, when it closed.

In 1967, efforts began to restart the mill, and skilled millwrights were brought in to rework the mill and bring back a historically accurate waterwheel. The mill began operating again in 1970, only to be damaged by tropical storms during that decade. The lowest level of the mill has flooded repeatedly over the years, and the water levels for some of the more recent floods are marked with dates on the wooden beams in the basement. During Hurricane Agnes in 1972, the water level reached the main milling floor, rising to a waist-high depth; the floodwaters essentially consumed the entire Rock Creek Park valley leading to the Potomac River. Fortunately, the mill was not washed away downstream and into the river.

After storm damages were repaired, the mill continued to grind corn intermittently until the main driveshaft failed in 1993. The structure and site were restored again beginning in 1997, and the mill reopened to visitors in 2011. Restoration is still ongoing. A highly capable National Park Service staff operates basic equipment in the mill for demonstrations; check the website for schedules.

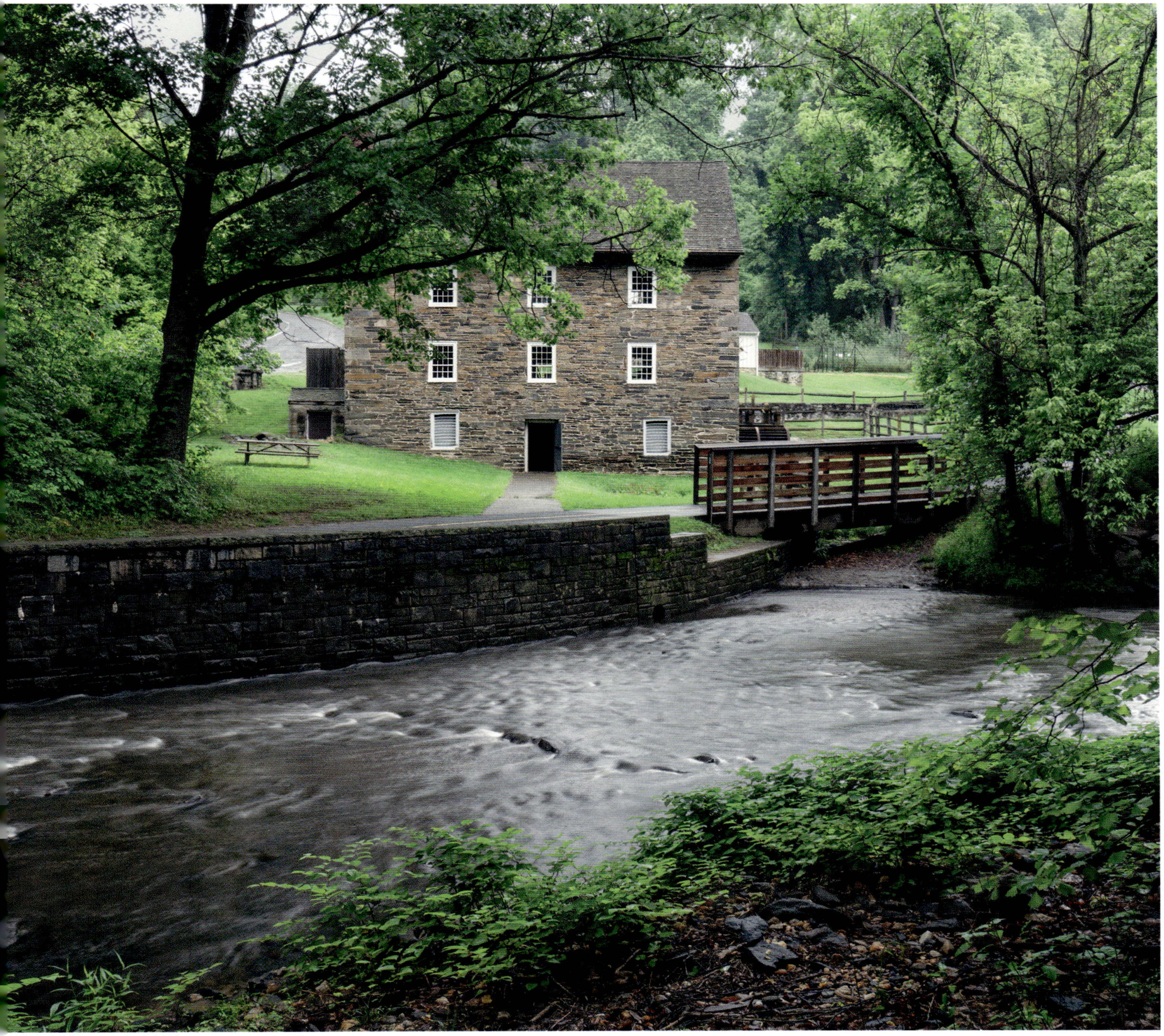

This recent springtime image of Peirce Mill offers a view of the lush site from across Rock Creek.

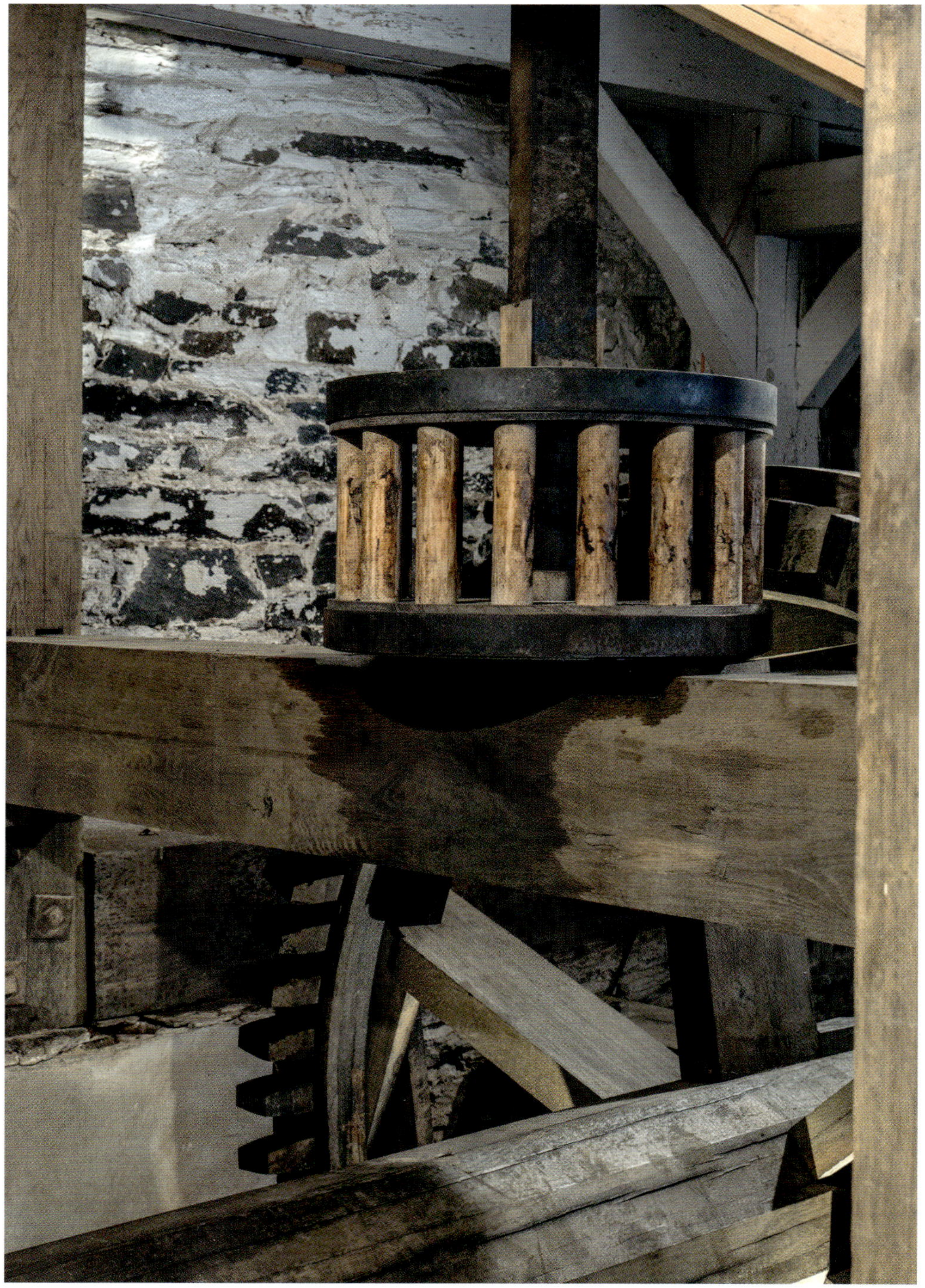

Peirce Mill is operated with classic traditional wooden gears, cogs, and pegs, much as the Peirce millers would have used in the early nineteenth century.

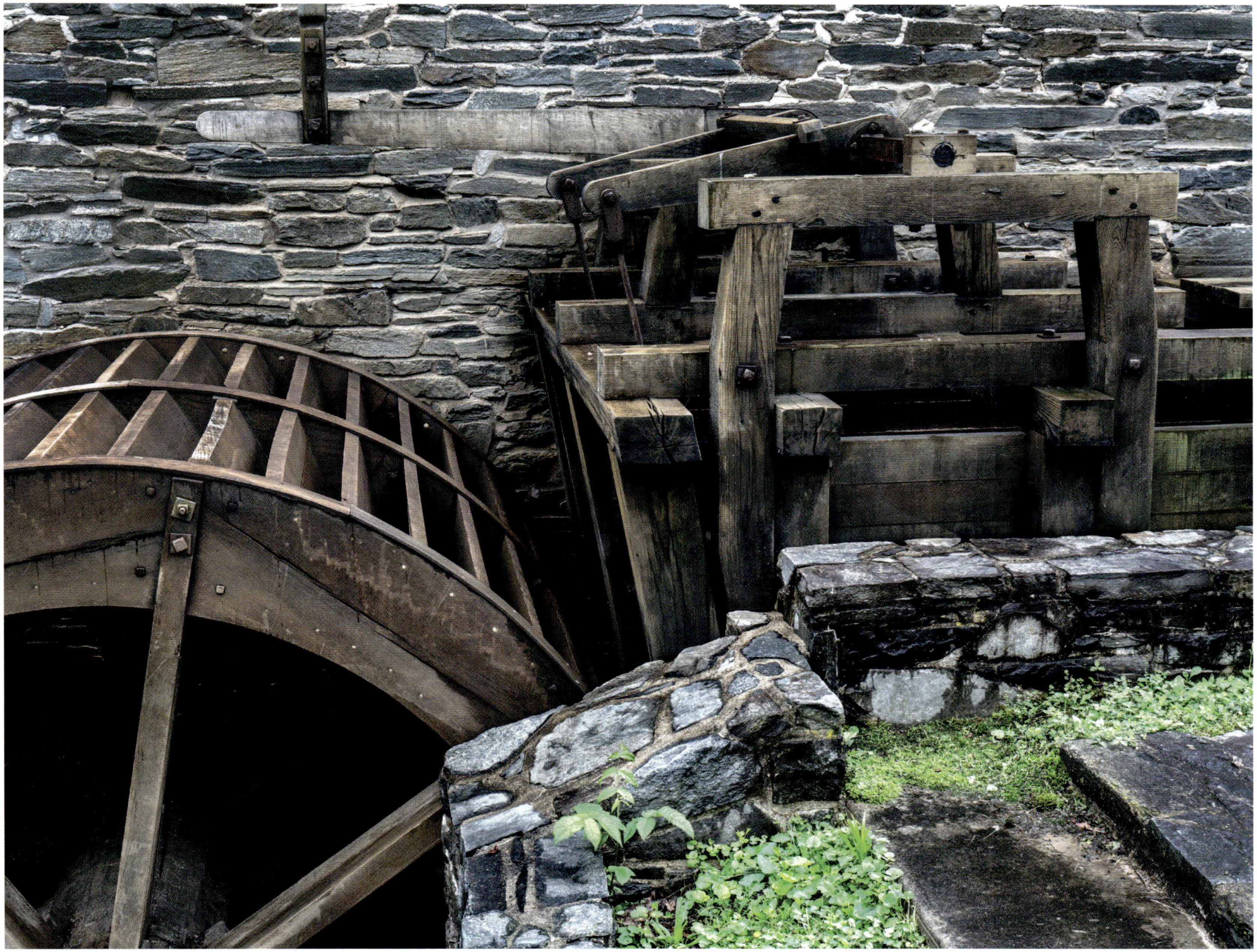

The buckets on the breastshot waterwheel and the water-delivery mechanism and flume can be inspected in this image. The quality and craftsmanship of both the current wheel and the nineteenth-century stonework on the mill's exterior are impressive.

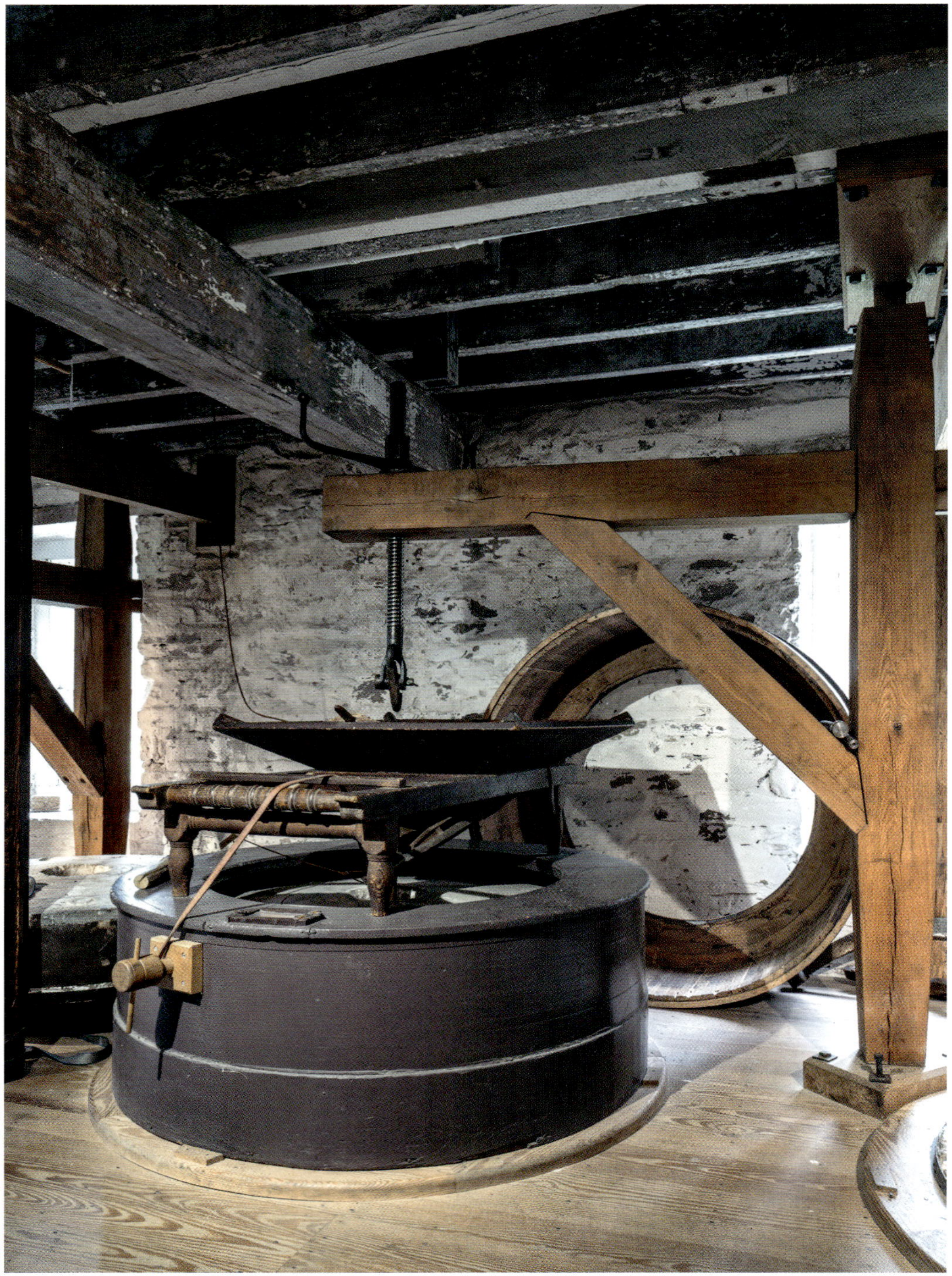

The gray circular housing contains millstones used to grind wheat into flour. The wooden mechanism to the right is a crane that can lift and turn the upper or "runner" millstone for inspection and sharpening.

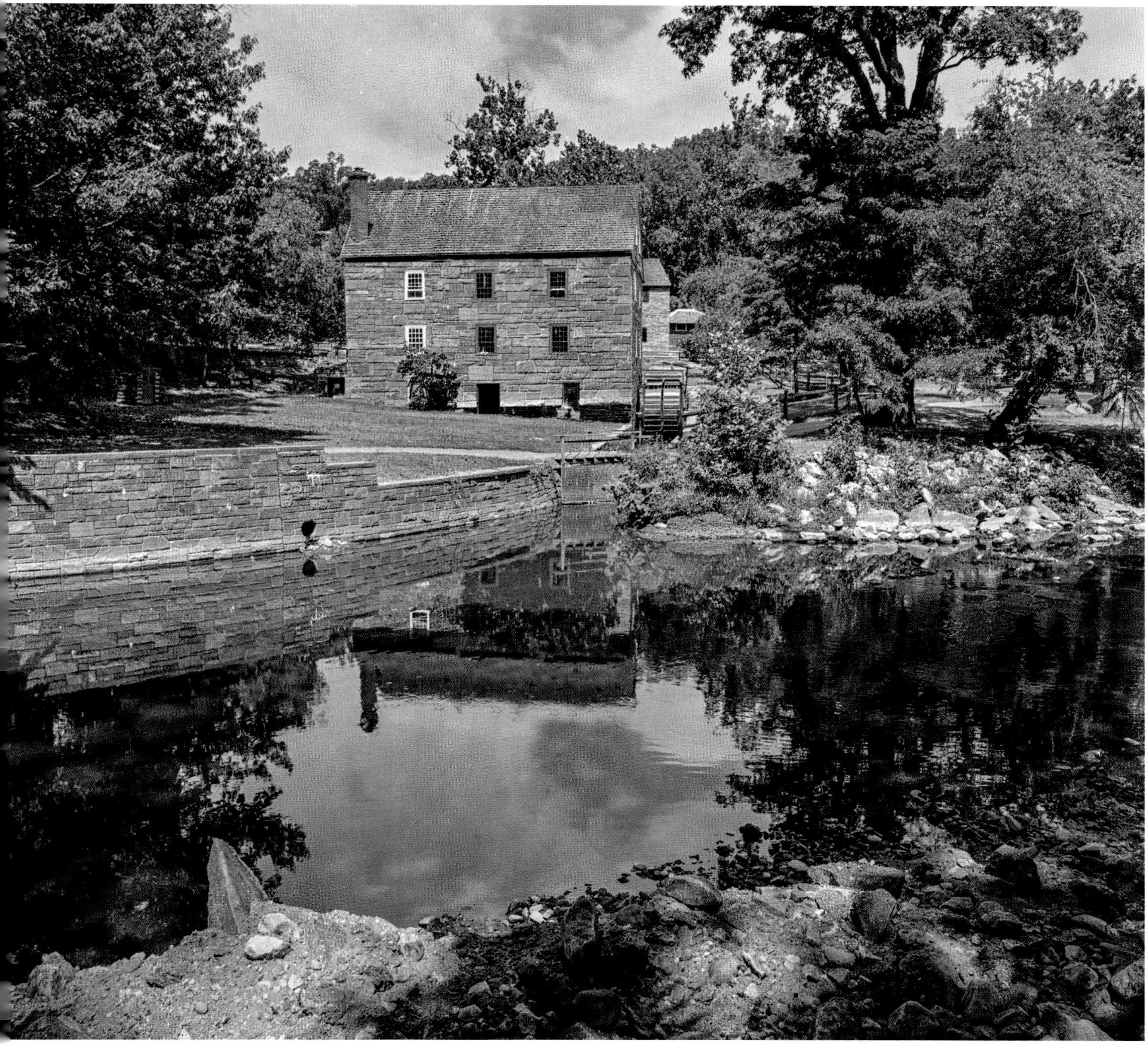

Peirce Mill is shown here in the late 1970s. The breastshot waterwheel is clearly visible to the right of the mill. The mill was not operational at the time this photograph was taken.

Columbus Historic Riverfront Industrial District

Columbus, Georgia, 1828 to the Early Decades of the Twentieth Century

BETWEEN COLUMBUS, GEORGIA, and Phenix City, Alabama, the Chattahoochee River crosses from the higher Piedmont geographical region to the lower Coastal Plain, and the water falls some 125 feet over about a two-and-a-half-mile stretch of the river. The transition is dramatic, with rapids and waterfalls that constantly produce an estimated 100,000 horsepower of energy. This area was recognized early in the Industrial Revolution as being ideal for nineteenth-century water-powered industry.

The modern city of Columbus was built up on the east bank of the river at this location, with Phenix City emerging later on the opposite side. The first project to harness power from the river came in about 1828, when a partial dam was constructed to channel rushing water to a gristmill. In this era, Columbus was still a frontier location on the site of what for years had been a Creek Nation village.

By the 1840s, several watermills were operating along the river. Eagle Mills began cotton mill operations in 1860 and, along with other Columbus industries, provided crucial materials, goods, and services to the Confederacy throughout the Civil War. Tragically, Eagle Mills was burned by the Union Army in 1865, *after* Grant and Lee had signed the peace treaty at Appomattox Court House but *before* the news had reached Georgia. Mill #1 was rebuilt by 1869, and Mill #2 followed in 1872. After the war, Eagle Mills was renamed Eagle and Phenix Mills to commemorate how the mill had risen from the ashes!

In the last half of the nineteenth century, five major water-powered industrial complexes were operating along the river. They were Bibb Mill, City Mills, Muscogee Mills, Columbus Iron Works, and Eagle and Phenix Mills. All these industries pulled power from the waters of the Chattahoochee.

Hydroelectric power came to Columbus in 1882, earlier than in most U.S. cities, as the mills installed electric generators to convert their water-powered machinery to electricity and sell power to residents and businesses in town. As evidenced by the tall stacks in the district today, steam boilers and engines were also used as the technology became available. The mills continued to expand and add new structures well into the twentieth century. Eventually, these operations fell victim one by one to competition from Asia, Mexico, Canada, and other lower-cost producers inside and outside the United States.

The first dam across the entire river, the Eagle and Phenix Dam, was completed in 1882. It was thirty feet high and twenty to thirty feet thick—the proportions necessary to hold back the power of the river—making it the largest masonry dam in the South at the time. In 1907, the City Mills Dam was completed further upstream to replace the 1828 partial dam. These dams were removed in 2012 and 2013 to restore the river's natural white water.

By the late twentieth century, economic markets and competition had evolved to the point that the old mills at Columbus were no longer competitive, and they all ceased operations. In the twenty-first century, work has been ongoing to restore, revitalize, and repurpose the district's historic mills. Commercial and residential real estate developers have repurposed much of the space as homes, hotels, offices, restaurants, and entertainment venues. The intent is to retain much of the historic character and flavor of the old industrial era while assuring a future for one of the South's greatest industrial sites.

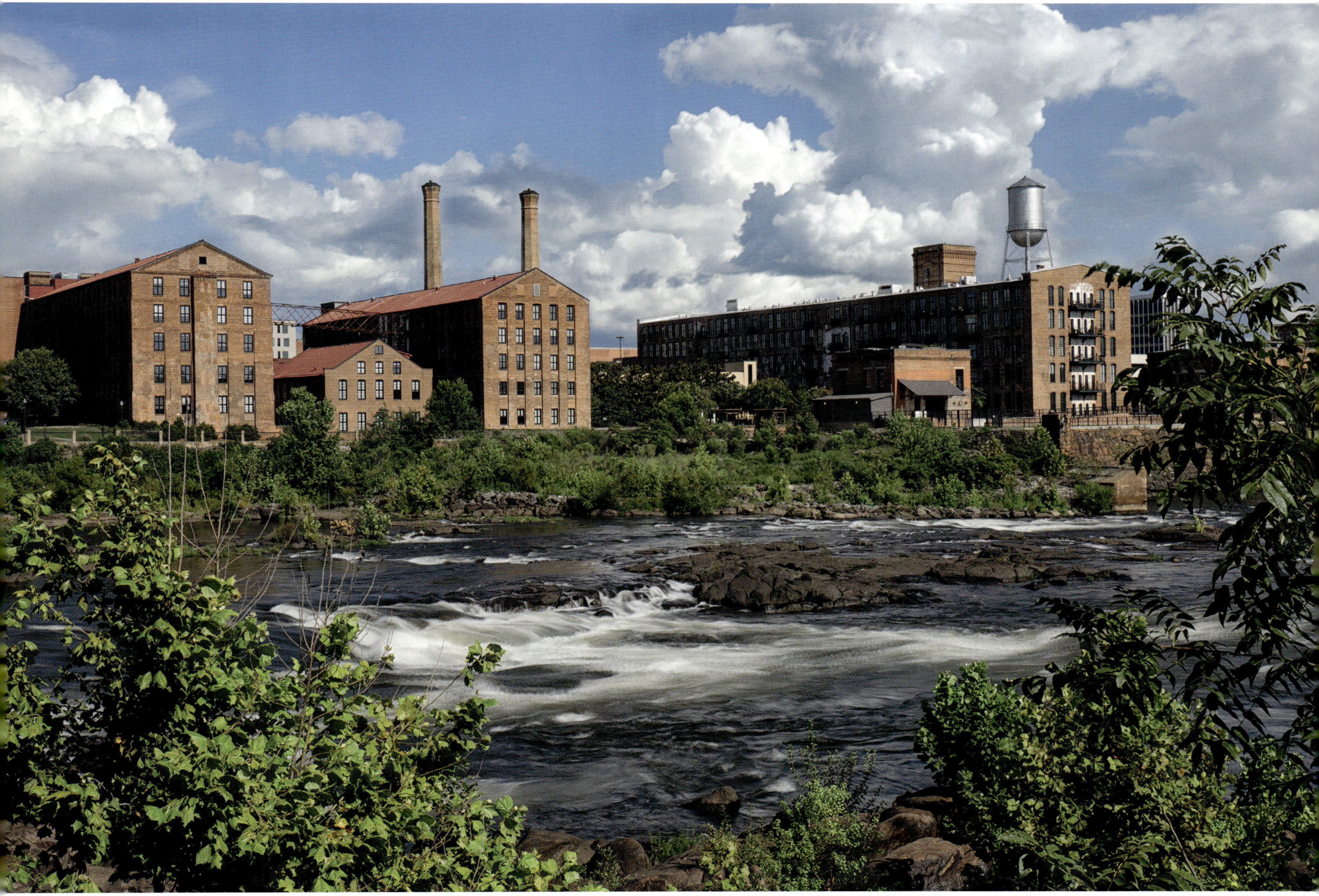

This is a partial view of the Columbus Historic Riverfront Industrial District, as observed from across the white water of the Chattahoochee River at Phenix City, Alabama.

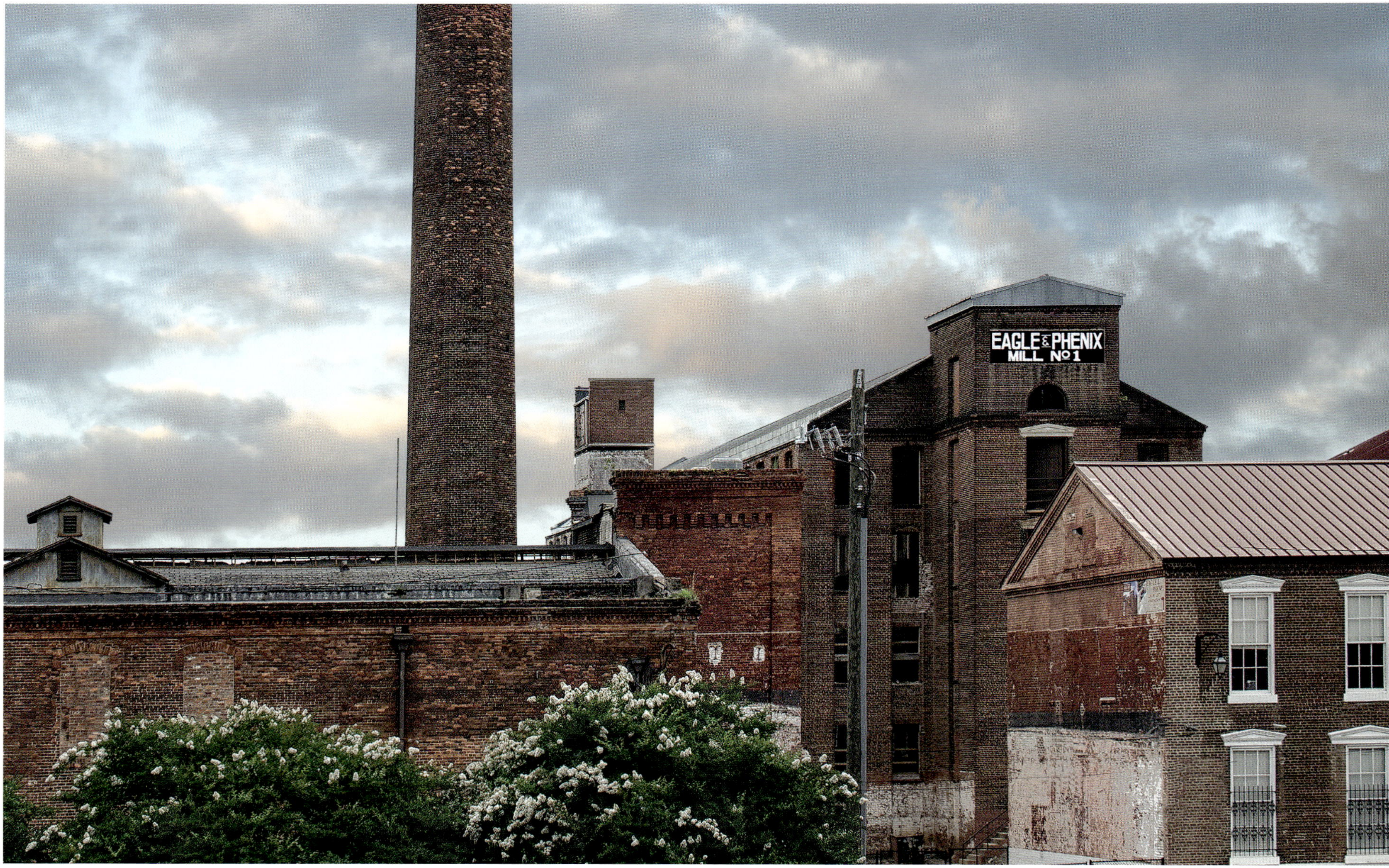

Eagle and Phenix Mill #1 dates to 1869 but was updated a number of times over the following century. Today this historic structure has been given new life through residential and commercial redevelopment projects.

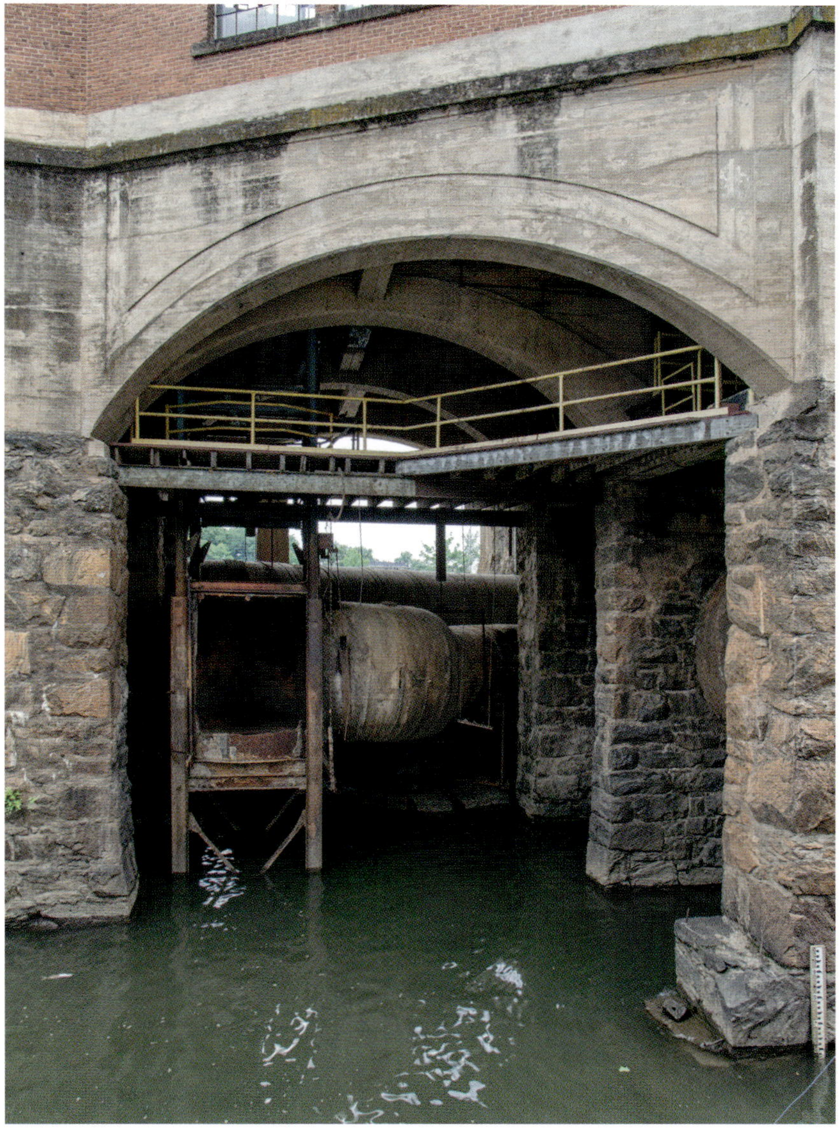

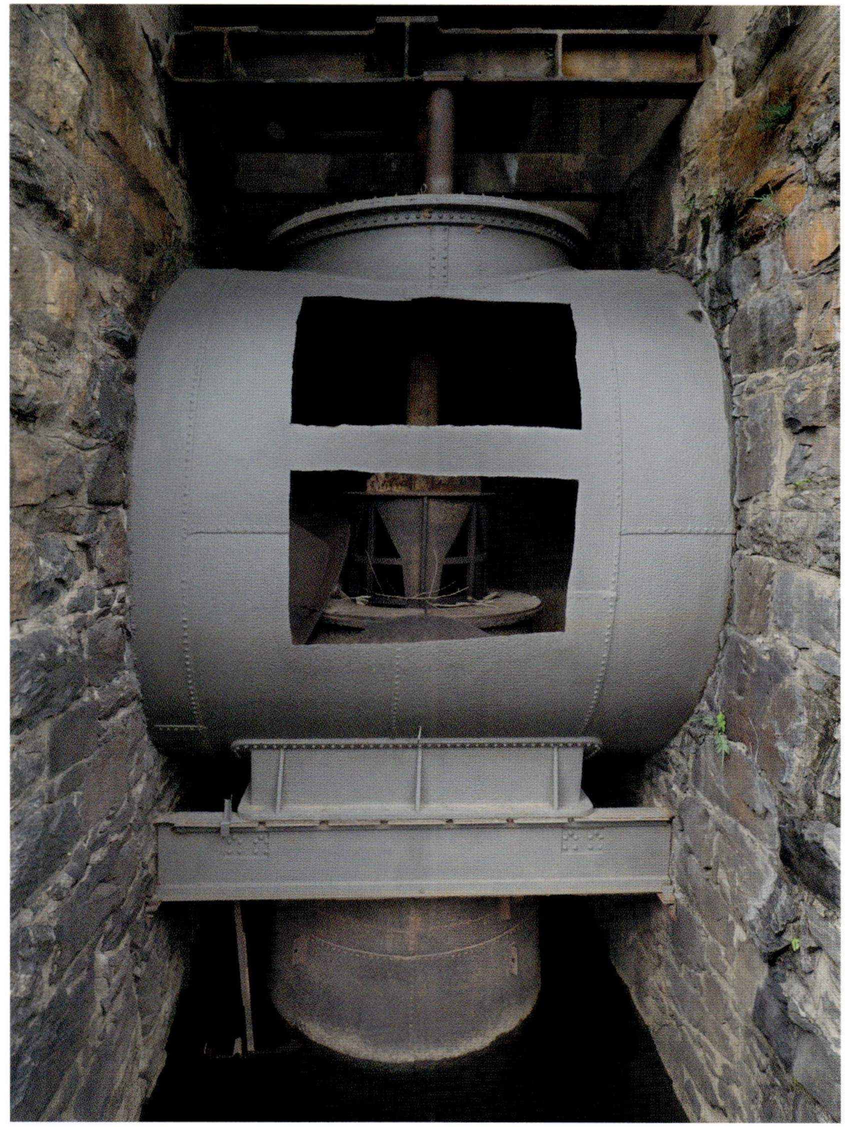

The water intake and handling systems at Eagle and Phenix Mills were massive and could harness enough power from the river for operation of numerous textile machines in the large mills.

Johnson Grist Mill

Cleveland, Georgia, Late 1880s

THE MOUNTAINS OF North Georgia are home to several charming watermills, and there is no better example than the old Johnson Grist Mill near Cleveland. This picturesque mill is set just above a landscaped receiving pond that catches water after it has passed over the waterwheel.

Perhaps the most notable feature of this mill site is the massive wooden overshot waterwheel with an equally substantial water delivery flume. The main driveshaft was carved from a single, select wooden log so that it could withstand the tremendous torque from the waterwheel. Using the waterwheel as a background, local high schoolers have taken their prom and homecoming photos here for generations!

Everything about the mill is welcoming, with a special Southern charm that, in some ways, watermills can offer best. The structure is a gabled-roof, wood-framed building with a main level and a basement. The front entrance has the look of a country store, complete with a porch, benches, a thermometer, an American flag, and other appointments. Interestingly, the basement below the main floor of the mill has two quaint and charming access doors that have weathered to take on an almost fairy-tale or hobbit-like appearance.

Operationally, the mill appears to be intact. Although it does not grind grain regularly and has not operated for some time, a skilled miller could probably have it running again fairly quickly. The interior is clean, orderly, and complete, and both the interior and exterior receive regular and ongoing maintenance and attention.

The mill shares the site with an adjacent cabin and a covered bridge just across the road. The bridge is thought to date to about 1940. The owner lives in a home beside the mill.

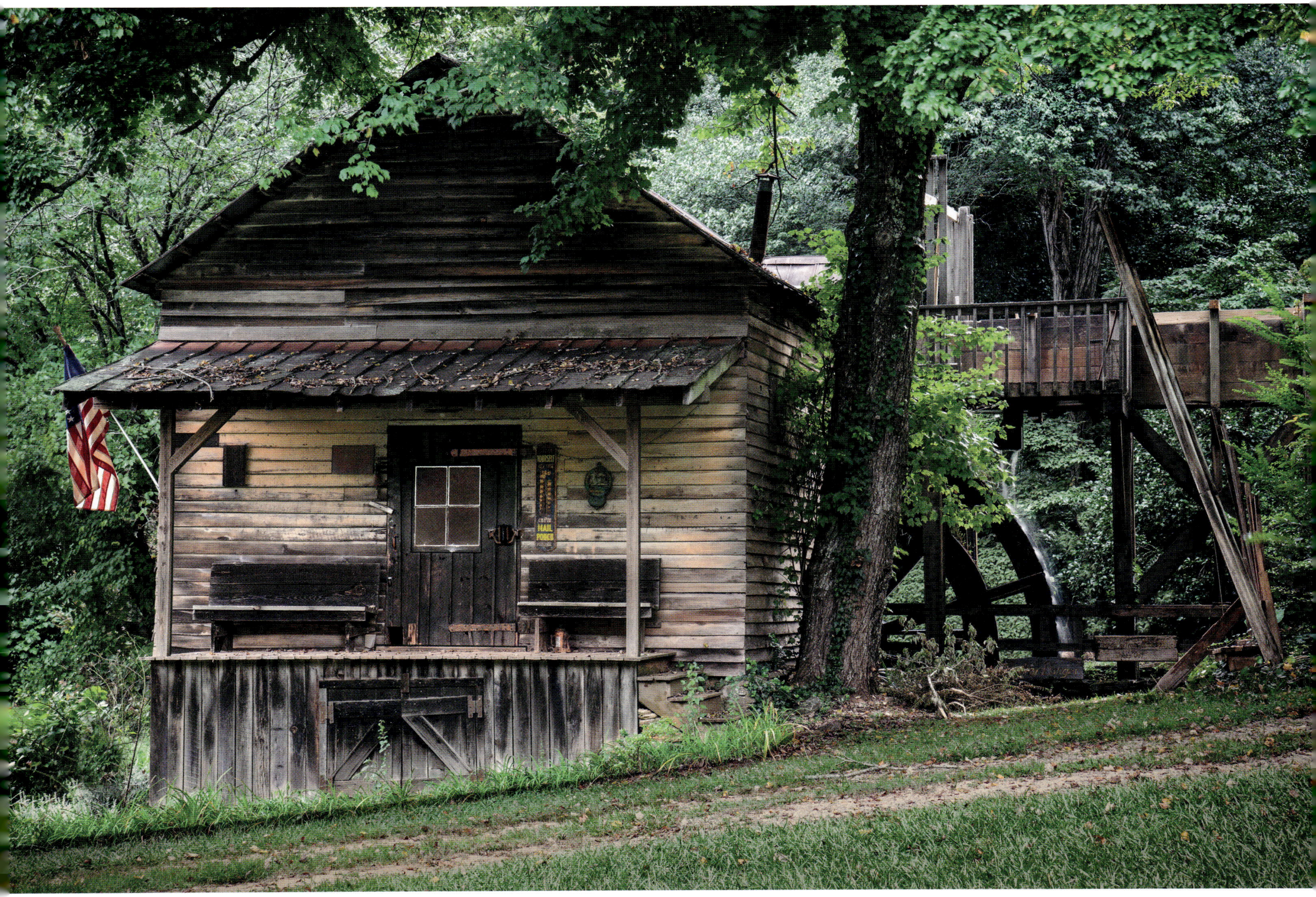

The waterwheel and flume at Johnson Grist Mill can be seen at right, just beyond the structure. The front porch features a pair of benches and is quite welcoming.

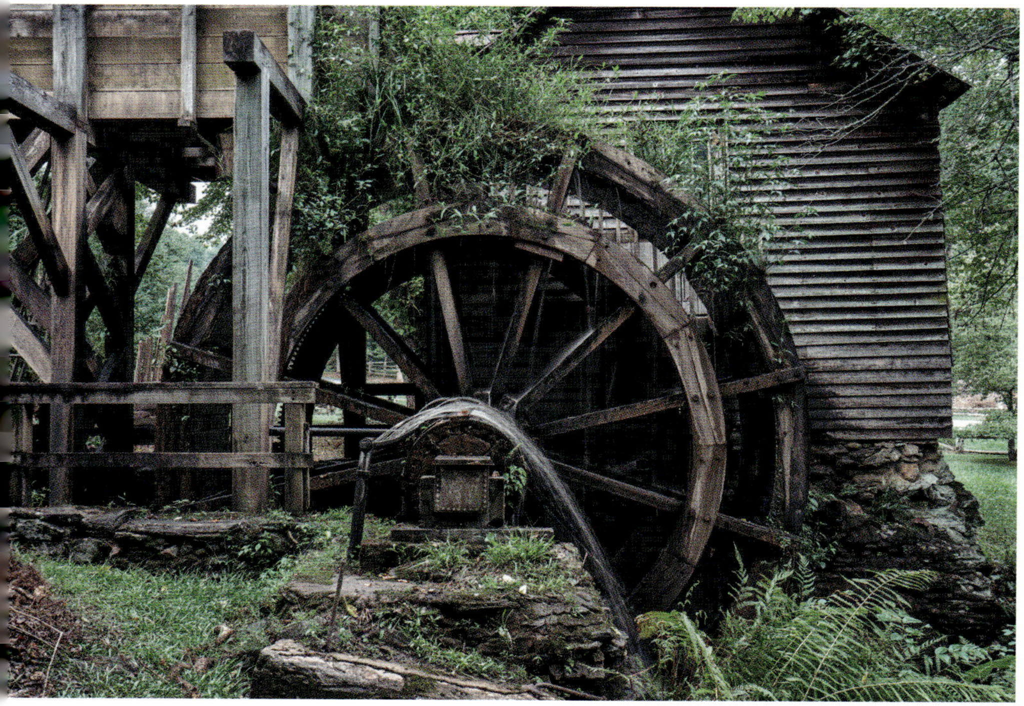

The impressive waterwheel is wooden and is substantially reinforced. Everything related to water handling at the mill seems to have been well designed.

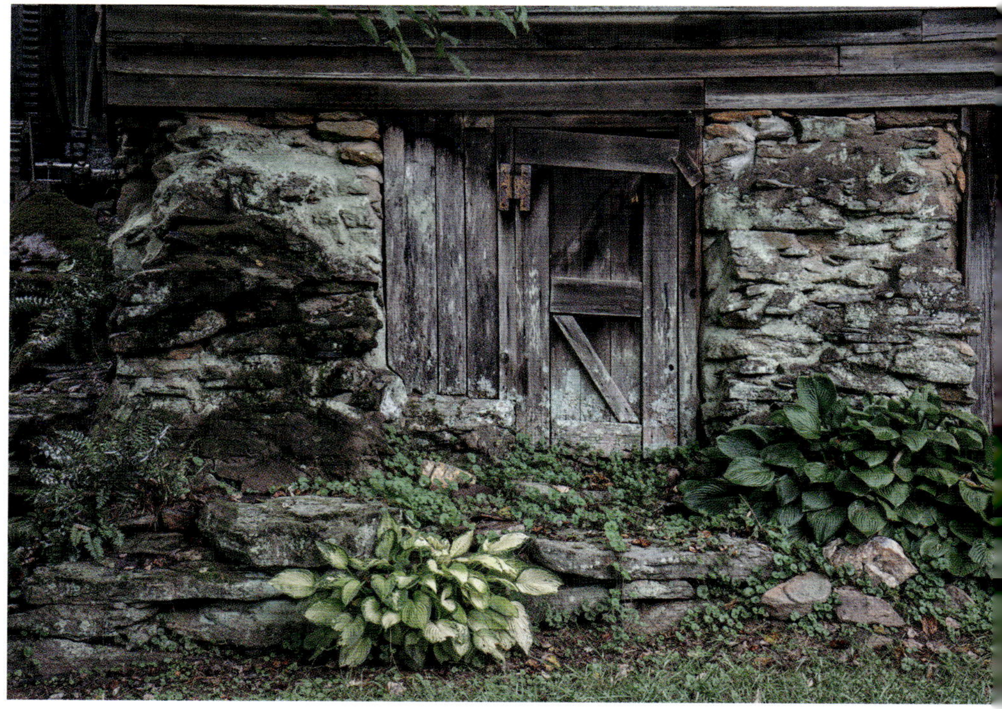

This is one of two quaint and charming access doors on the basement level of the mill.

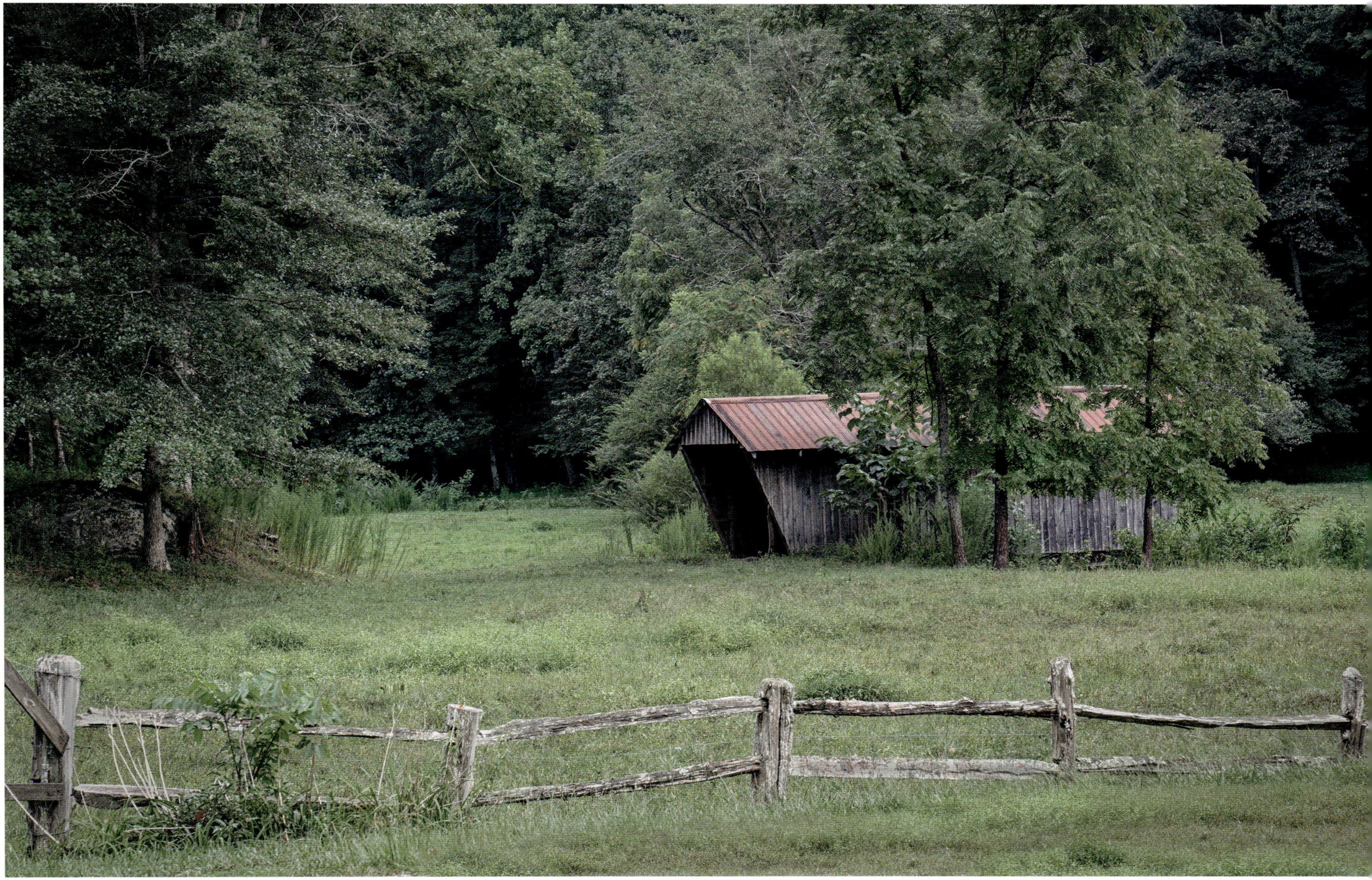

In a classically pastoral scene, the tin-roofed wooden bridge across the road spans the creek downstream of the mill.

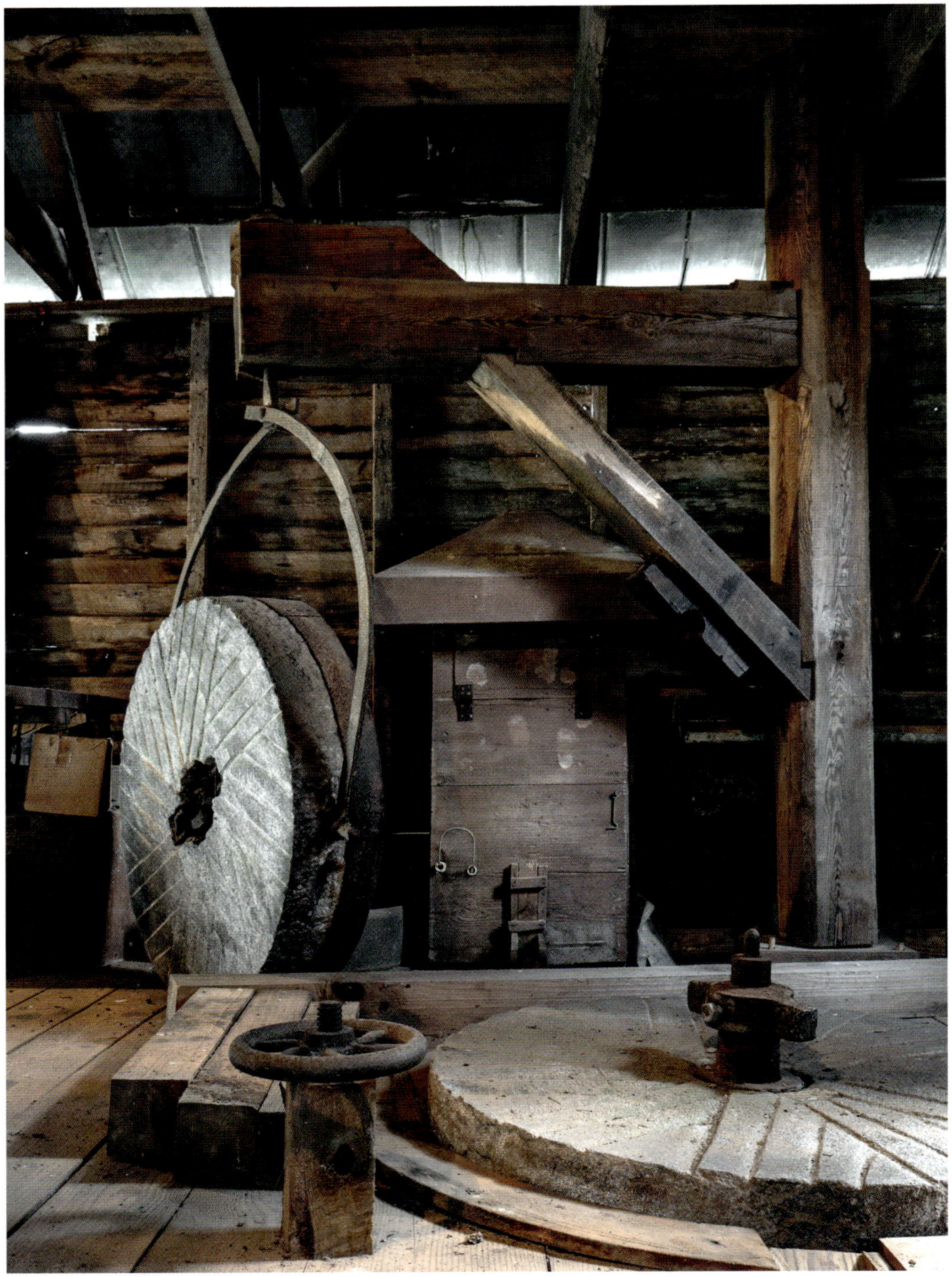

The interior of Johnson Grist Mill is clean and orderly. In this image, the lower or "base" stone can be seen in the foreground. The stone at left in the background is the upper or "runner" stone. The small metal wheel at lower left is used by the miller to adjust the spacing between the stones when the runner stone is installed above the base stone for grinding.

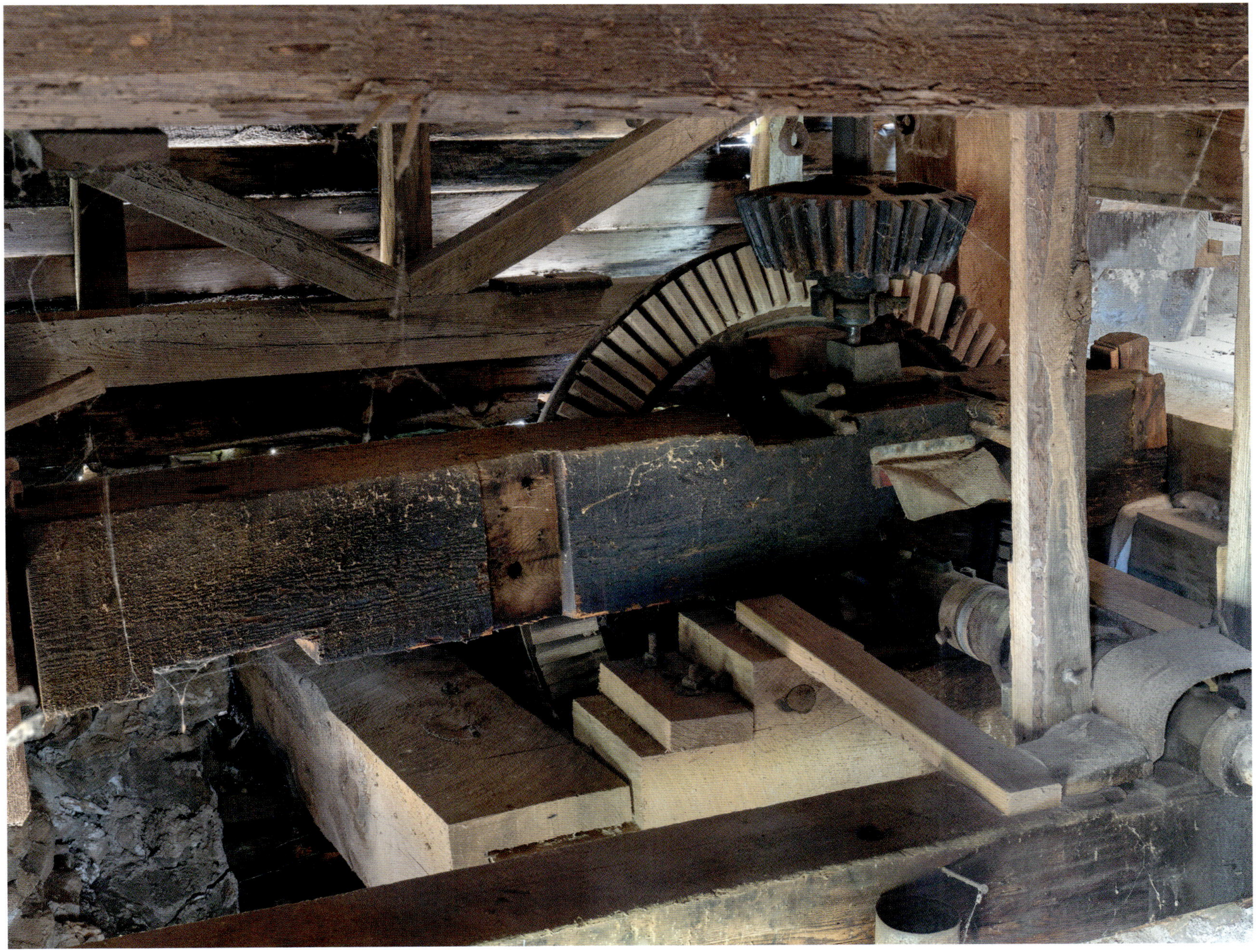

This is a view inside the basement beneath the mill showing the massive wooden structural beams and the beveled gears that link the waterwheel to the millstones.

Loudermilk Mill / Short's Mill / Tumlin's Mill

Clarkesville, Georgia, 1850s or 1880

LOUDERMILK MILL is a popular and well-known landmark among locals and visitors in Habersham County around Clarkesville. The mill is frequently photographed and painted because of its setting, with the idyllic rapids of Hazel Creek flowing through and across the scene. The mill is on posted private property, but it is clearly visible from Highway 197. In fact, the elevated view from the highway is probably the best view and is well away from the property owner's home at the creek across the valley.

Conflicting information shows the mill as dating from the 1850s or early 1860s or perhaps from as late as 1880. Such uncertainty is fairly common with rural country mills; little documentation exists, and tax records may not be available. Adding to the confusion, the mill has also been known by various names over the years.

Loudermilk Mill ground corn, and for a time it also included a shingle mill. The seventeen-foot metal wheel was installed in the 1920s, replacing a wooden wheel. Grinding was done with a set of forty-eight-inch millstones.

There is a story that Tom Loudermilk was greasing the wheel one day when his foot slipped, and he lost a leg in the powerful rotation of the wheel. After that, his daughter took over as miller, and she operated the mill for the next eighteen years. As time passed, the mill ran less frequently, and it finally closed in the mid-1970s. The demand in North Georgia for mills like this had passed by then.

Although still picturesque, the mill is clearly not operational today. Hopefully this mill will not be lost to history.

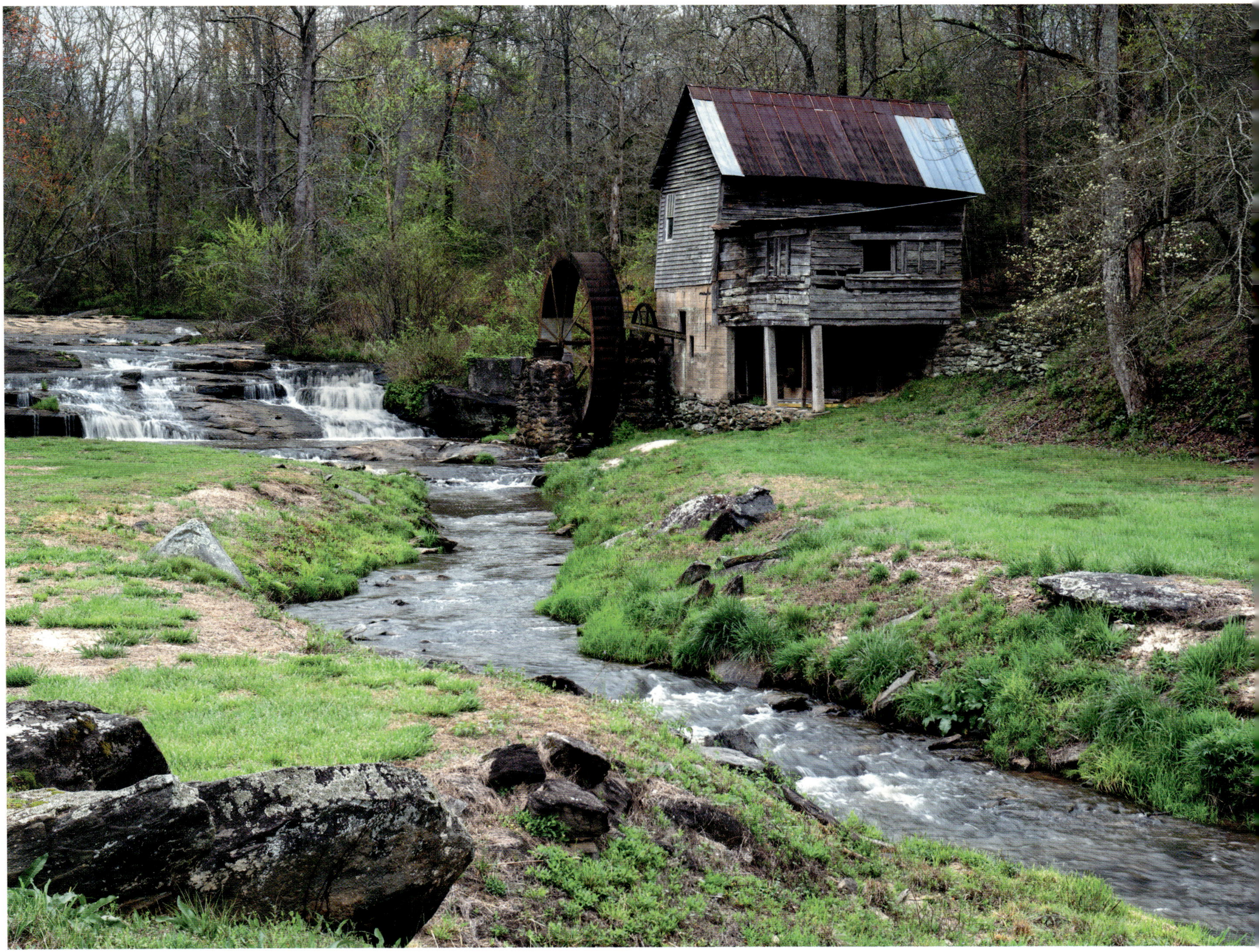

Loudermilk Mill is located in a scenic North Georgia valley along the rapids of Hazel Creek.

Nora Mills Granary

Helen, Georgia, 1876

FOR ANYONE with an interest in nostalgia and a yearning for the good old days, Nora Mills Granary is the perfect destination. Take just a few steps inside the entrance doors, and you are transported to that time when life seemed a little simpler, friendlier, and slower in pace. All the vintage milling equipment is operational and on display for perusal.

The mill grinds a full range of products for customers. In addition, the mill is a classic general store selling thousands of old-fashioned items, such as candies, jams and jellies, cast-iron and speckle ware, honey, and vintage-style kitchen items. Like a few other active mills, Nora is equipped with a kitchen for demonstrations and samples.

This mill has its roots in the gold rush days of North Georgia. Located mere feet from the Chattahoochee River, the current mill dates to 1876 but is on the site of an earlier mill. A turbine rather than a conventional waterwheel powers the machinery. The original millstones weigh in at 1,500 pounds apiece and are still in regular service.

Lamartine Hardman, who would later serve two terms as governor of Georgia, purchased the mill in 1902 and named it in memory of his sister, Nora. The Hardman family owned the mill until 1998, when Nacoochee Village, Ltd., bought the property. The mill is operated today by the Fain family, who came to the mill in the 1980s and worked with the Hardmans.

Because Nora Mills Granary is located along Georgia Highway 75, the main route into Helen—one of the most popular visitor and tourist destinations in the Peach State—the mill's future looks bright. Many guests to the mill are repeat customers, and quite a few make the trek from Atlanta just to enjoy and shop at the mill.

In recent years, the dam, raceway, and penstock have been rebuilt, and a new porch and deck have been added over the river. The porch and deck are especially popular overlooks for the river. In addition, the Hardman Heritage Trail winds its way along the opposite bank of the river for anyone interested in a pleasant walk along an old railroad bed. A panoramic view of the mill opens up after just a short, low-impact walk from the Helen parking area. This is one of the nicest walking trails in the state.

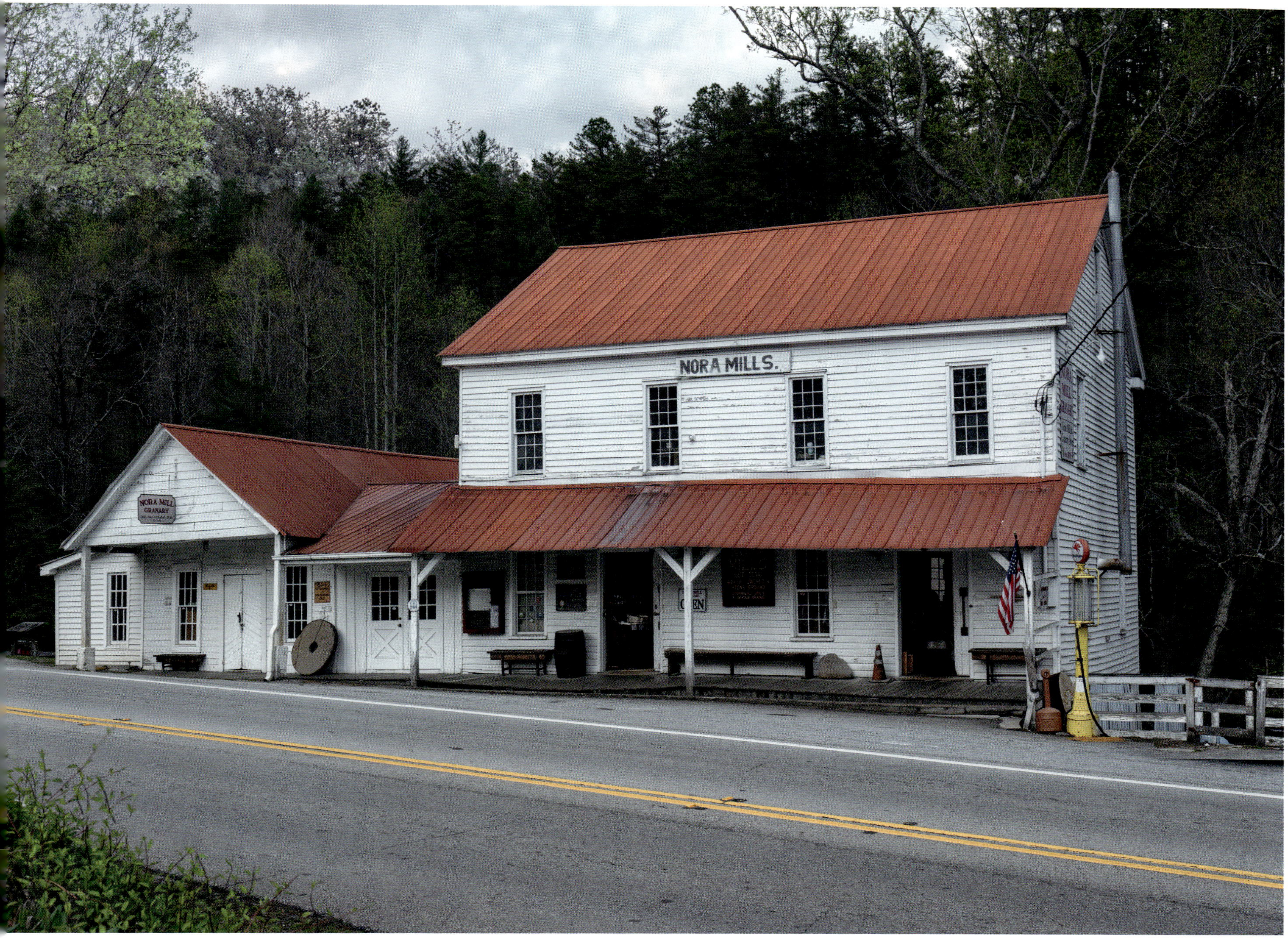

With its distinctive red roof and white siding, Nora Mills Granary has long been a popular stop along Highway 75, just a few miles south of the Bavarian-themed town of Helen. The Chattahoochee River flows a few feet behind the mill. In the old days, steam locomotives operated along tracks on the opposite bank of the river.

Stone Mountain Grist Mill

DeKalb County, Georgia, 1867 / 1965

THE OLD MILL at Stone Mountain Park east of Atlanta was moved to its current location from Ellijay, Georgia, in 1965. It was originally built in 1867 or 1869. The mill features a thirteen-foot wooden overshot waterwheel. The mill is located adjacent to a lake at the popular park in what is often described as the most attractive setting on the expansive property.

When planners working to develop Stone Mountain Park discovered the mill, it was basically abandoned, in disrepair and in peril. In all likelihood, its relocation to the park saved the mill from complete loss, the fate of so many Southern watermills. Currently, the mill is not thought to be operational, but it is relatively well preserved.

The historic 151-foot Washington W. King Bridge, a covered bridge designed and built by King in 1892, was moved to the park from Clarke County and reassembled near the watermill in 1965.

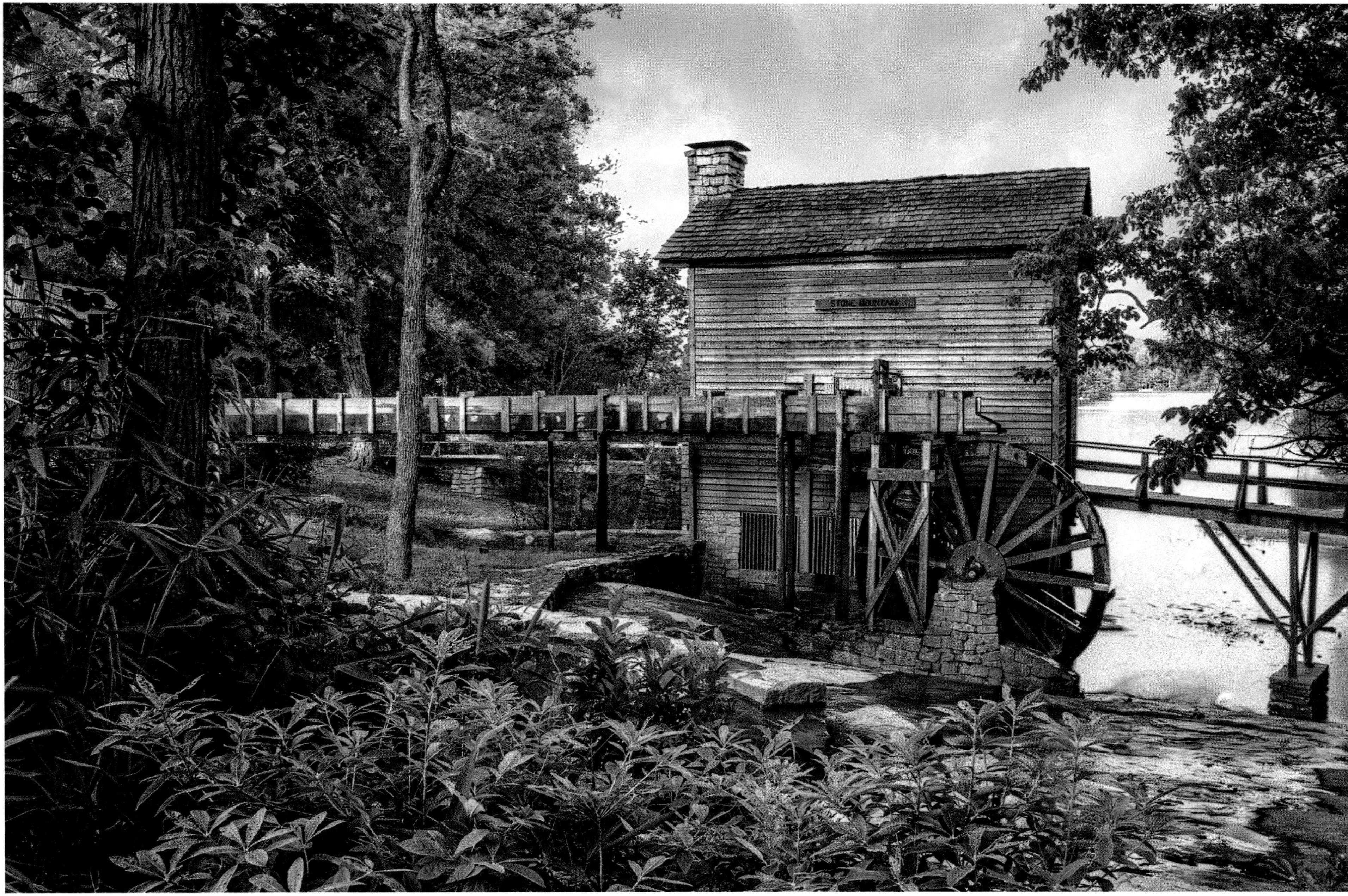

This grist mill was relocated to Stone Mountain Park from North Georgia and authentically reassembled. As is evident from this highly detailed black-and-white image, the mill has provided a lovely destination for many park visitors since the mid-1960s.

The Old Mill at Berry College

Rome, Georgia, 1930 / with Several Restorations

A VISIT TO THE famous old mill on the campus of Berry College begins with a drive through the main academic campus at one of the nation's most beautiful colleges. Classic examples of Gothic, Colonial Revival, and neoclassical architecture can be seen among the college buildings along the drive. From this area of the campus, the main road next enters a vast woodland area where deer and other wildlife are abundant. As the drive continues, visitors will begin to appreciate that this is the largest college campus in the United States and one of the largest in the world. Continuing, the road emerges into the school's mountain campus and winds around several quiet swan ponds, classroom and residential facilities, and open fields. Finally, a narrow gravel road heads through a gate and up a hillside through a dense forest, before the mill finally comes into view.

For first-time visitors, seeing the watermill can be breathtaking. The iconic image of the mill reflected in a small pond is the view upon arriving at the mill. The forty-two-foot-high steel and wooden waterwheel is three feet in width, making this, by some definitions, the largest waterwheel in the country. The mill building itself is a classic stone structure with a small entry porch, but everything is dwarfed by the scale of the waterwheel.

This mill's actual history reaches back much further than 1930. Some parts of the large waterwheel and much of the mill machinery date to the 1800s. Part of the waterwheel had previously been used in a gristmill and cotton gin at a mining community between Rome and Calhoun, Georgia. Henry Ford arranged for the waterwheel hub to be moved to Mount Berry, and other components of the older mill were also relocated. Students at the school then worked to reassemble the parts and build the rest of the structure.

Through the Great Depression era, the mill was operated to grind corn grown on the campus into meal and grits. Much of this grain was eaten by students during this difficult period.

As World War II consumed the nation, the mill was basically neglected, and it fell into disrepair. It was not until 1977 that the wheel was rebuilt by staff, students, alumni, and other supporters of Berry College. In 1985, the mill received further mechanical restoration to bring it to full operating condition. Today the mill runs on special occasions.

Interestingly, the tall waterwheel is overshot. To deliver water to the top of the wheel, the pipes are primed and the water finds its own level above the wheel by an inverted siphon. A nearby reservoir provides a source of water for operation. In theory, a tall waterwheel can provide power to operate a mill with minimal water supplied. A shorter, wider wheel can be used to produce the same power with more water flow, if the necessary volume of water can be supplied.

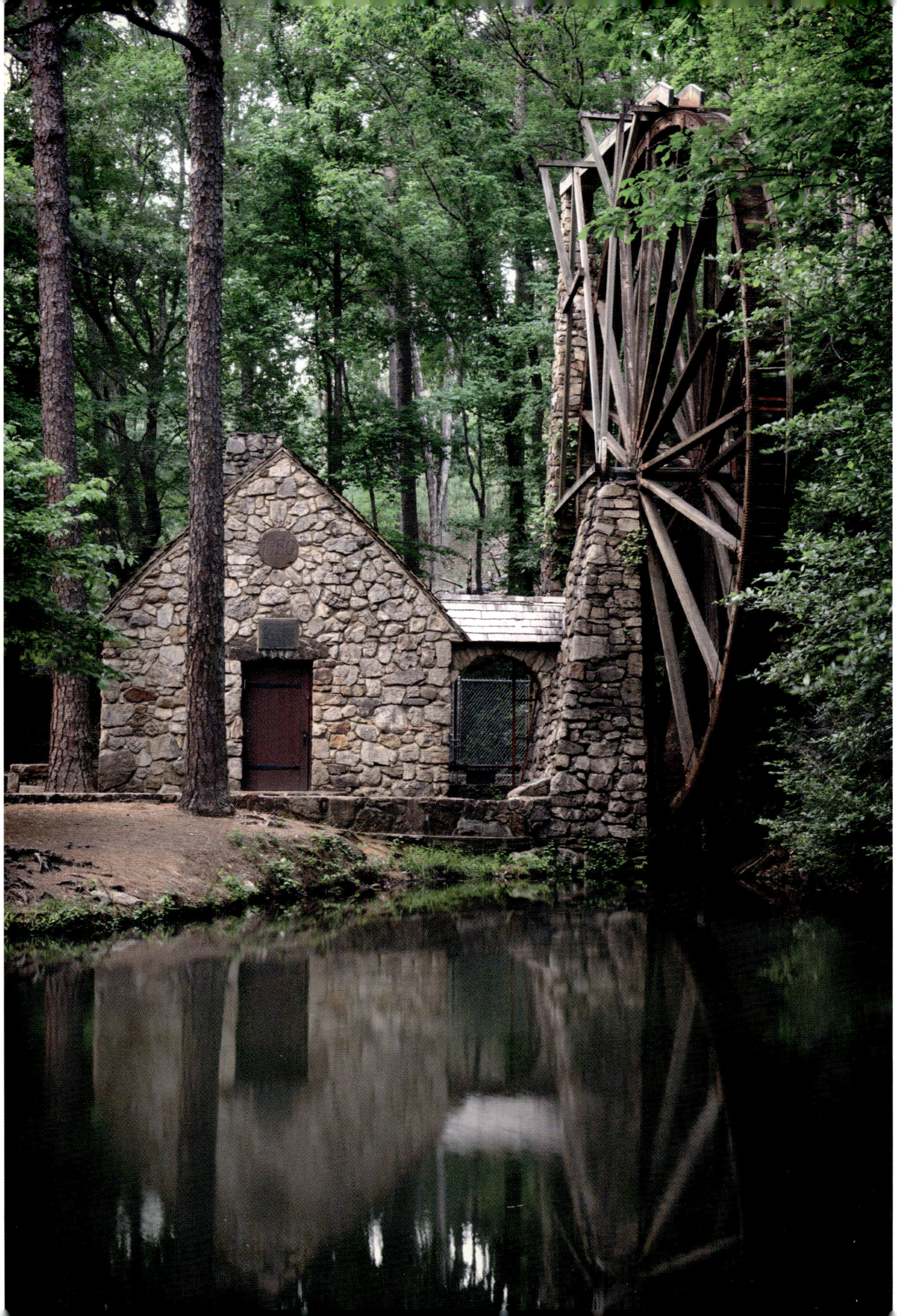

Unlike many early country watermills across the South, the Old Mill at Berry College was designed with visual impact as a consideration. This is the view that greets visitors as they arrive at the mill.

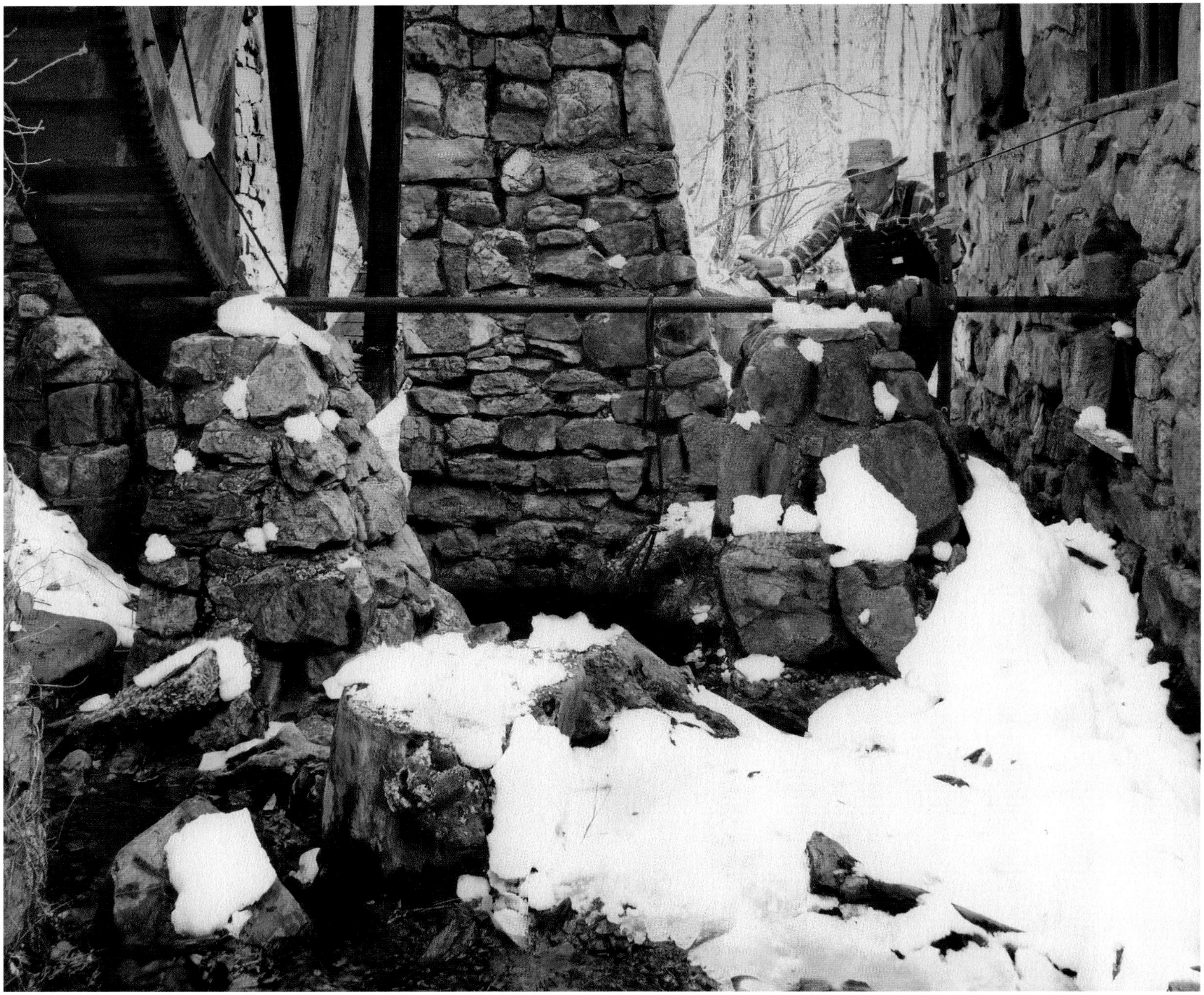

The massive scale of the waterwheel is apparent in this 1990 photograph of the author's father, Wesley Boyd, posing in the snow. Boyd passed away in 1998.

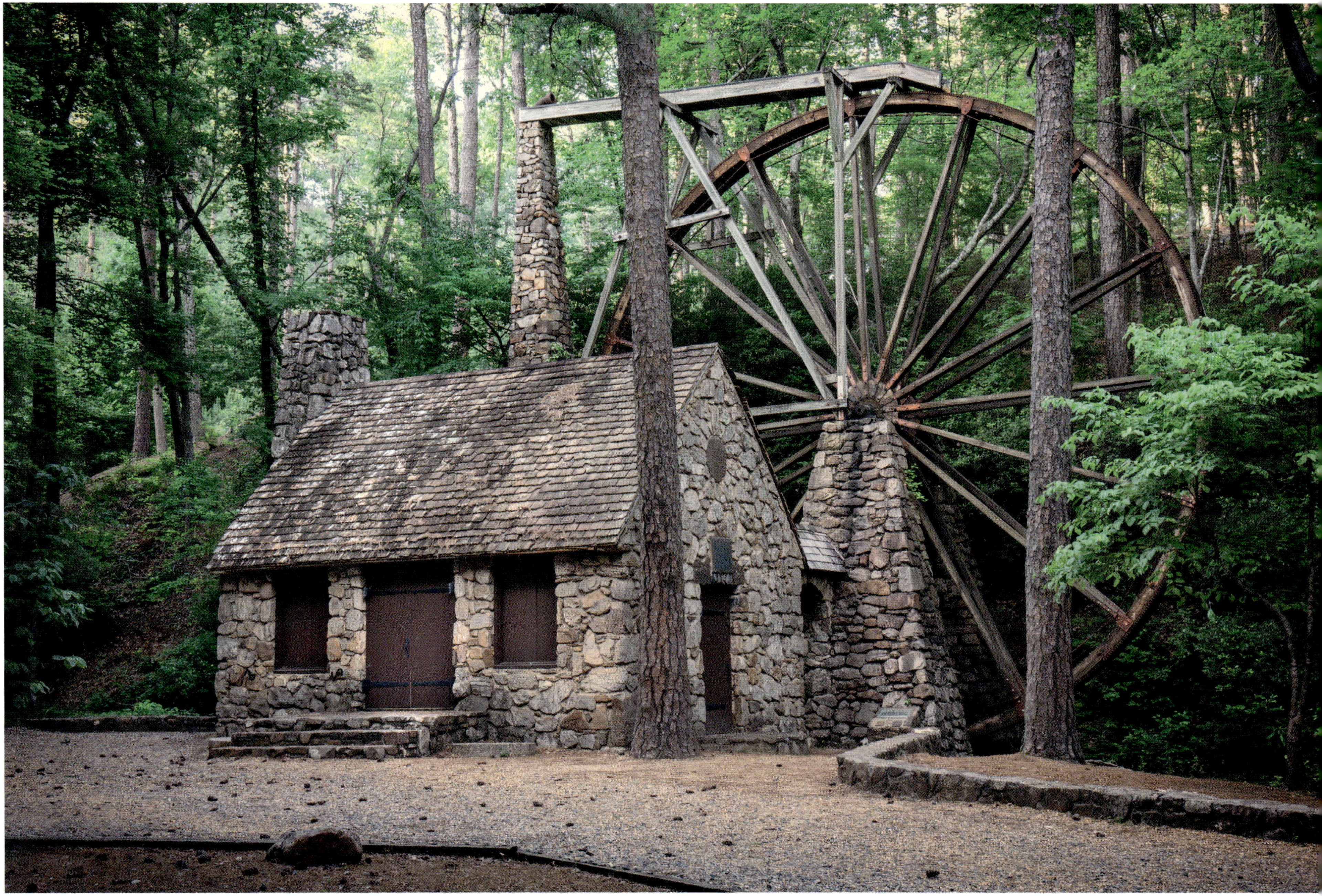

This image of the Old Mill at Berry College provides a view from the front of the mill.
The mill is set in a beautiful forest.

Ragsdale Mill

Homer, Georgia, 1863 / with Restorations in Recent Decades

IN 1863, at the height of the Civil War, the Reverend Francis Ragsdale built Ragsdale Mill. An earlier mill had been built here on Nail's Creek sometime before 1837. The current mill is set at the base of a steep hill and along a dramatic rocky shoal and gorge.

French millstones housed inside the mill arrived through the Union blockade of Savannah during the Civil War. The stones were used to grind local grains for area farmers, and a vertical saw was installed on the lower level below the mill to cut boards from logs. The overall operation also included threshing equipment to separate grain from stalks and heads of straw. The mill ceased operation in 1955.

In 1967, Dr. James Cantrell and his wife, Nancy, bought the mill. Dr. Cantrell was a math professor at the University of Georgia in Athens. The couple worked tirelessly to restore the mill, and it was made operational again for grinding cornmeal.

Doc and Nan Sisk purchased the mill in 2003, and by 2007 they had restored the metal overshot waterwheel and rebuilt the wood and stone flume and trestle; unfortunately, within ten years the flume had already fallen into disrepair again. The Sisks have also worked to maintain and preserve the other historic structures near the mill.

Because of the colorful granite shoal, the long wooden flume, and the large overshot waterwheel, Ragsdale Mill is often considered to be one of the most scenic mills in Georgia. The site has been used for special events and festivals. Mount Olivet Methodist Church, just up the hill from the mill, has also been restored. It was organized in 1868, and a large cemetery is located on the church property.

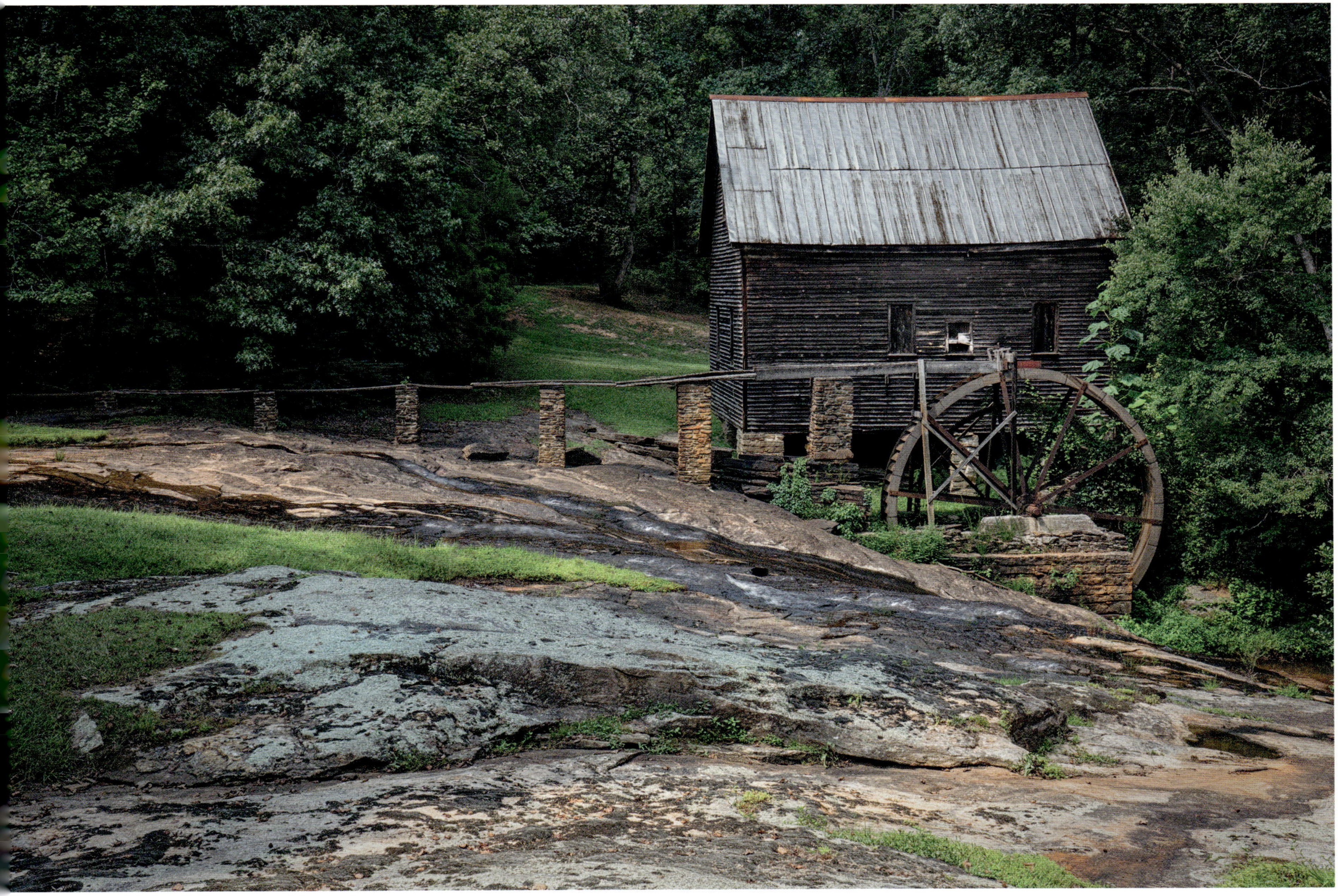

Ragsdale Mill near Homer was not operational when this photograph was taken, but it is in solid structural condition, except for the wooden flume, which seems to require regular reconstruction. The rocky shoal provides an interesting visual setting for the watermill.

Starr's Mill

Senoia, Georgia, 1825 / 1907

ANYONE WHO HAS SEEN the 2002 Reese Witherspoon movie *Sweet Home Alabama* may find Starr's Mill familiar, because the mill was featured prominently in that film. The first watermill at this site was built in about 1823 to 1825. By that time, the land for the mill had sadly been awarded to White settlers as part of the 1821 lottery of Creek Nation lands in Georgia.

The current mill is actually the third watermill on the site and is named for the owner of the second mill, Hilliard Starr. Starr owned that mill from 1866 to 1879, when it burned. Prior to his ownership, the name of the mill had changed several times, probably each time the mill was bought and sold.

The mill is a wood-framed structure and has been painted red with white trim for many years. As the paint fades in the sun and weathers between paintings, the hues of red gradually change.

Although the current mill looks as if it might have been powered by a waterwheel, it has always been turbine driven. With a good water supply from Whitewater Creek, the turbine operated a gristmill, a sawmill, and a generator to supply electricity to Senoia. The mill did not cease operation until 1959. A separate turbine-powered cotton gin was located on the opposite side of the creek and dam.

Today, the watermill, dam, and surrounding lands belong to the Fayette County Water System, and the site is popular for fishing, picnics, gatherings, and events. A restored old schoolhouse, also painted red, is located just behind the mill.

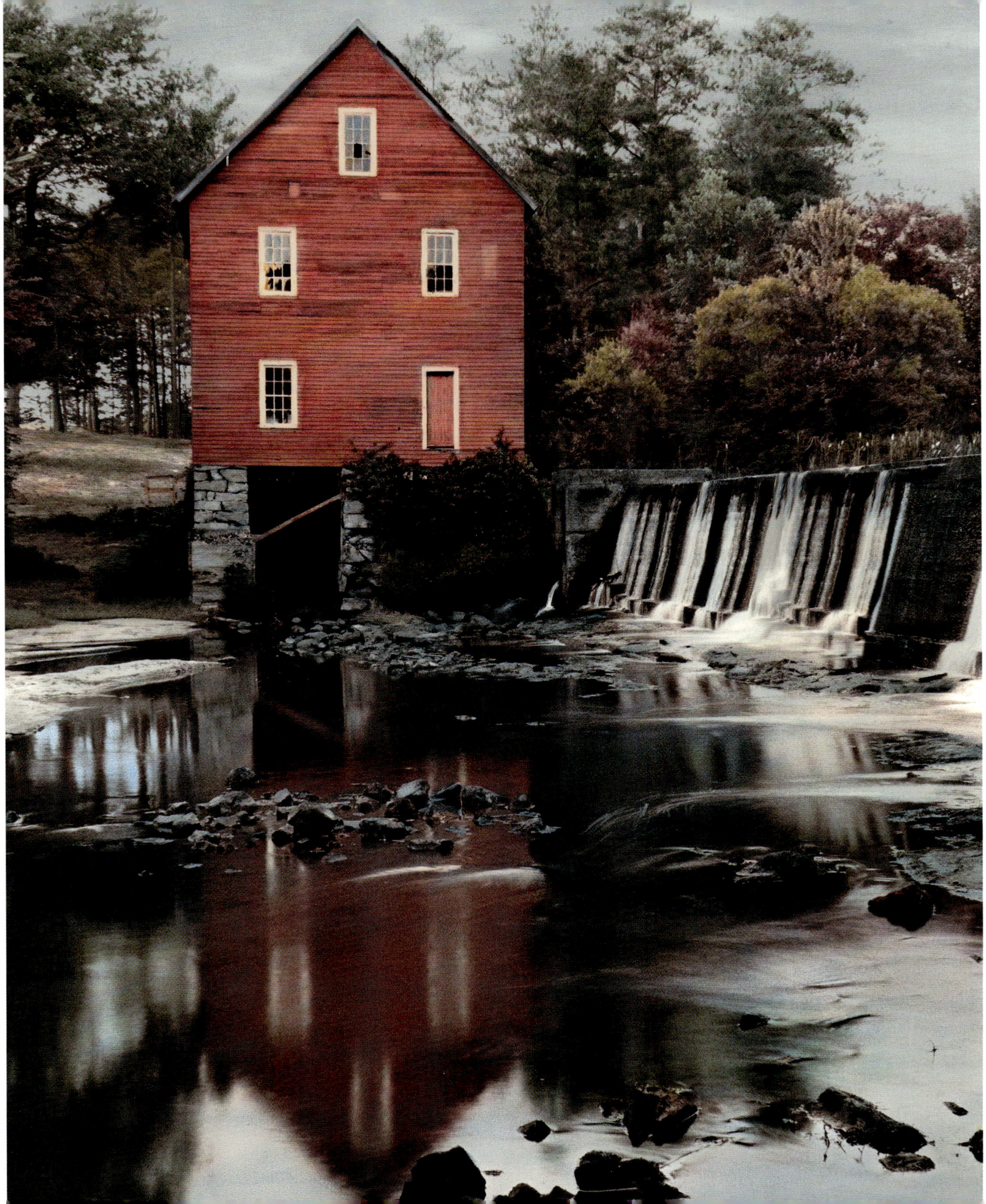

In this classic hand-tinted image, Starr's Mill is viewed from across Whitewater Creek, in the area where a cotton gin once stood opposite the watermill.

Historic Augusta Canal and Industrial District

Augusta, Georgia, 1845 to the Present

AUGUSTA IS LOCATED near where the Savannah River crosses the fall line between the Piedmont Plateau region and the Atlantic Coastal Plain. Early developers and industrialists in the city recognized that the energy of the Savannah River rapids could be harnessed to power mills, much as had been done with similar hydrologic features in England and New England.

In 1845, workers began to construct a canal that would deliver the hydropower of the river water from an elevation above the city into an Augusta industrial area. Enslaved laborers dug the canal by hand, with some help from hired Black and White workers. The canal soon proved its value by providing drinking water for the growing city, as well as power for mill operation and a waterway for shipping and transportation. By 1847, the first gristmill and sawmill were operating on the energy from the water in the canal, and a textile mill was soon to follow.

For mill operations, water from the canal was routed through each mill and past waterwheels and/or turbines and then released behind the mills at lower elevations and back into the Savannah River. This same basic concept had worked for centuries at industrial sites around the world.

When the American Civil War broke out, the Confederacy selected Augusta as the site for major industrial operations. The Confederate Powder Works became what was said to be the second-largest gunpowder plant in the world at the time. The powder works spread out for some two miles along the canal. Operations continued throughout the war without assault by Union forces. However, it was a highly dangerous working environment; in August 1864 some eighteen thousand pounds of gunpowder exploded at the facility, killing eight men and a boy and destroying one of twenty-eight buildings.

The powder works continued to operate until the end of the war, but the facility was dismantled soon after and ultimately was mostly demolished, as the canal was widened and textile mills expanded along the canal waterfront. The obelisk chimney was preserved from the powder works and has been restored as a war memorial.

In 1880–82, the water-turbine-powered Sibley Textile Mill was established on the former Confederate site. Bricks and other building materials from the powder works were reused for the construction of a massive four-story mill that continued to operate until about 2006, though it was long past its operational peak by then. The facility still generates electricity today with waterpower from the canal.

Work on the adjacent and even larger King Mill began in 1882, with expansion projects spanning the next few decades. The hydro powerhouse, operated by canal turbines, was erected in 1886 and also continues to generate electricity today. Both the Sibley Mill and the King Mill sell power into the local electricity grid operated by Georgia Power Company.

As of the twenty-first century, the canal is about thirteen miles long and operates on three levels with locks. It essentially forms the centerpiece of the city as the Historic Augusta Canal and Industrial District, which was named a National Historic Landmark in 1977. The Sibley and King Mills are preserved and are currently undergoing continued redevelopment. The water-powered mill turbines will continue to crank out energy for the area. Canal boat tours are offered on a regular schedule and depart from the Augusta Canal Discovery Center at nearby Enterprise Mill.

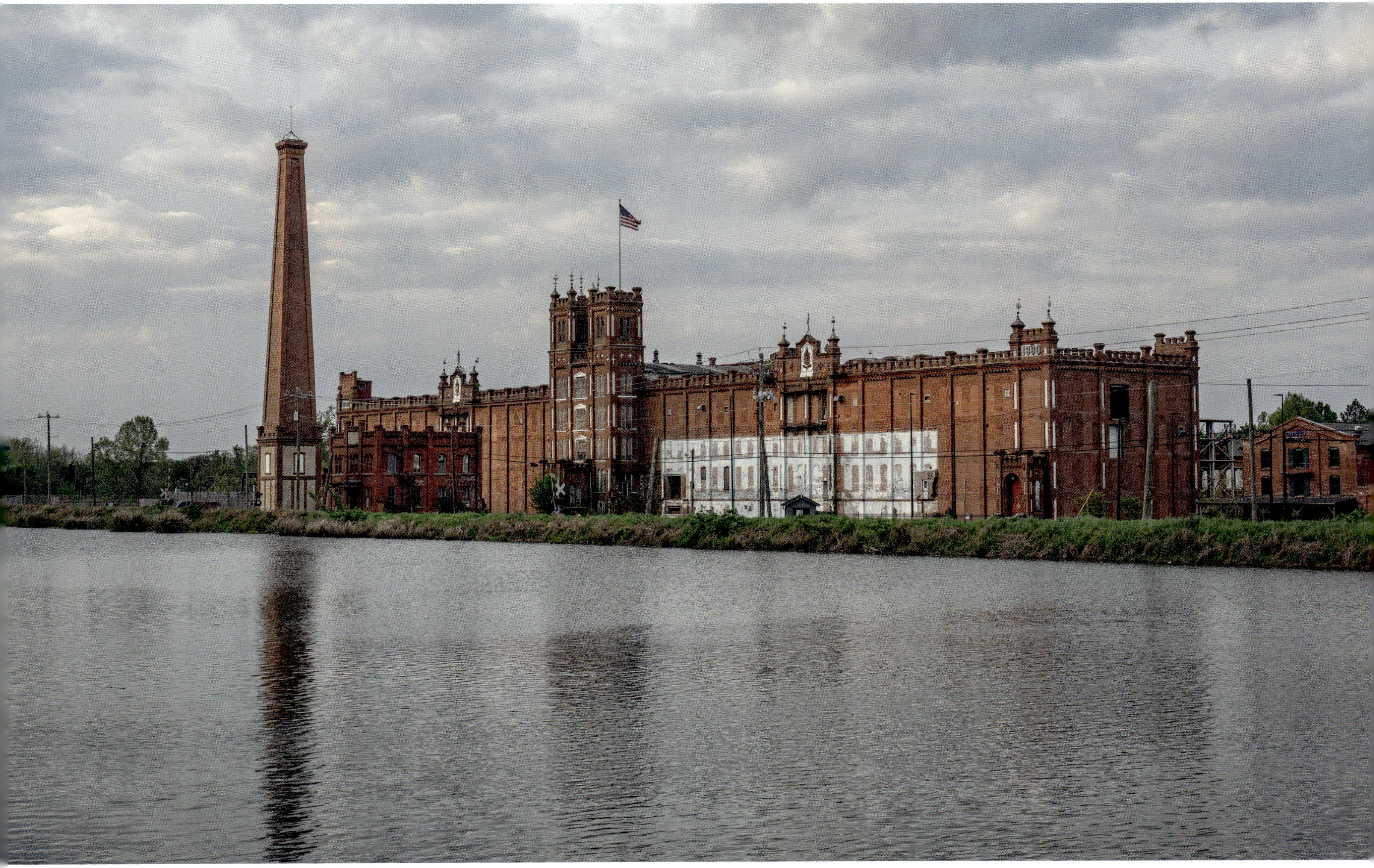

The Sibley Mill is somewhat dominated by the Confederate Powder Works chimney that has been restored and preserved as a Civil War memorial.

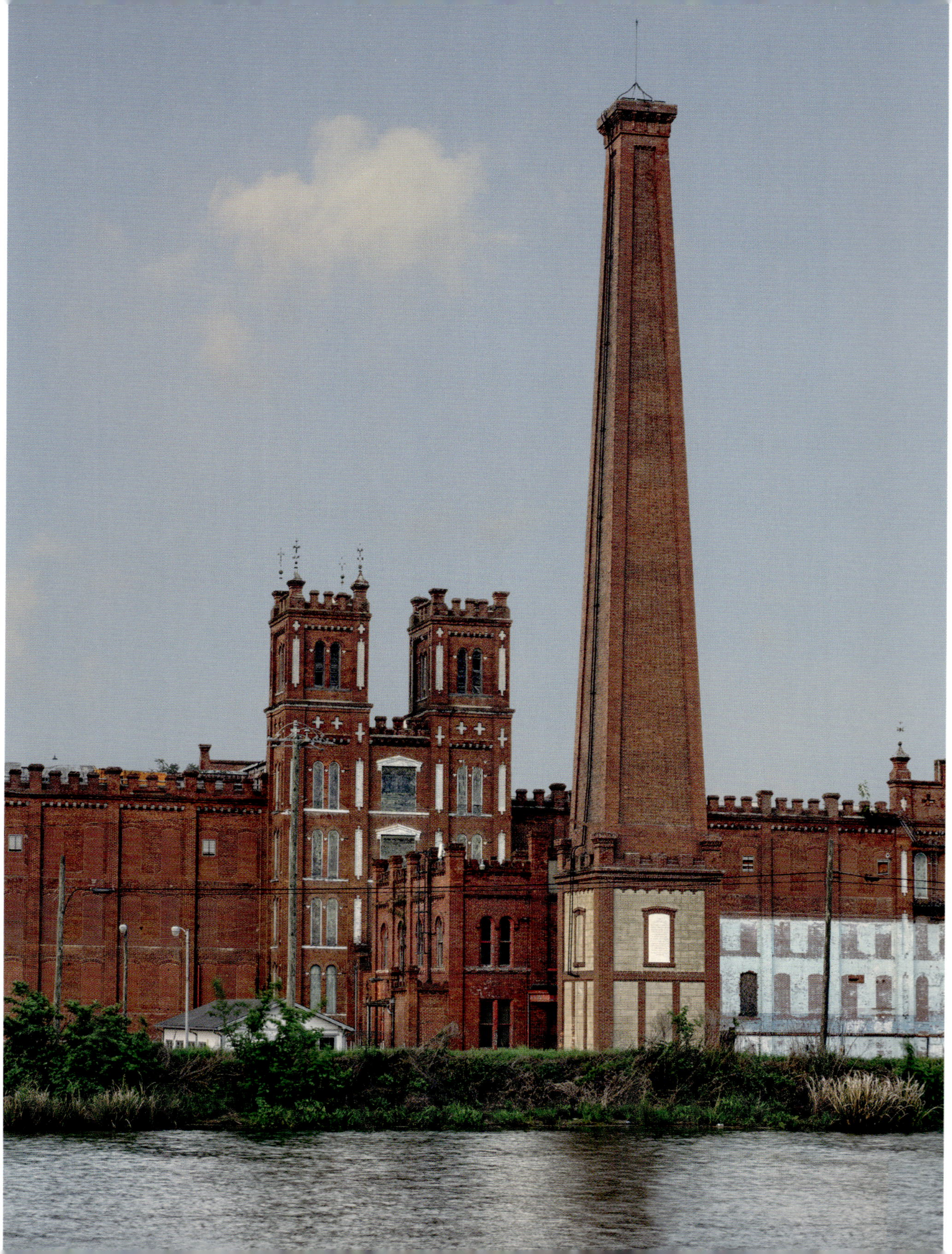

The massive scale of the Sibley Mill and the 150-foot Confederate Powder Works chimney is apparent in this image.

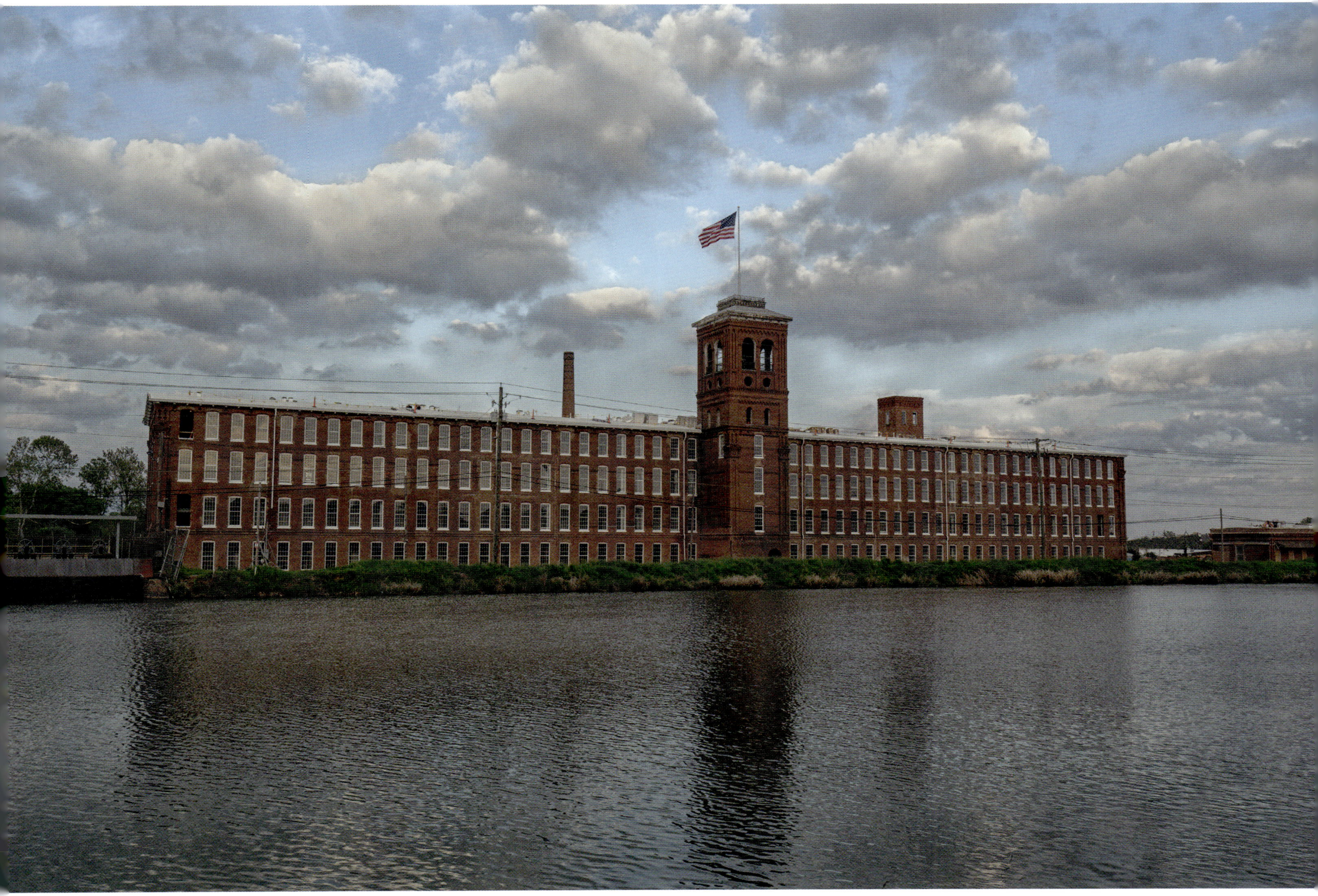

The King Mill, viewed here from across the Augusta Canal,
is a strikingly impressive industrial-era structure.

Pardue Mill

Cleveland, Georgia, 1930s

THE PARDUE MILL is like hundreds of similar small and simple country watermills that once dotted the Southern landscape. The equipment in the mill included a single set of millstones and a corn sheller. Power came from an eighteen-foot waterwheel that was about four feet across. A miller's house and a barn were also originally part of the property.

Situated along a spectacularly high sixty-foot rocky shoal and just below a seven-foot-high stone dam that is forty-eight feet across, the mill must have been something to see from along the opposite bank of the creek. The wheel and much of the flume are now long gone, likely washed down the shoal and out along the Pardue Mill Creek, en route to the Chattahoochee River.

In the years after World War II, a young couple fixed up the mill and began grinding corn for the surrounding community. Operation continued for a total of thirty years or so.

After operation ceased in the mid-1960s, the mill deteriorated quickly. The building's rusty tin siding and tin roof blend naturally with the colorful exposed stone along the cascading creek. Several smaller shoals characterize this area along the creek, but this is the largest shoal, and it is almost as steep as a cliff.

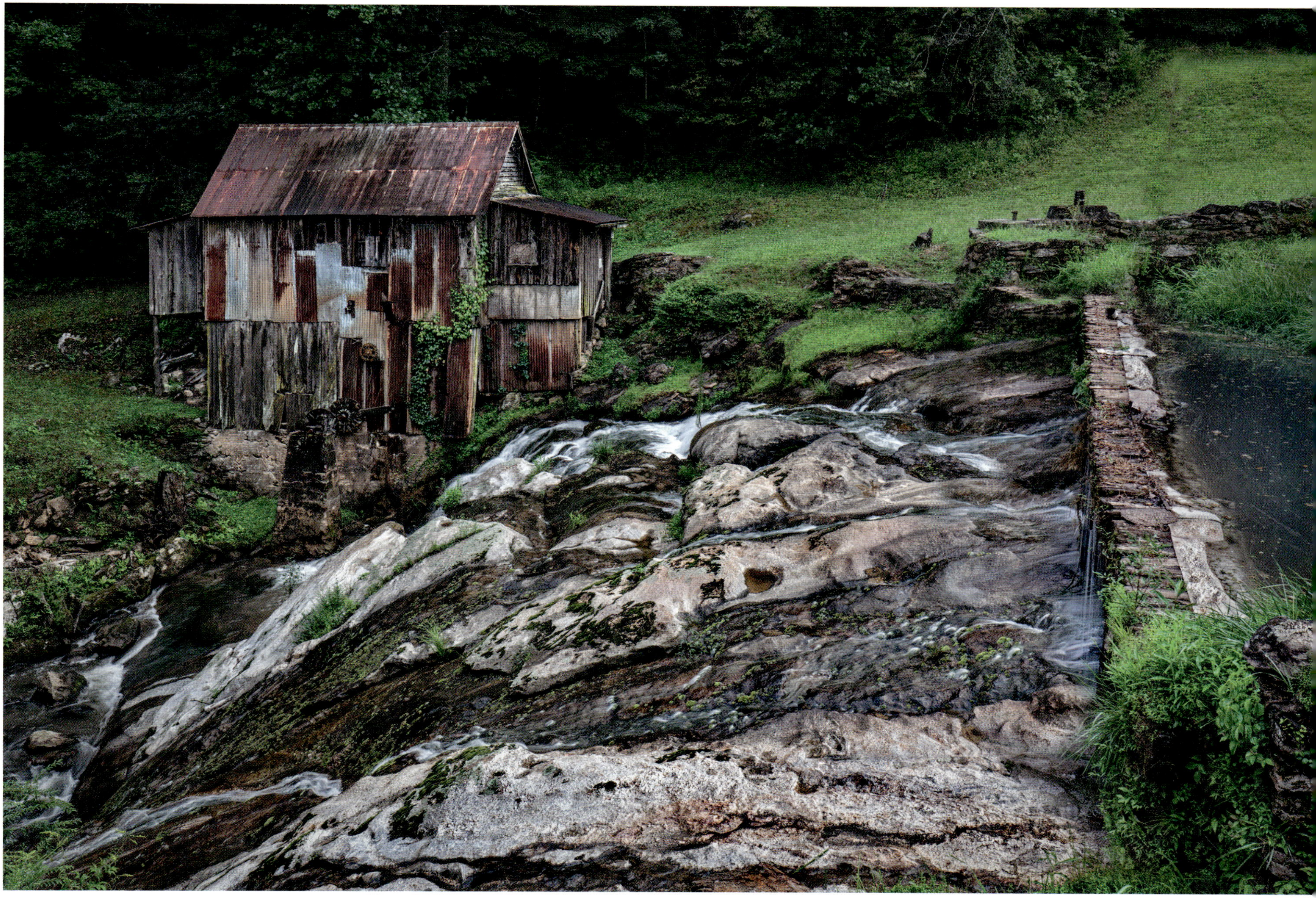

Pardue Mill is one of several surviving but threatened watermills in the historic Cleveland and Clarkesville, Georgia, area.

Sells Mill

Hoschton, Georgia, 1914 / with Recent Restoration

LOCATED NEAR Hoschton in northeast Georgia, and adjacent to one of many impressive rocky shoals in the hills and mountains across the region, Sells Mill ground corn into the 1990s. Power was provided by a sixteen-to-eighteen-foot overshot wheel driven by water from a millpond on Indian Creek. The first mill on the shoals dated to the late eighteenth century. The present structure is from 1914.

In the early decades of the twentieth century, Sells Mill also generated electricity and was a popular community meeting place. For a time, the site was also supposedly used for auto repair.

Today the mill is the featured attraction at a small park near the rapidly growing community of Braselton. Since the site was purchased by the county in 2001, the mill has been undergoing restoration and is in seemingly excellent structural condition. A high dam and picturesque millpond are located just above the mill, and water is channeled to the wheel through a large pipe.

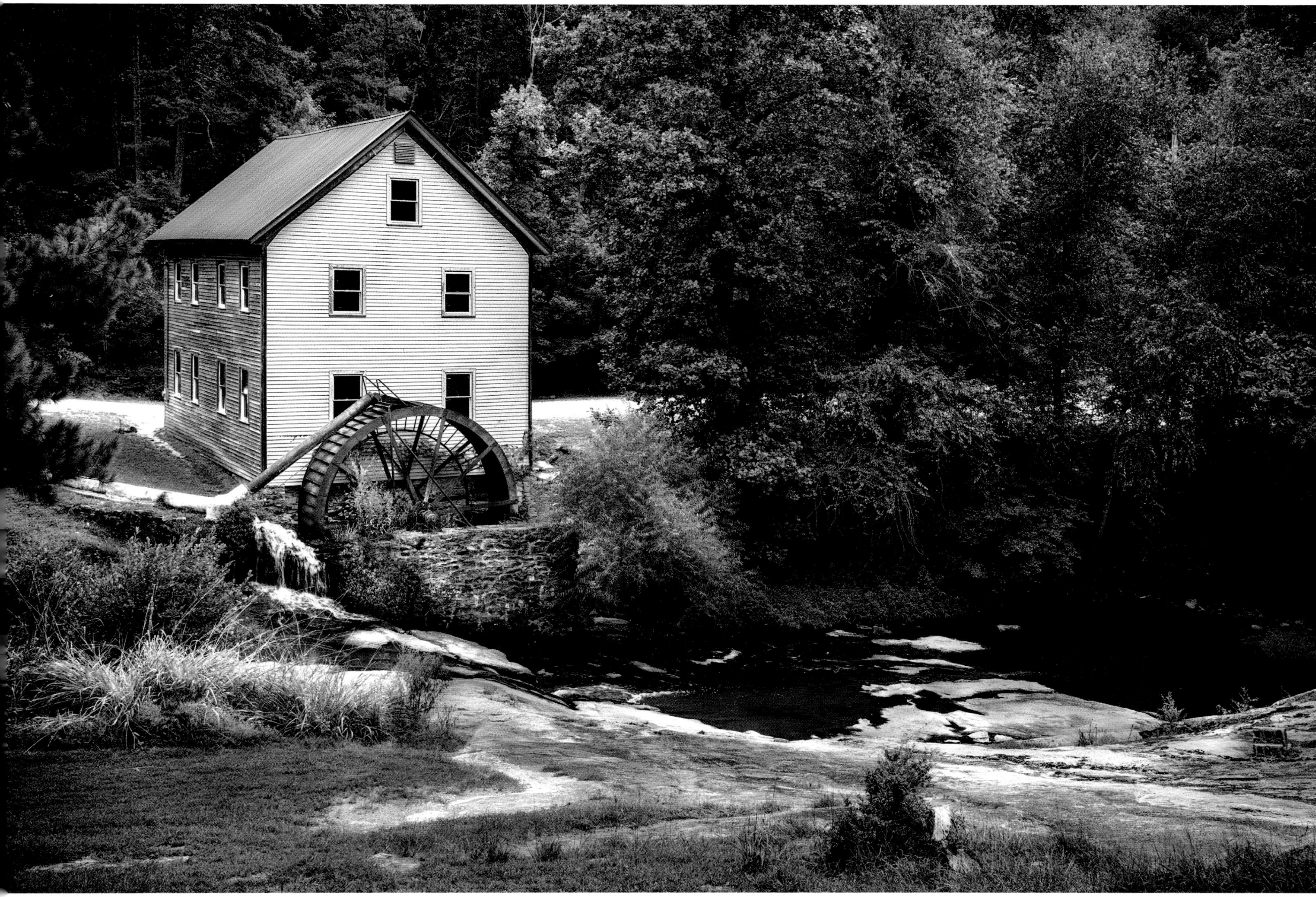

Sells Mill has been undergoing restoration since it was acquired by Jackson County in 2001, and today the structure and grounds appear to be in excellent condition, as evidenced in this photograph.

Prater's Mill

Dalton, Georgia, 1855 / 1970s

SOON AFTER THE railroad opened access to farmlands in northwest Georgia, Prater's Mill began 108 years of operation along flood-prone Coahulla Creek, near Dalton. As the area population grew, farmers would line up early with their wagons for a turn at the Prater millstones.

Benjamin Franklin Prater built the mill on land that had belonged to the Cherokee people. The Indigenous people called the site Fish Trap Shoals. The construction of the mill was typical for the South at that time. The timbers were hewn by hand, and the building was assembled with wooden pegs.

Both Union and Confederate armies made use of the mill and spared it from the torch during the Civil War. Over time and after the war, the Praters added a cotton gin, sawmill, woolen carding mill, syrup mill, general store, and blacksmith shop at the site.

The Prater family sold the mill in 1953, and a nonprofit, volunteer-based foundation took over and began restoration work at the historic site in 1973. The turbine-powered mill is operational, but it has been somewhat inaccessible due to structural issues. The mill site is the focal point of a well-maintained heritage park, now owned by the county. An arts and crafts festival has been held at the site in October most years. The park grounds around the mill and along the creek are popular picnicking areas in good weather.

Two arsonists were convicted of attempting to burn the mill in 1995. Most of the destruction occurred inside, though windows were also blown out and the roof was damaged, but the mill was repaired.

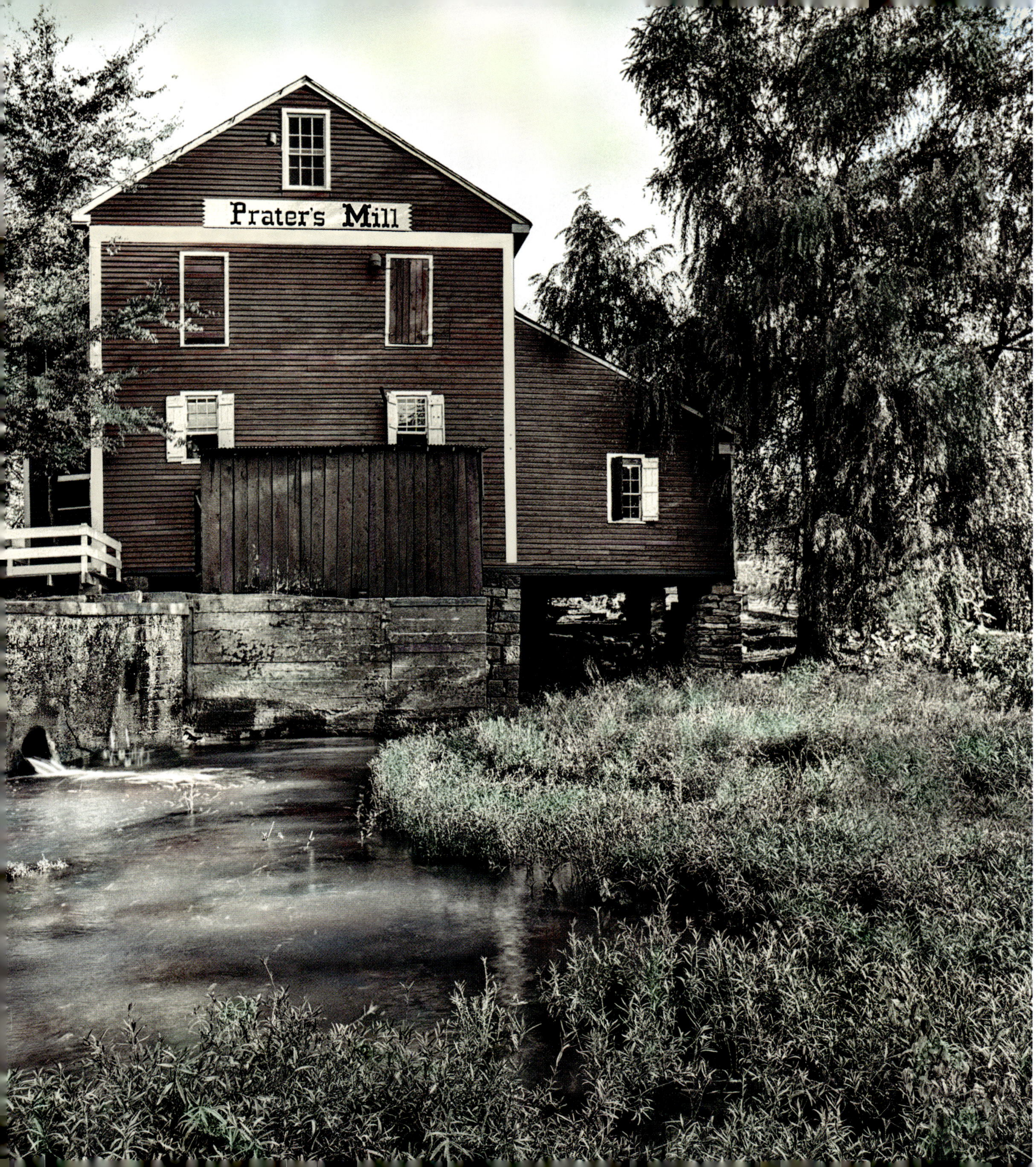

This vintage, hand-tinted image of Prater's Mill from across Coahulla Creek shows the mill as it looked some forty years ago. Since that time, the mill has endured numerous floods as well as an attempt to burn the structure.

Gulf Coastal Plain, Mid- and Late 1800s

Shepard's Mill, Greensboro, Florida, 1875; Rikard's Mill, Beatrice, Alabama, 1845; Parrish Mill, Twin City, Georgia, 1880s

WATERMILLS DEPEND on the energy of flowing water to operate. As water flows and falls from a higher level to a lower level, the energy can be captured by a waterwheel or water turbine. The rotating wheel can then be applied to do useful work like grinding grain or sawing timber.

Many regions of the South are ideally suited to watermill operation. Water races down the hillsides of the Appalachians, gushes freely from large springs in the Ozarks, and roars over falls where the Piedmont Plateau meets the Atlantic Coastal Plain.

Yet in South Georgia, southern Alabama, and the Florida Panhandle, the terrain is almost flat. Water still flows, of course, but it spreads out and moves more slowly, often through marshy and swampy areas. Such a slow, lazy flow is obviously not good for watermills, but the settlers and farmers of these areas still needed their corn ground and their timber sawed; after all, the crop-growing season in this area is long, and timber is abundant along the Gulf Coastal Plain.

Thus watermills had to be cleverly designed to extract what little energy was available from the lethargic water flow. The millpond was often set up to be much larger in area and might cover many acres. The mill itself was sometimes constructed directly over the stream and the dam to maximize the extraction of whatever energy the water could provide. By the last half of the nineteenth century, turbines were used because they were more efficient than waterwheels with limited millpond height and water head.

These watermills were not the high-horsepower plants found under better waterpower conditions, but when skillfully conceived, they could be made to work. As transportation and energy options evolved, milling was done elsewhere or with electricity, and the products were delivered back to the area. Only a few of these coastal plain mills remain today, but three good examples are shown here: Shepard's Mill in Florida, Rikard's Mill in Alabama, and Parrish Mill in Georgia.

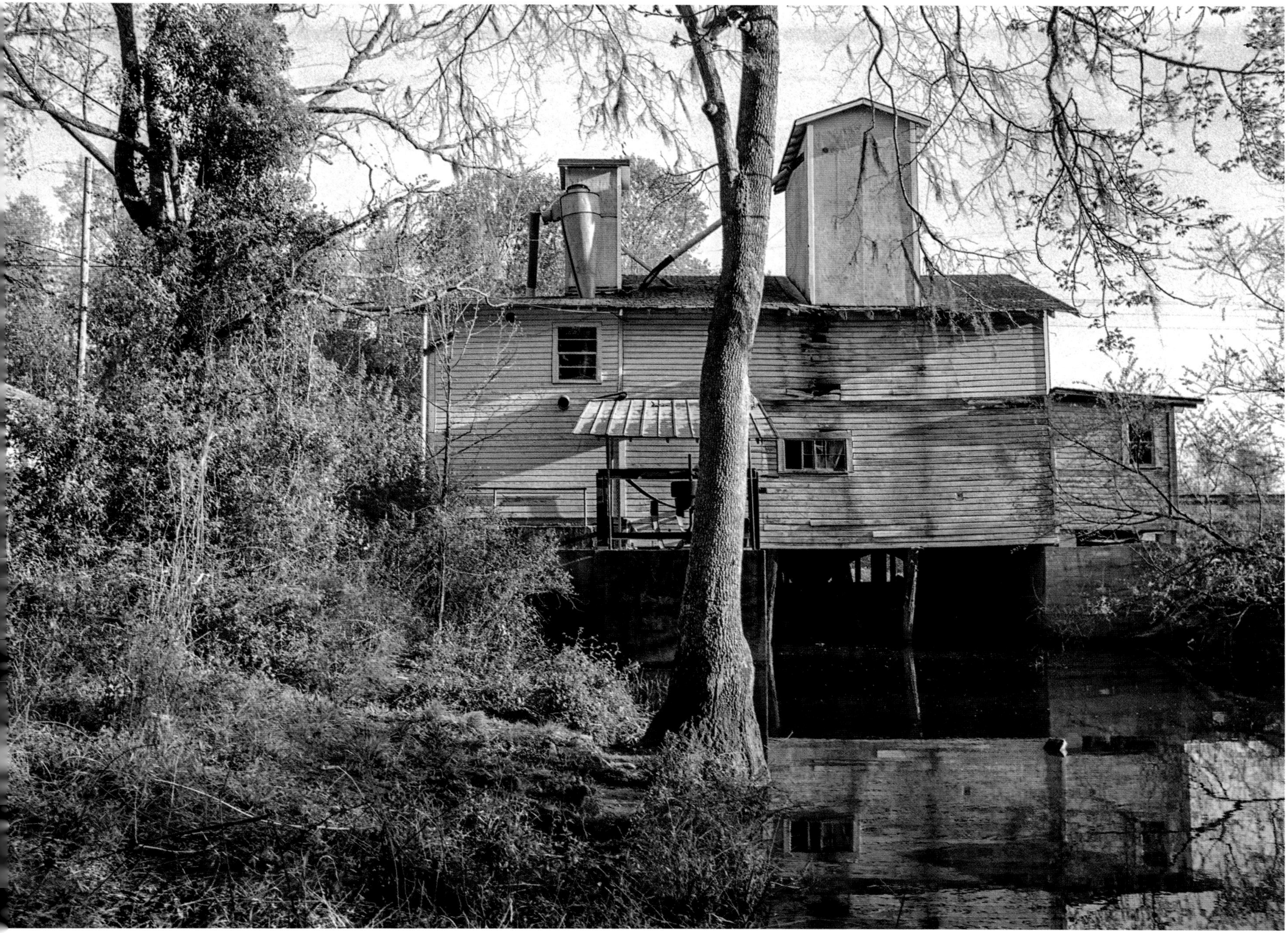

Shepard's Mill is located in the Florida Panhandle, an area that has endured many powerful hurricanes over the years. The terrain at the site is almost flat and leads down to the Gulf of Mexico. The mill straddles Telogia Creek, so the turbine below the mill can take full advantage of the low head of water pressure from the millpond.

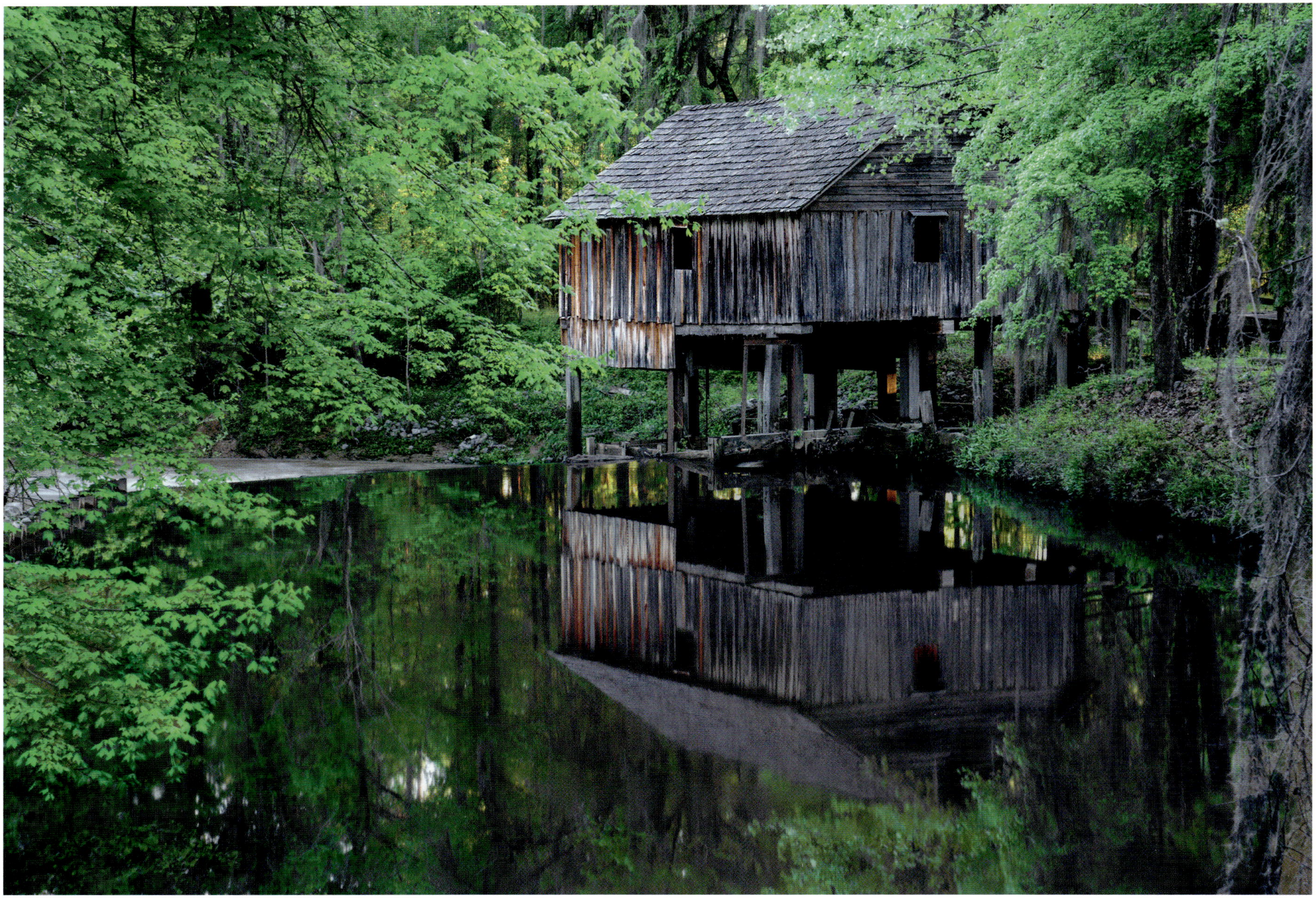

Rikard's Mill stands a few miles north of Monroeville, Alabama, in the land of Truman Capote, Harper Lee, and *To Kill a Mockingbird*. The mill dates to the antebellum period and stands on tall stilts at one side of the millpond. Water from Flat Creek flows through the turbine under the mill. This area of southern Alabama is a classically beautiful, swampy coastal plain.

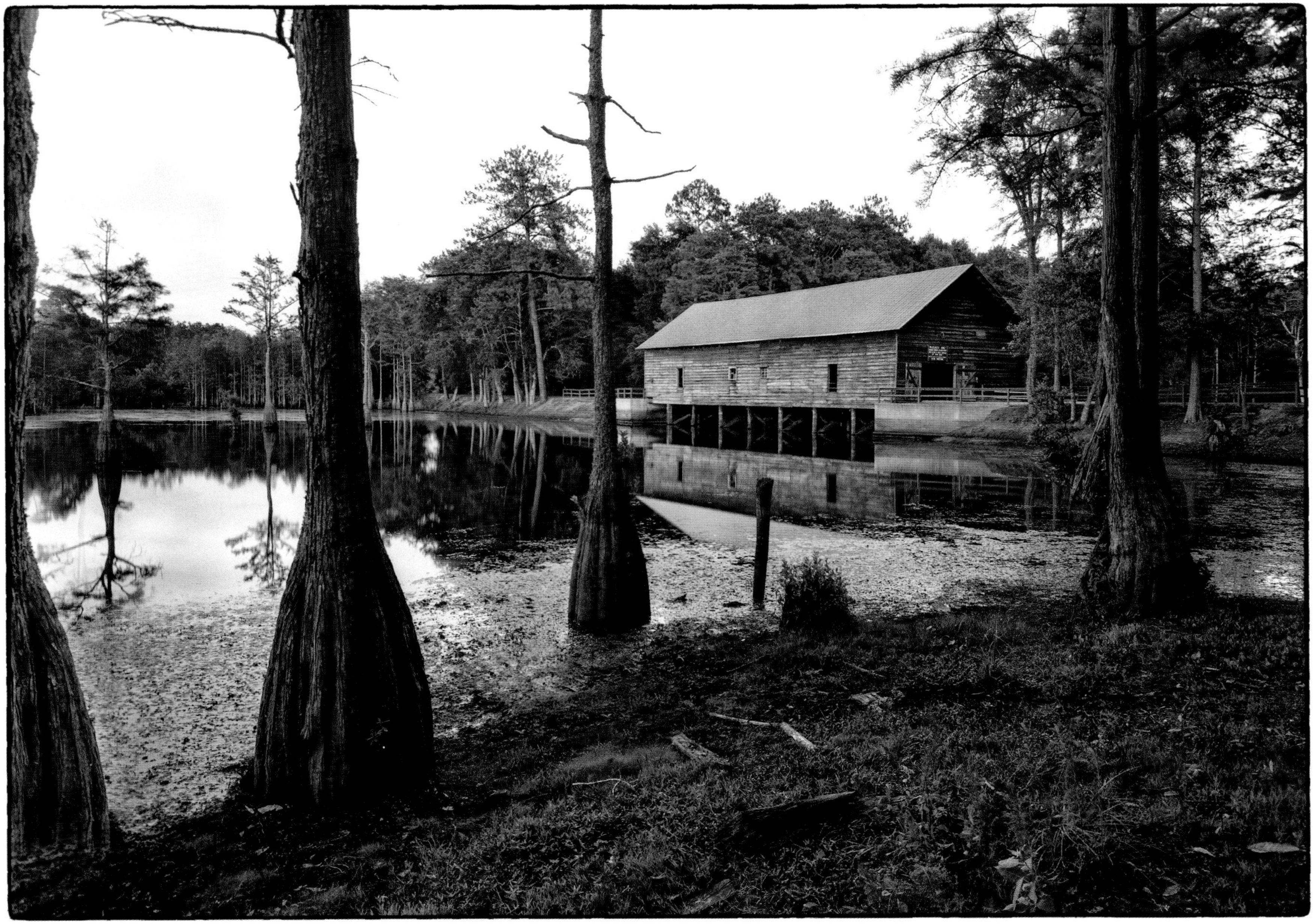

Parrish Mill in southeastern Georgia, known as Watson Mill to some in the area, is set in a cypress swamp complete with alligators, snakes, and Spanish moss. The structure also serves as a covered bridge, and over the years it has operated as a gristmill, a sawmill, and a cotton gin. Like Shepard's Mill, this structure straddles the dam and the water source, Fifteen Mile Creek, and the turbine is positioned under the mill.

Mill Springs Mill

Monticello, Kentucky, 1877 / 1970s

AS THE STORY GOES, in 1774 Daniel Boone came upon an area in central Kentucky with thirteen springs on a steep slope above the Cumberland River. With the springs representing a potentially abundant source of waterpower, he noted the area as a good location for watermills. A succession of settlers and then several watermills would soon follow. The area came to be known as Mill Springs, and today the closest city is Somerset.

A gristmill was first erected at the site in 1817. This early mill was well equipped with a pair of forty-eight-inch French millstones known as buhr stones, or just "burrs." As occurred all too commonly, the mill burned in 1824; it was not replaced with another operating mill until 1839. At one point during this era, the steep hillside was actually home to three mills—a carding mill, a sawmill, and a gristmill.

Historically, the area around the mill was the site of the 1862 Battle of Mill Springs, sometimes known as the Battle of Fishing Creek. Remarkably, unlike many other mill sites during the Civil War, the Mill Springs Mill survived the conflict.

In 1877, the current mill was built on the steep slope below the springs to replace the earlier Civil War–era structure. New owners in 1907 replaced the original and quite large twenty-eight-foot waterwheel with an even larger forty-foot metal wheel that was three feet in width, making the wheel one of the largest in the United States. Even today, the wheel can hold two tons of water during operation and can grind up to eight hundred pounds of corn hourly on each of two sets of millstones. Because of its large diameter, the wheel turns at the relatively slow rate of about 2 revolutions per minute, while the millstones turn at about 150 revolutions per minute with step-up gearing and linkage. (Smaller waterwheels at other mills often rotate at a rate of 10 to 12 revolutions per minute.)

Improvements continued at the mill in 1910, when a long wooden flume that had delivered water from the springs to the mill was replaced with the current lower-maintenance metal pipe. In the 1920s, a fifteen-horsepower, oil-fired engine was added under the mill building to supplement power during dry weather.

Confusingly, the area under the mill is busy with a maze of wooden spouts, shafts, and elevators that move grain. The explanation is that this mill is ingeniously configured to continuously circulate grain; this keeps the grain dry and avoids fermentation. Thus, when the mill is not grinding, grain can be kept in good condition despite the humid conditions along the Cumberland River, in an area where spring water constantly flows around the mill and down the steep hillside.

The main mill structure is a three-story wood-framed and wood-sided building supported by twenty-foot-high stone columns, which provides room under the mill for the grain circulation equipment. Access to the mill is by a steep path from a parking area above the mill. With its forty-foot waterwheel and its location on such a steep slope, the mill looks incredibly tall when viewed from the low side.

Around the time of the American Bicentennial, a number of mills across the South were restored. In preparation for the Bicentennial and as part of the national celebration, the U.S. Army Corps of Engineers completely renovated the gristmill at Mill Springs in the 1970s. The corps still operates the mill, which is open for tours seasonally, on weekends and holidays. The grounds and the path down to the river are open every day.

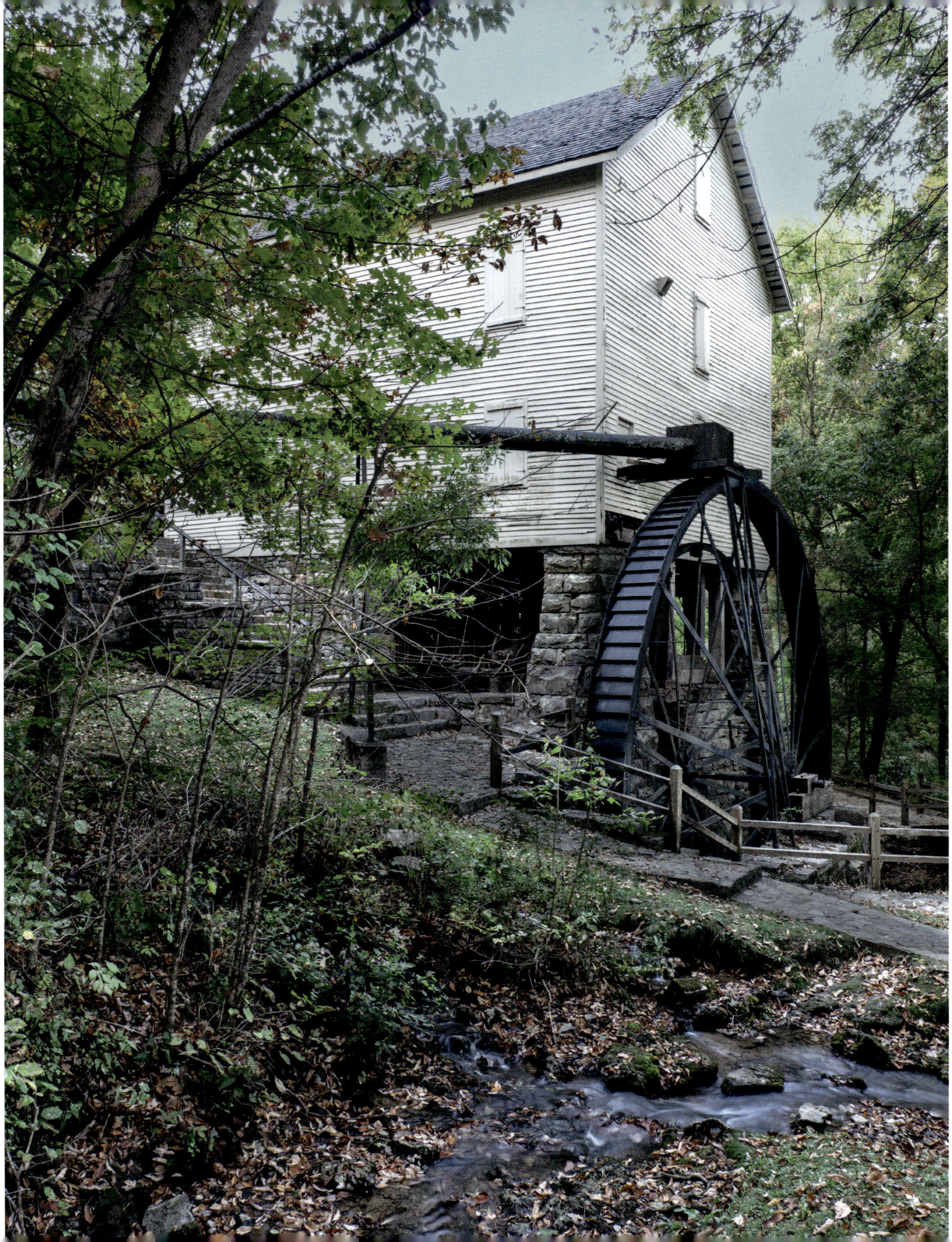

This image of Mill Springs Mill might deceive some viewers because the forty-foot waterwheel may be perceived as smaller in diameter than it actually is.

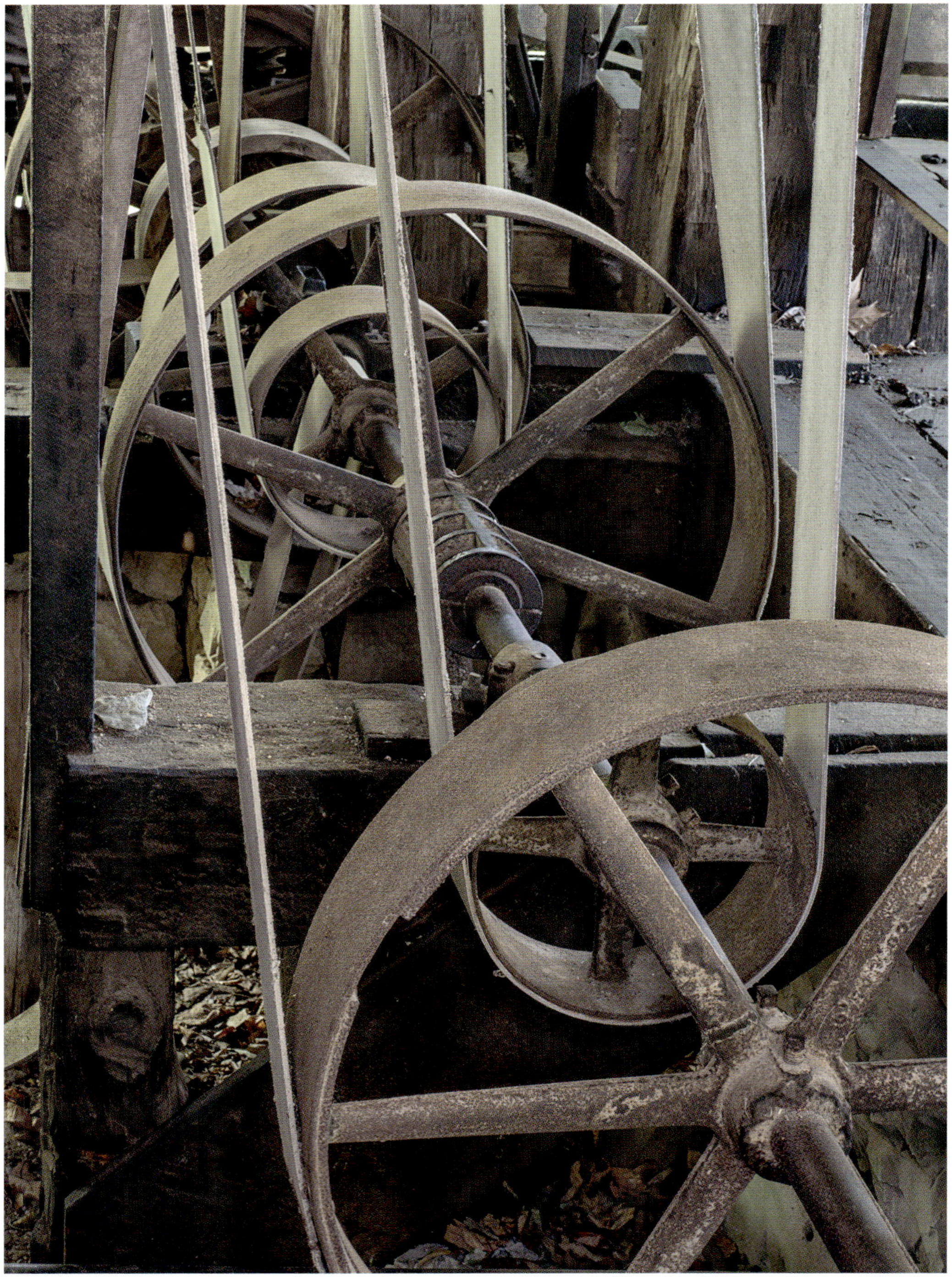

The machinery at the mill is powered by a complex combination of metal gears, belts, and pulleys.

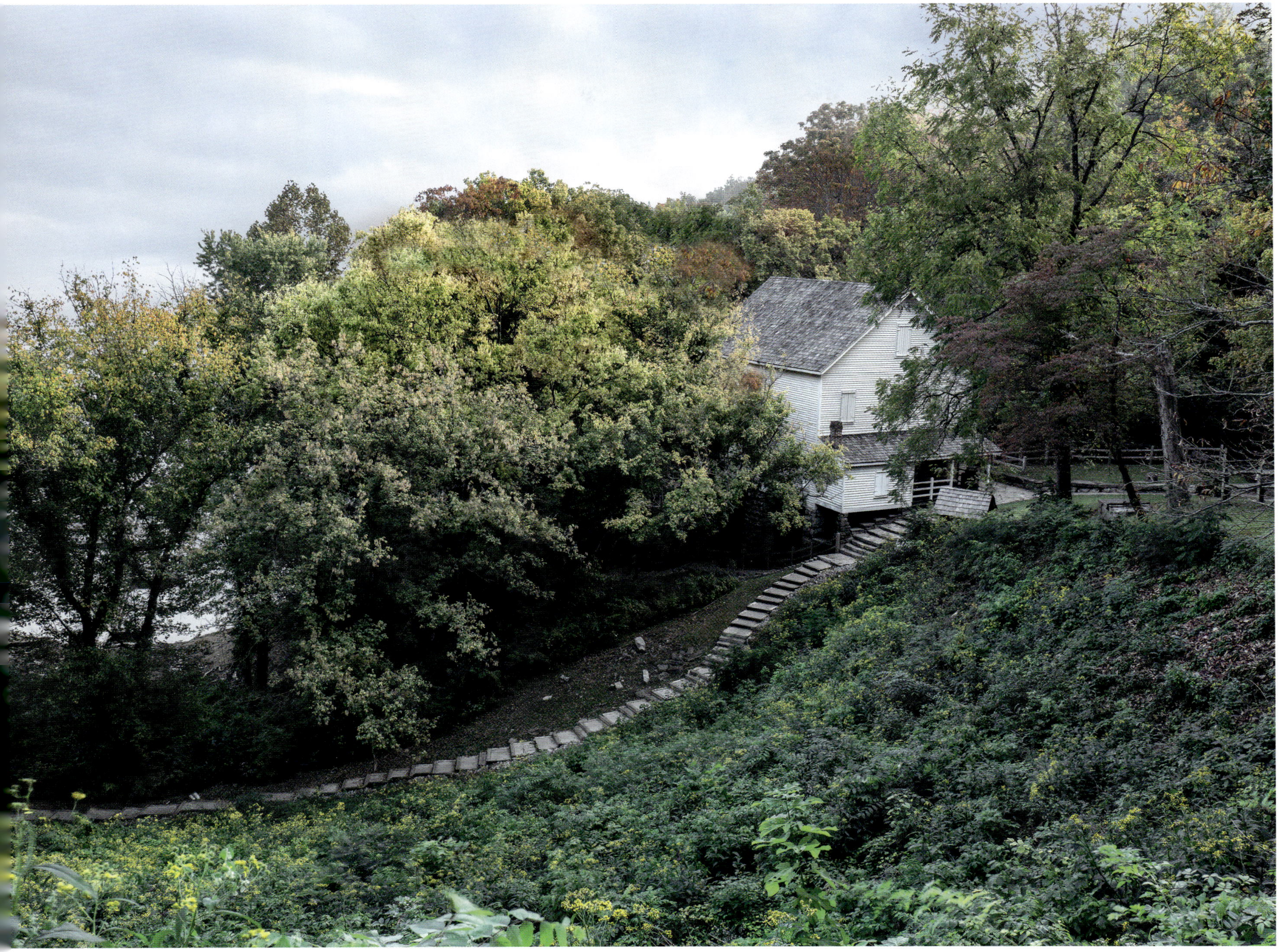

A steep pathway leads past the mill and down to beautiful Lake Cumberland just below. Power for the mill is provided by the generally abundant flow of spring water from sources up the hillside, above the mill.

Kentucky's Early Frontier Watermills

Bardstown Mill, Bardstown, Kentucky, Circa 1830; McHargue's Mill, London, Kentucky, 1812 / 1939

SOON AFTER Daniel Boone passed through the Cumberland Gap and into the frontier that would become Kentucky, settlers began to make the trek. These rugged individuals and families traveled over the Great Valley Road, the Wilderness Road, and trails forged by Indigenous Americans to a new life beyond the Appalachian Mountains.

Historians tell us that settlers journeyed to the frontier in waves or "migrations," usually traveling in groups for safety. Following Daniel Boone's trip in 1775, the first wave of pioneers began by 1780. Before long, thousands of settlers were spread across the territory.

That first wave included not only settlers but also adventurers and explorers who continued west across the great rivers and beyond to Missouri, the Great Plains, and the Rocky Mountains. The second wave, sometimes called "the Great Migration," occurred just after the War of 1812. These settlers came to clear land, build homes, and farm the soil. Early life in this area was a constant struggle. These hardy pioneers designed and constructed bridges, churches, and schools, and they built watermills. Their goal was to achieve a civilized quality of life for their families. (Modern studies show that about 25 percent of all Americans today can trace their heritage back to these settlers.)

Some documents from this period indicate that as early as the 1780s, a few crude watermills, probably small tub mills, were already being set up in Kentucky. The farmers needed mills to grind corn, wheat, and other grains.

Interestingly, and in what is now a famous Kentucky tradition, some innovative and creative souls early on began using watermills to make whiskey! Millstones could be used to grind corn mash, a process that was quite messy if done by hand. After a few years, whiskey was being shipped down the Kentucky, Ohio, and Mississippi Rivers to markets in New Orleans.

Unfortunately, the earliest watermills in Kentucky no longer exist, and even the state's watermills from the nineteenth century are mostly long gone. However, a few good examples of watermills from the early frontier in Kentucky have been preserved, restored, or recreated. The Bardstown Mill and McHargue's Mill, shown here, provide some insight into watermills from the frontier era in the Bluegrass State.

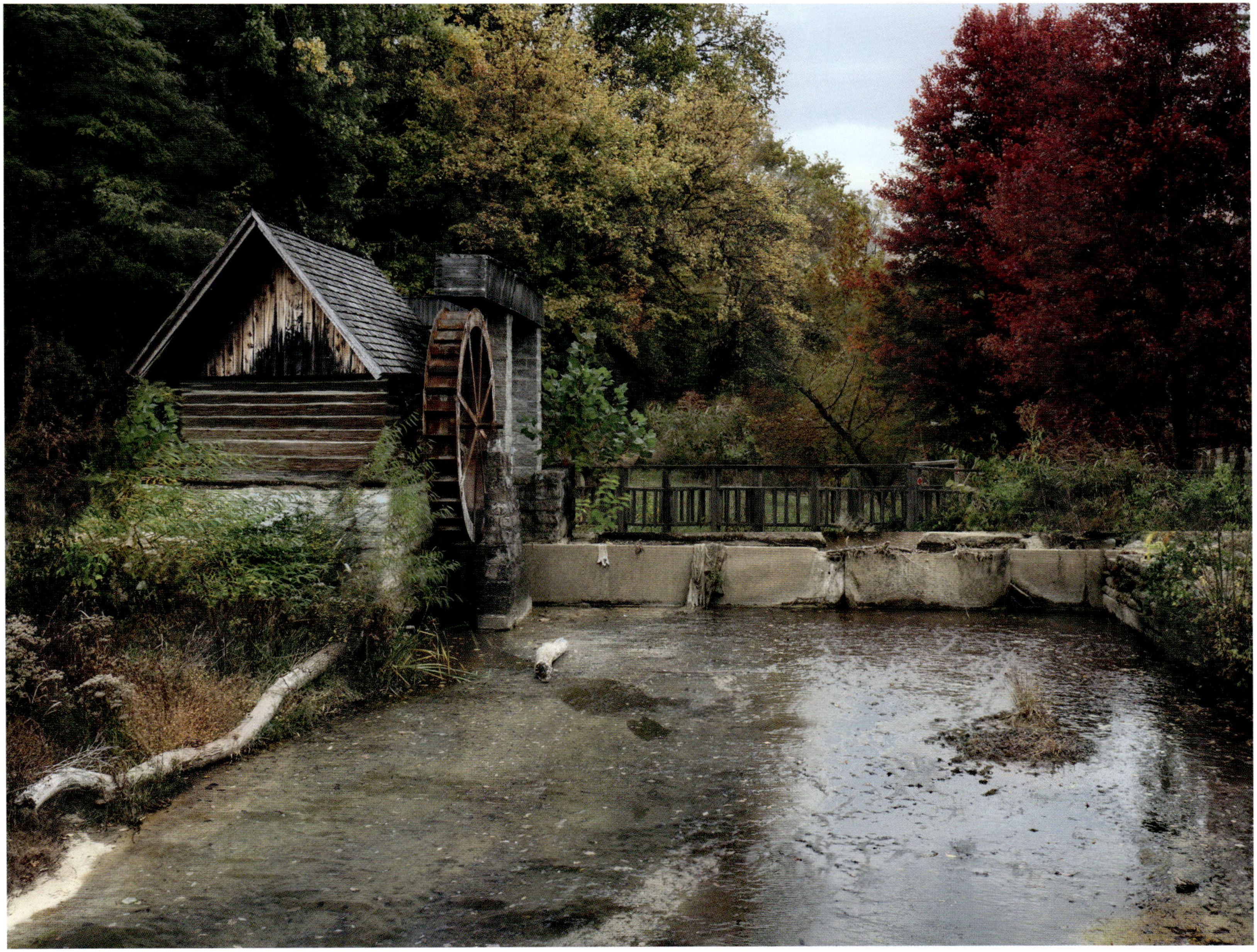

The Bardstown Mill, known by some as Brown's Mill, was originally built near Sand Gap, Kentucky. The log-constructed mill was carefully moved to Bardstown, Kentucky, in 1980. The millstones in this early mill are only two feet in diameter. The tall waterwheel is of the overshot design, though the mill dates to a time when many other mills in the area were still using small horizontal waterwheels.

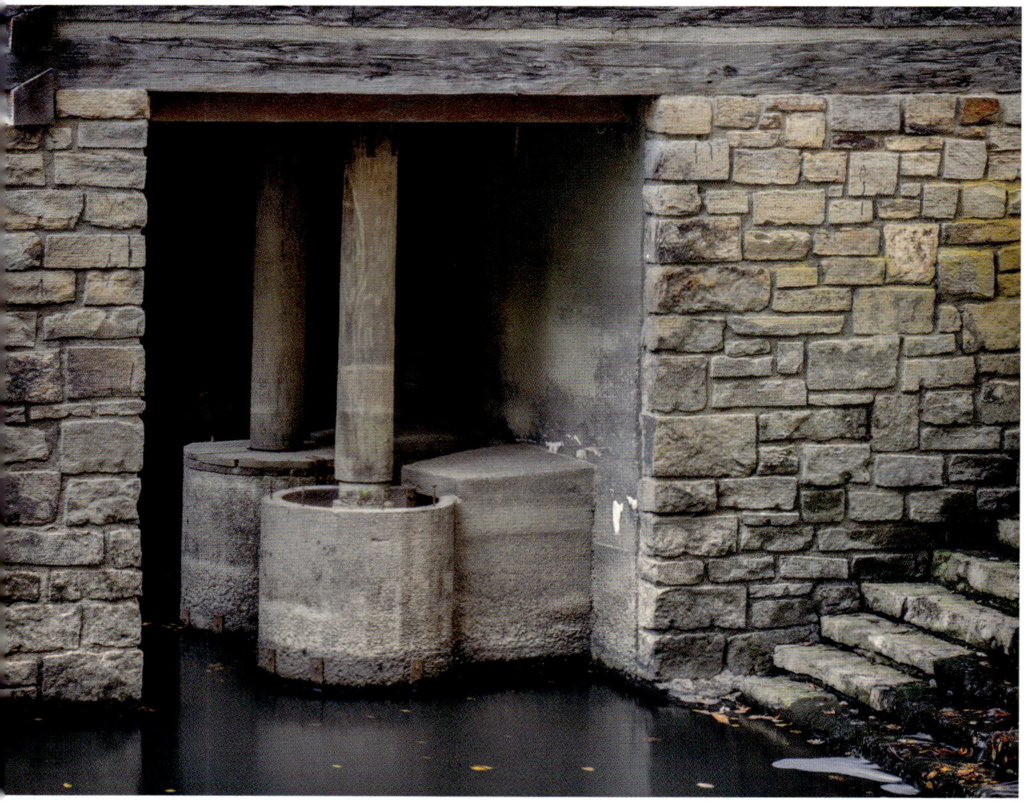

The “tub-type” turbines and vertical driveshafts are clearly visible in an open area below McHargue’s Mill.

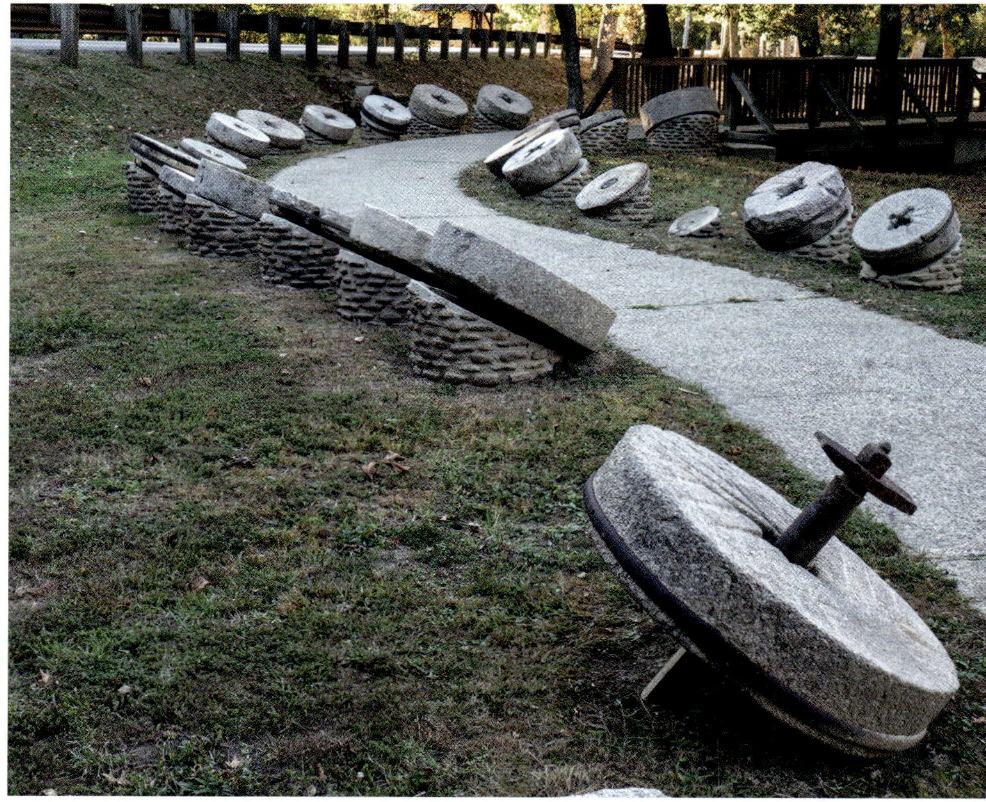

These are a few of the many millstones on display at McHargue’s Mill.

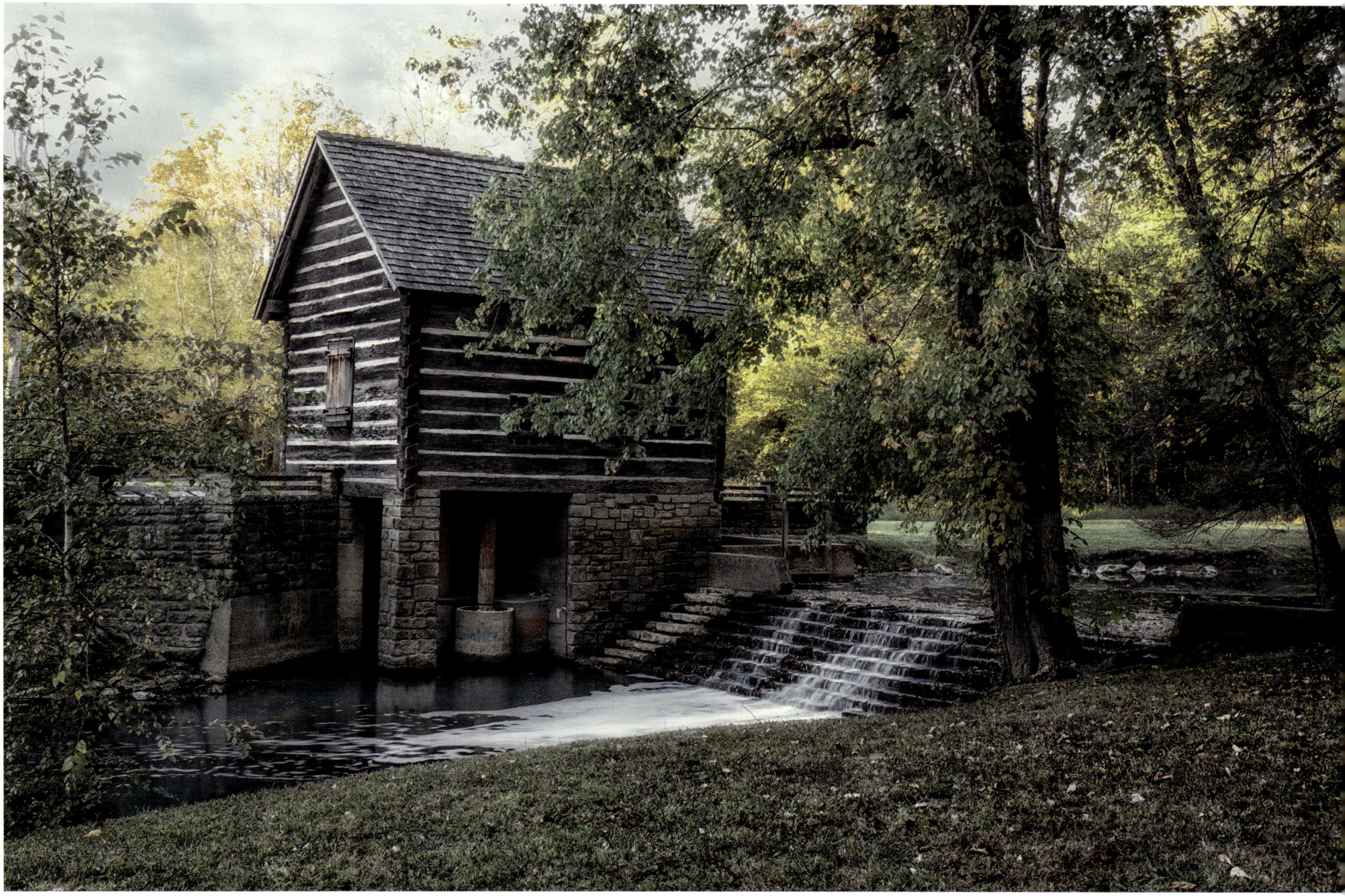

McHargue's Mill is primarily a 1939 reconstruction by the Civil Conservation Corps of an 1812 frontier watermill. Simple turbines can be seen below the mill. The mechanisms inside the mill are mostly authentic.

Weisenberger Mill

Midway, Kentucky, 1865 / 1913

THE WEISENBERGER FAMILY has operated a watermill at the same location near Midway since 1865. Before that time, an old stone mill had been located on this site along South Elkhorn Creek since about 1818. The Weisenbergers purchased the property in 1865 and replaced the old mill with the current mill in 1913. Reflecting the technology of the early 1800s, the old mill used a traditional waterwheel and millstones, but twin turbines were installed in the new 1913 mill, along with roller mills and an automated milling system.

Today, in the twenty-first century, the family still grinds, with a current daily production of approximately 150 bushels of local wheat and about the same amount of white and yellow corn. From 150 bushels of wheat, five thousand pounds of fine white flour are produced.

Weisenberger Mill products are in demand across the central Kentucky area by hotels, restaurants, and households that appreciate their high quality. The family remains motivated to continue improving its products and services.

The mill has also been involved in research projects with the University of Kentucky and the U.S. Department of Energy to evaluate water turbine optimization and hydroelectric generation at the site.

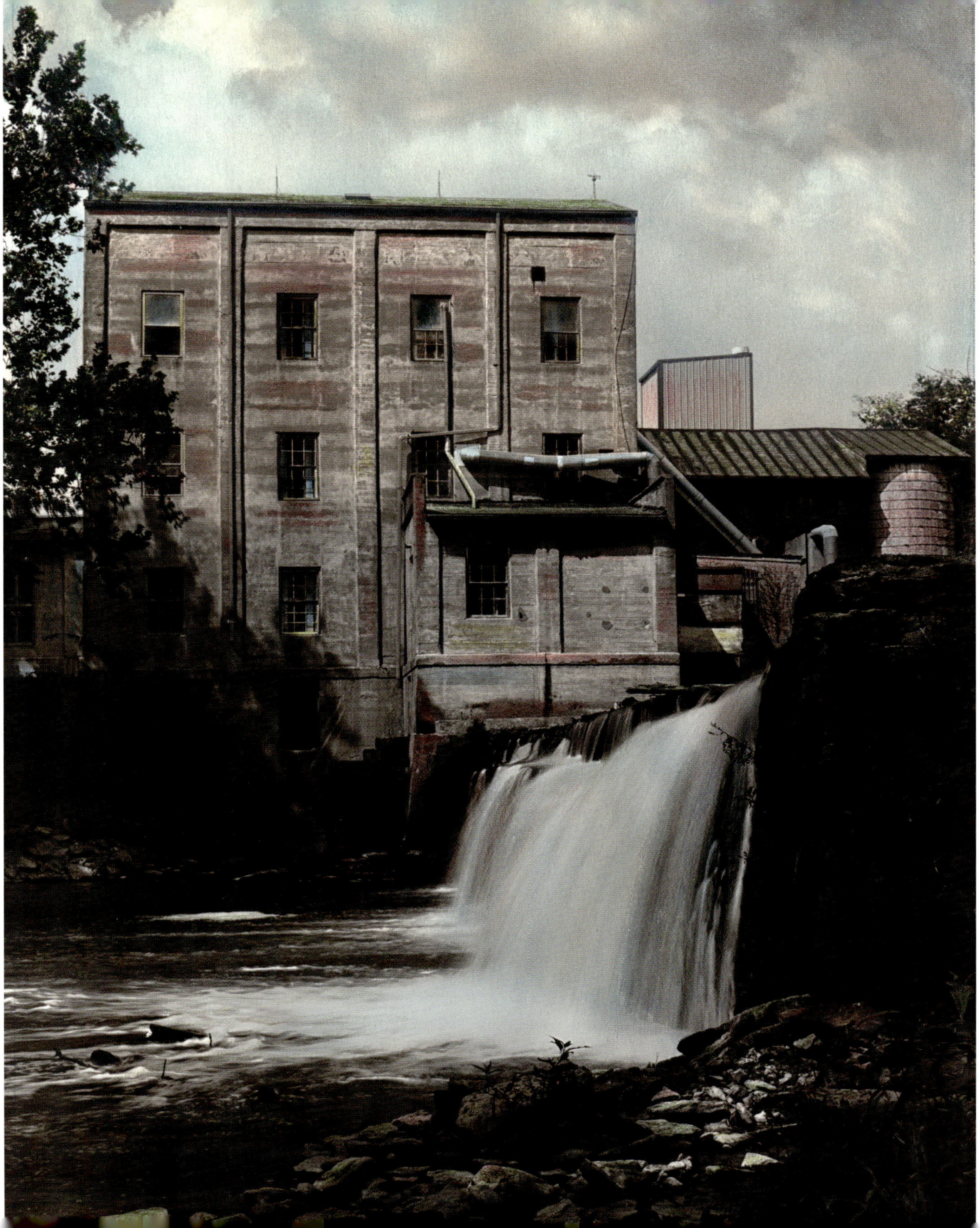

Weisenberger Mill, situated on South Elkhorn Creek, has been operated by the same family for six generations. This hand-tinted image dates to the early 1990s.

Linchester Mill

Preston, Maryland, 1681 / 1827 / Recent Restoration

ALONG WITH VIRGINIA, Maryland was the location of some of the earliest European settlements in the mid-Atlantic region of the South, and with those settlements came watermills. At one time, Linchester Mill was said to have been the oldest operating business in the nation. In 1827, the current mill replaced an even older seventeenth-century mill at the site, and it did not end commercial operations until the late 1970s, when a flood breached the dam.

Long before Europeans arrived in Maryland, this area along Hunting Creek was home to the Choptank Indigenous people. The area where the mill is located was active for centuries, and a Choptank trail crossed the creek here.

Washington's Continental army operated in the area during the American Revolution. The mill provided flour for the troops during the war.

Over many years of commercial operation, the site included the mill, a general merchandise store, the miller's house, the assistant miller's house, and other homes. It was an area in which Blacks and Whites, the free and the enslaved, and people of various religious faiths came together. As it did for the Choptank people in earlier centuries, the creek near the mill served as an overnight crossing location for the Underground Railroad movement in the period leading up to the Civil War.

Of concern to escaping enslaved people, Hunting Creek could be difficult to cross, especially at night. The water could be dangerously high or swift, and some did not survive their attempt at crossing. In addition, bounty hunters sometimes caught unlucky freedom seekers as they tried to cross the creek and dragged them back to their enslavers.

It is said that, as a result of this sad and frightening history, Linchester Mill experiences paranormal activity! Investigators and psychics report high energy levels around the mill. Similar stories persist at other watermills where people have been killed in the mill machinery or died mysteriously.

In 2003, the Caroline County Historical Society purchased the mill property, and since that time the society has worked carefully to restore the site. The mill restoration has been thoughtfully done, and the remarkable and classic 1917 Fitz metal waterwheel has been preserved as a link to the past. Inside, the mill houses an exceptional collection of milling machinery.

The site includes several other restored historic structures. The 1879 one-room Hog Island School with bell cote was moved to the property in the mid-2000s and restored. The grounds are lovely and include beautiful trees, picnic areas, tables, and walking and braille trails.

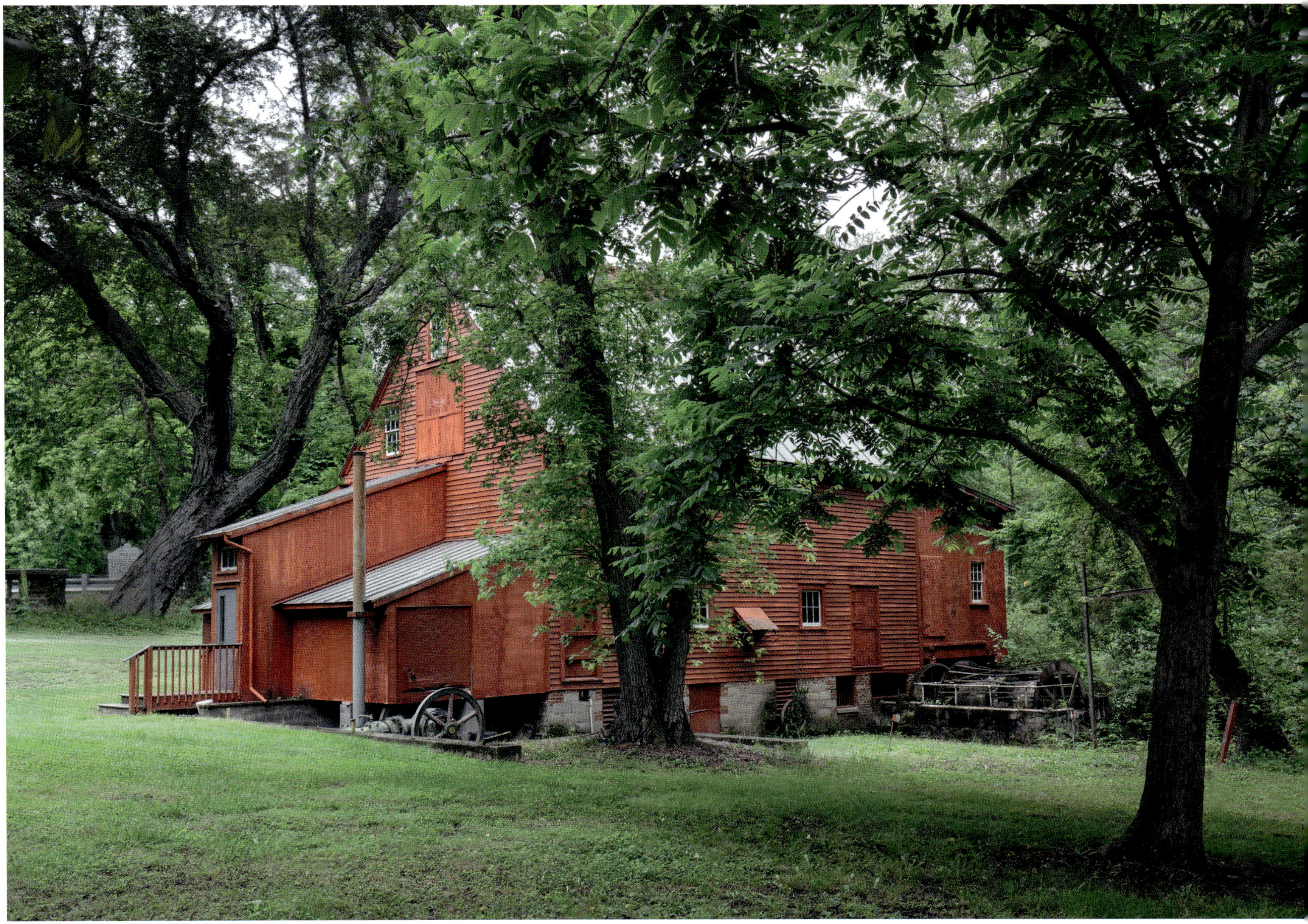

This view of Linchester Mill shows the back side of the building but nevertheless conveys the charm and character of the restored site, with an old engine at left and the classic waterwheel on the right.

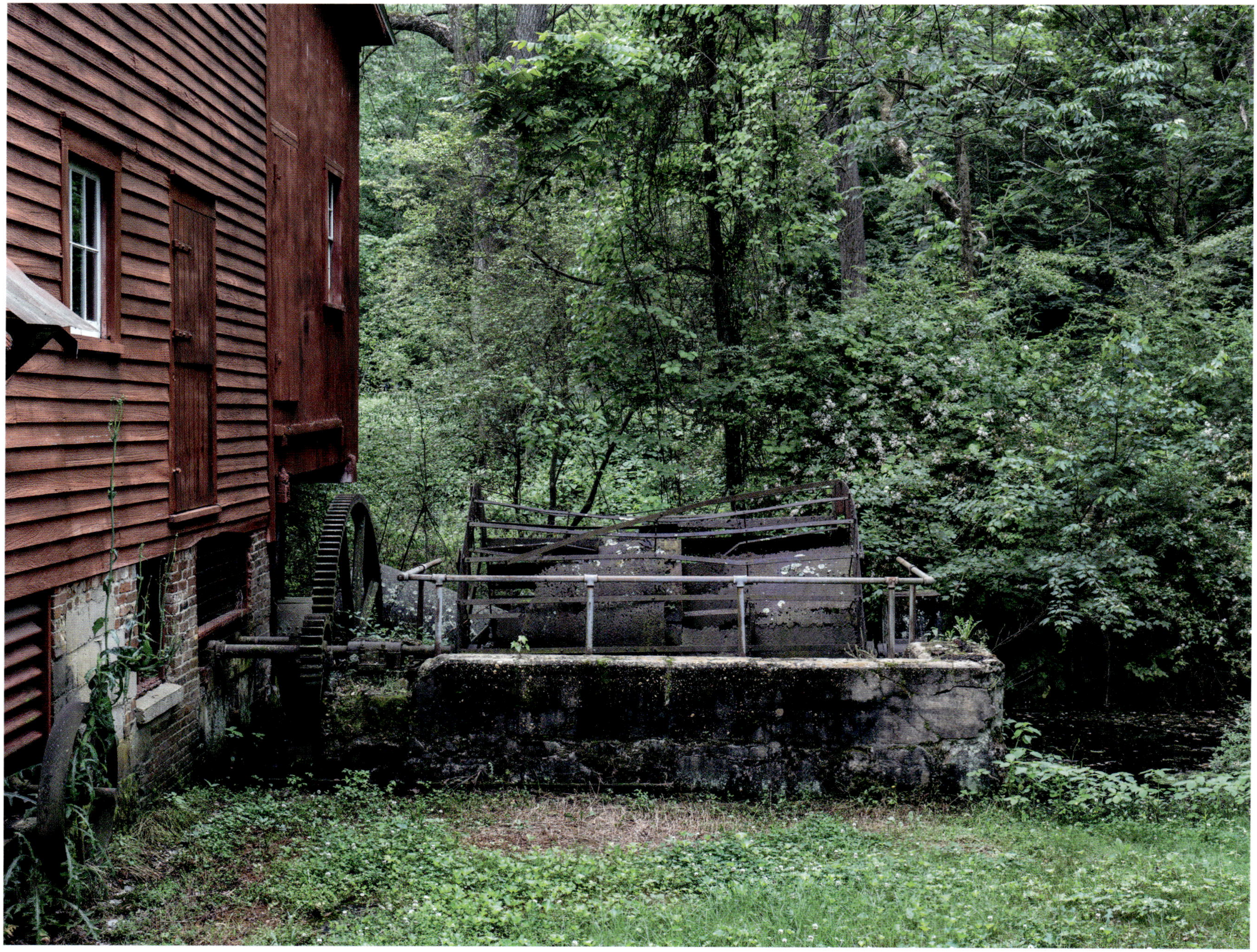

Restoration efforts at Linchester Mill included the successful preservation of the old 1913 Fitz metal waterwheel, and the wheel is now one of the most fascinating exhibits at the site.

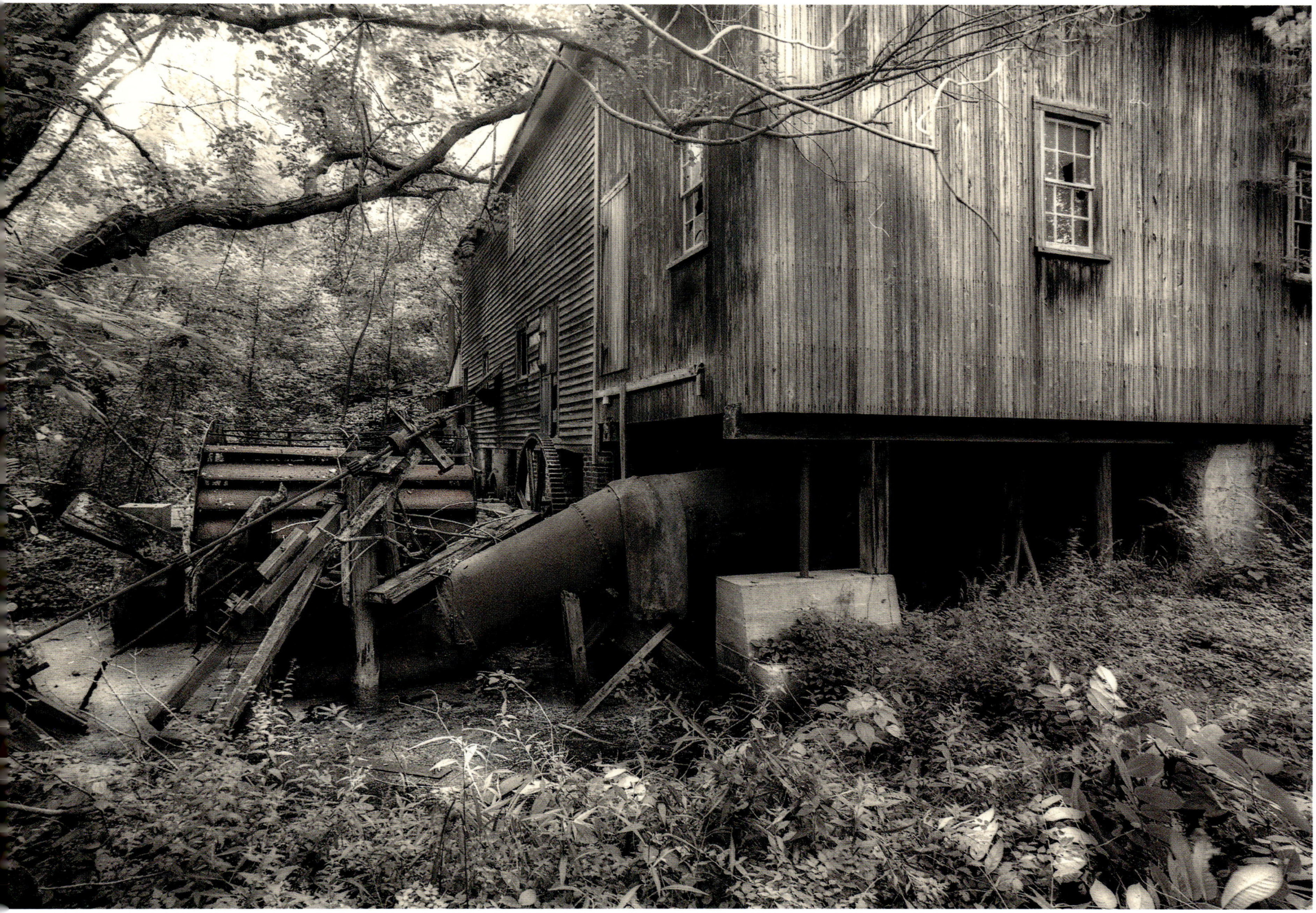

This vintage image by the author shows Linchester Mill in the late 1970s, when the site was mostly abandoned and in peril, before its restoration by the Caroline County Historical Society.

Old Wye Mill

Wye Mills, Maryland, 1682 / with Ongoing Maintenance

FOR ANYONE with a special fascination for the colonial and early Southern era, a visit to Old Wye Mill is akin to stepping onto hallowed ground. This mill is literally as old as anything in the former Southern colonies, and its history is as colorful as any landmark in the region.

Located in the community of Wye Mills on the Eastern Shore, Old Wye Mill is the oldest continuously operating watermill in the United States and one of the oldest commercial businesses in Maryland. The mill is near other historic sites, including Old Wye Church (built ca. 1721), Wye Mills Methodist Church (ca. 1793), an eighteenth-century school, and Wye Oak State Park.

In the early years of European settlement, the Eastern Shore in what would become Maryland was tobacco-producing country. During and after the American Revolution, farmers switched to growing wheat, and by the late 1700s gristmills had sprung up across the countryside to support the area farms.

Old Wye Mill is actually the second mill built on the site, with the first mill dating to about 1668. As happened at other mills across the South, the area around Old Wye Mill soon became a social and cultural center, and other services, churches, and schools soon opened nearby.

In this era, small boats would arrive in the creek below the mill and take bags of flour downstream to larger boats. During the American Revolution, much of the flour from the mill went to feed the troops of the Continental army. It is a relatively small mill, but this has never discounted its important history.

The structure is wood-framed and painted white. The millstones are said to be local and not exotic European rocks as at some mills, but they do the job and do it well. The millpond is just across the road, and today water passes under the pavement and onto the ten-foot-diameter Fitz metal overshot waterwheel. From there, the waterpower is transmitted by shaft to a belt and then to metal and wooden gears, which convert the hydro energy to mechanical energy that drives the stones. The current twenty-six-horsepower metal waterwheel improved the efficiency of the mill relative to the wooden wheels that had been used historically.

The mill applies the Oliver Evans automatic system for handling and processing grain, with buckets, lifts, and belts that reduce the strenuous lifting and manual labor involved in operating a gristmill. Without the Evans system, the mill would require about five workers, but with the system installed, it can be run by two people. The mill's production would also be considerably lower without the Evans system.

The State of Maryland purchased the mill in 1953. In 1955, the mill was badly damaged by Hurricanes Connie and Diane. A community effort, with help from Preservation Maryland, had the mill up and running again by 1959.

Today the mill provides educational demonstrations, and the site and the mill's interior serve as a museum of early American industrial technology. The Friends of Wye Mill, a volunteer nonprofit organization, cares for and operates the mill with funding from local and state grants, visitor contributions, and retail sales of stone-ground grain products and gift items.

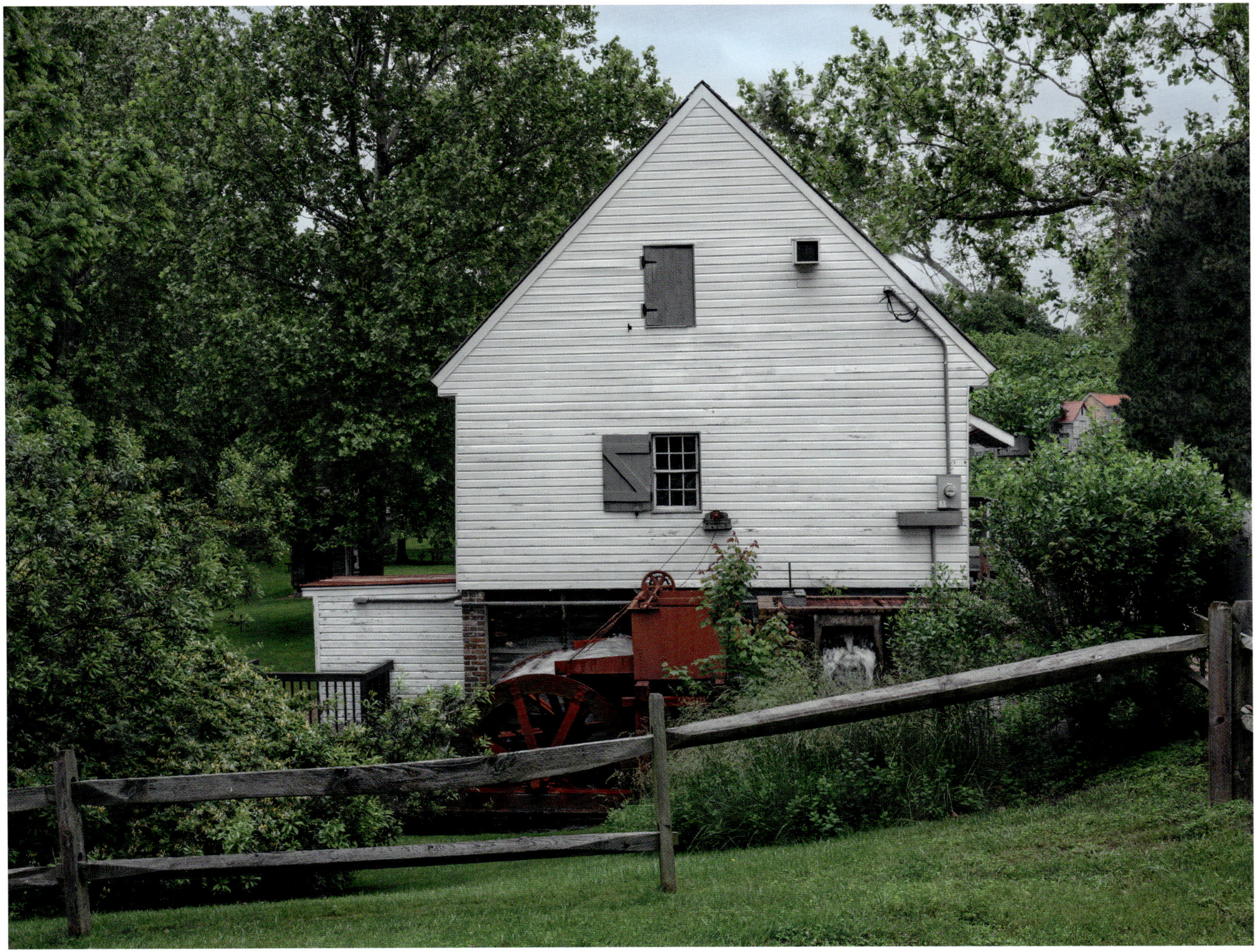

In this image of Old Wye Mill, the ten-foot-diameter, twenty-six-horsepower waterwheel is rotating under waterpower, while just inside the mill, the millstones are turning with a steady rhythm as they grind grain.

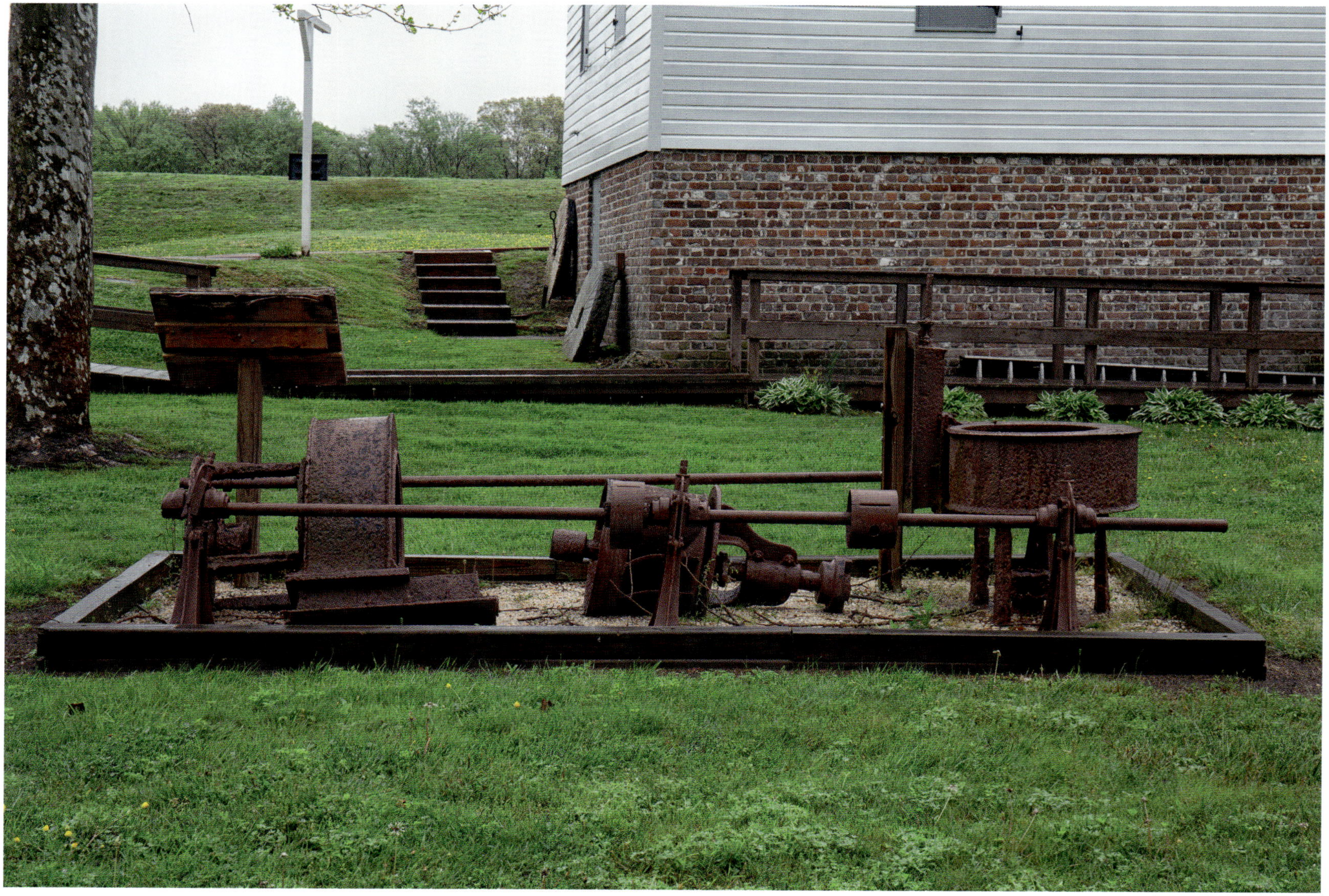

This display of strange metal pieces preserved on the grounds behind the mill features parts from a pair of turbines that a slick salesman sold to the miller in the late 1800s or early 1900s. This salesman convinced the miller that the turbines would work better than the waterwheel being used at the time. In reality, the turbines needed to be twenty-five to thirty feet below the level of the water in the millpond for the blades to spin fast enough for good performance. At Old Wye Mill, the water falls only about twelve feet from the millpond elevation to the mill, so the turbines never worked as promised. The turbines likely were removed shortly after installation and replaced with an old-fashioned waterwheel!

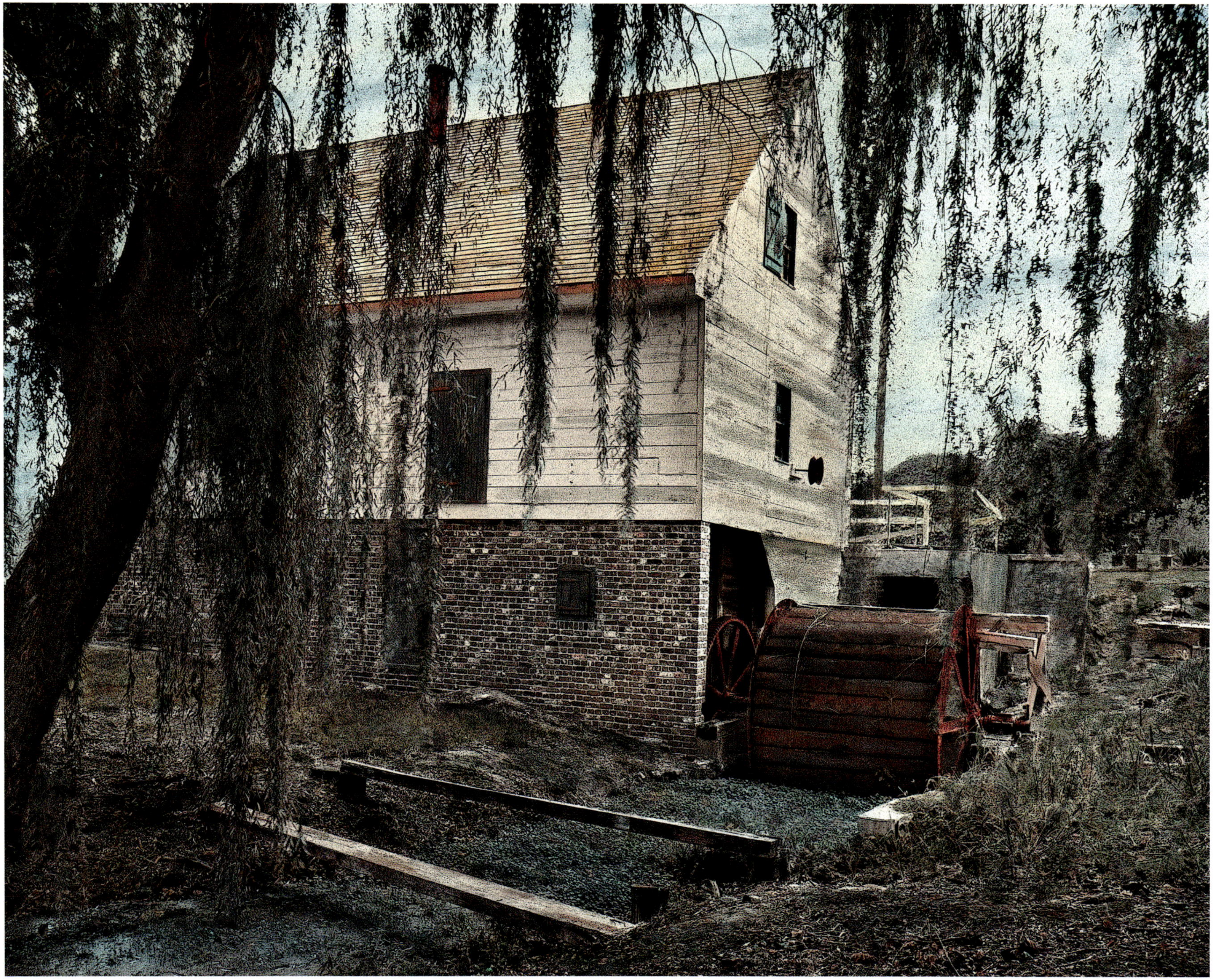

This hand-tinted image of Old Wye Mill shows the site as it looked in the early 1980s before modern landscaping by the Friends of Wye Mill.

Lake Bounds Mill

Stafford Springs, Mississippi, Before 1850

LAKE BOUNDS MILL IS—or more accurately *was*—located on a spring-fed millpond lake in southeast-central Mississippi just north of Laurel. The lake is fed by three large springs that produce fifty thousand gallons of water per hour, much like the more famous springs in Missouri. The log-framed mill was built to take advantage of the substantial water flow.

According to an article in *The Clarion-Ledger* (Jackson), when Theodore Bounds bought the property in about 1850, the mill was already there, along with several other log buildings. Bounds became a well-known miller in the area and, ever resourceful, soon added a blacksmith shop to his operation. In addition to grinding corn and wheat, Bounds traded with farmers in salt, bacon, tobacco, snuff, and canned tomatoes at his general store. Some references also mention a sawmill and sorghum mill on the spring-fed lake.

This is a valuable Southern historical site, but the structure has fallen into disrepair in an area where invasive kudzu can grow a foot a day during the summer months. The mill is currently in peril, if it even still exists.

The nearby Stafford Springs community is named for Edward Stafford. Stafford learned that the Choctaw Indigenous people used the local spring water to treat their sick and wounded because they believed it possessed healing powers. By 1892, a hotel had opened at the spring, and a bottling works began shipping the spring water all across the United States as a medical treatment! The hotel remained a popular destination until the Great Depression, when guests and patients stopped coming, and the building was demolished in 1952.

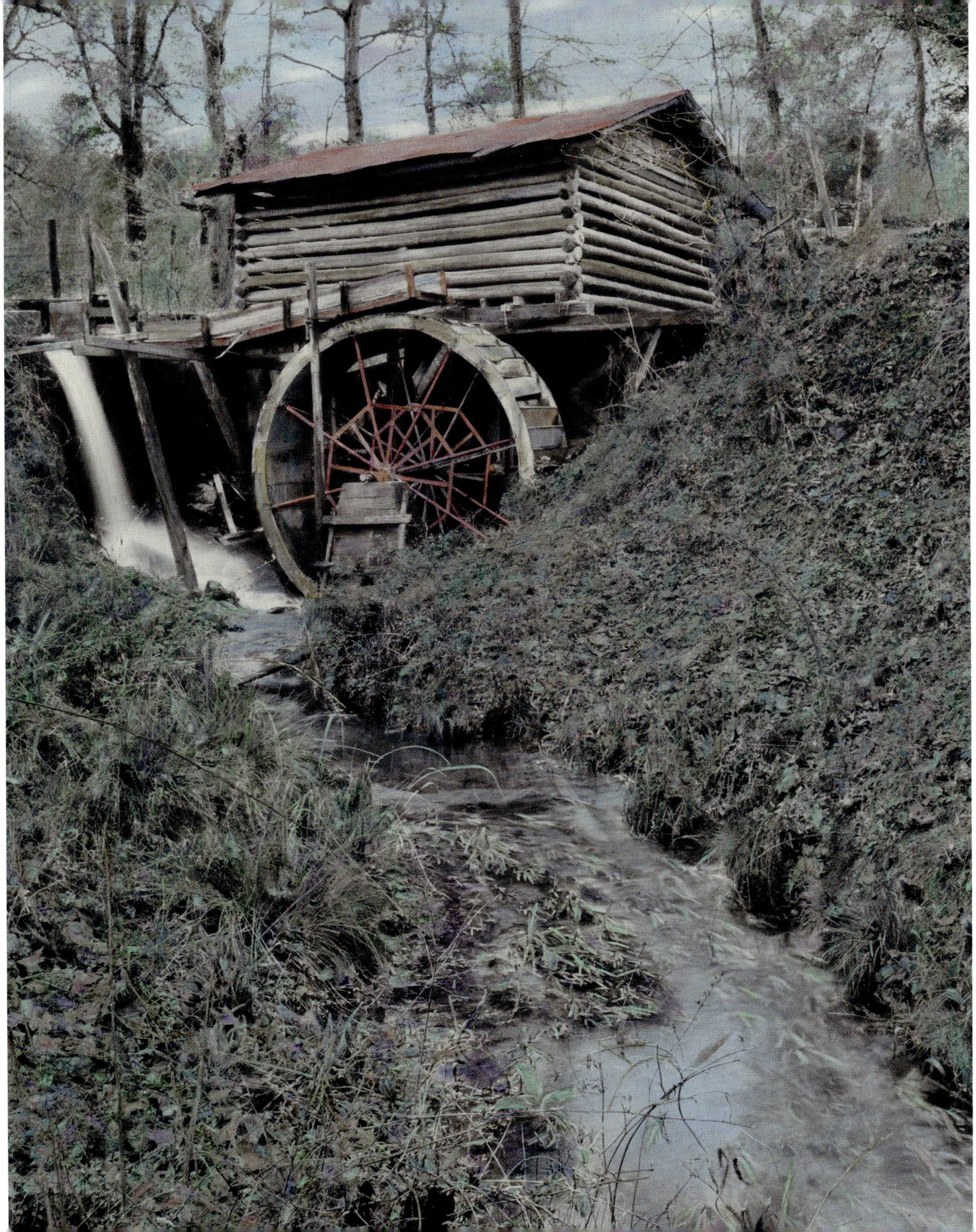

In this 1987 image, Lake Bounds Mill exudes a pleasing old-world charm. The log cabin construction and the large overshot wheel highlight an intriguing backdrop for southern Mississippi.

Carroll-Richardson Grist Mill

Enterprise, Mississippi, 1857 / 1987

THE CARROLL-RICHARDSON GRIST MILL was relocated from Cave Spring, Georgia, in 1987 and set on the precipice of a sixty-to-seventy-foot cliff at Dunn's Falls, overlooking the Chunky River. The stream over the falls provides a natural source of waterpower for the waterwheel without the need for a dam or millpond, although a small dam and millpond have been constructed just above the mill.

The waterfall is only somewhat natural and authentic. Apparently, in the mid-1800s the streambed was relocated a short distance to have the water flow over the steepest part of the ledge, improving the site's desirability as a resource for waterpower.

The mill was moved from Georgia to Mississippi to replace an earlier gristmill that had been located at the same site. The original mill was a cotton mill constructed by John Dunn at about the time the Civil War broke out. During the war, the mill produced blankets, knives, and other items, but it was especially known for making Stetson hats after they were introduced in 1865. A blacksmith shop, carding machinery, and a distillery were also part of the extensive operations at the falls.

Despite the original mill's importance to the Confederacy, it was not burned as General Sherman marched his Union army forces from Vicksburg to Meridian and across Alabama to Georgia. Historians believe Sherman was unaware of the mill, and his troops did not find it.

When the war ended, the mill continued to make hats and also produced cornmeal and flour. Somewhat surprisingly, the story is that the mill also printed money! As with other Southern mills, the site became a social gathering place and even hosted dances on the mill floor. The mill changed ownership several times over the years before eventually washing into the river below.

The current, relocated gristmill dates to the 1850s, making it about the same age as the original mill that had washed away. The building is open for self-guided visits, and throughout much of its three stories, machinery is labeled with names and descriptions to explain how a gristmill works.

The Carroll-Richardson Grist Mill is in a privately managed park that is popular for hiking, swimming, picnicking, and camping. Cabins are available for rent. The area is a natural wildlife refuge.

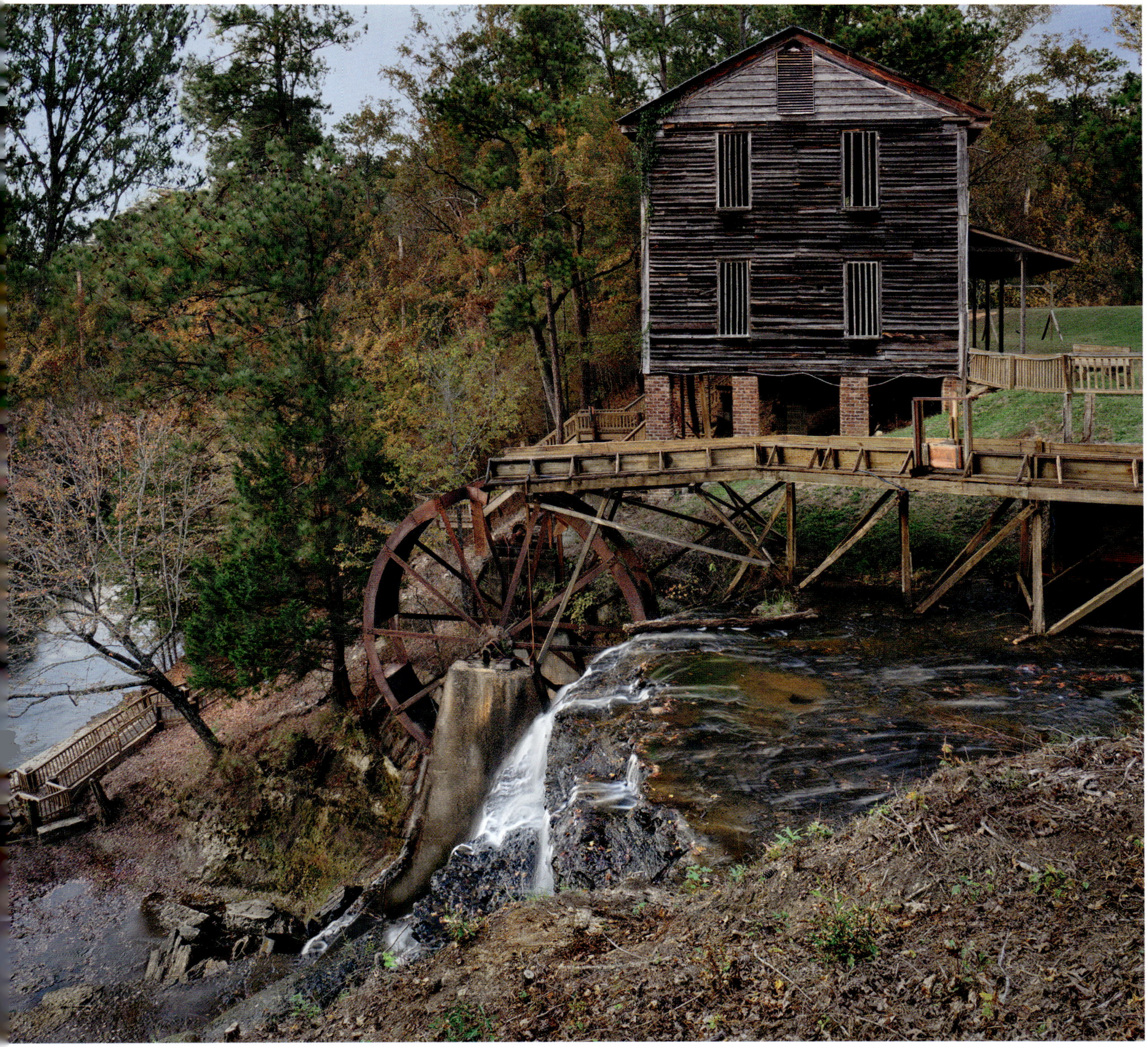

The Carroll-Richardson Grist Mill is set at the top of Dunn's Falls, with the waterwheel suspended over the ledge of the steep cliff. The Chunky River can be seen below and in the distance.

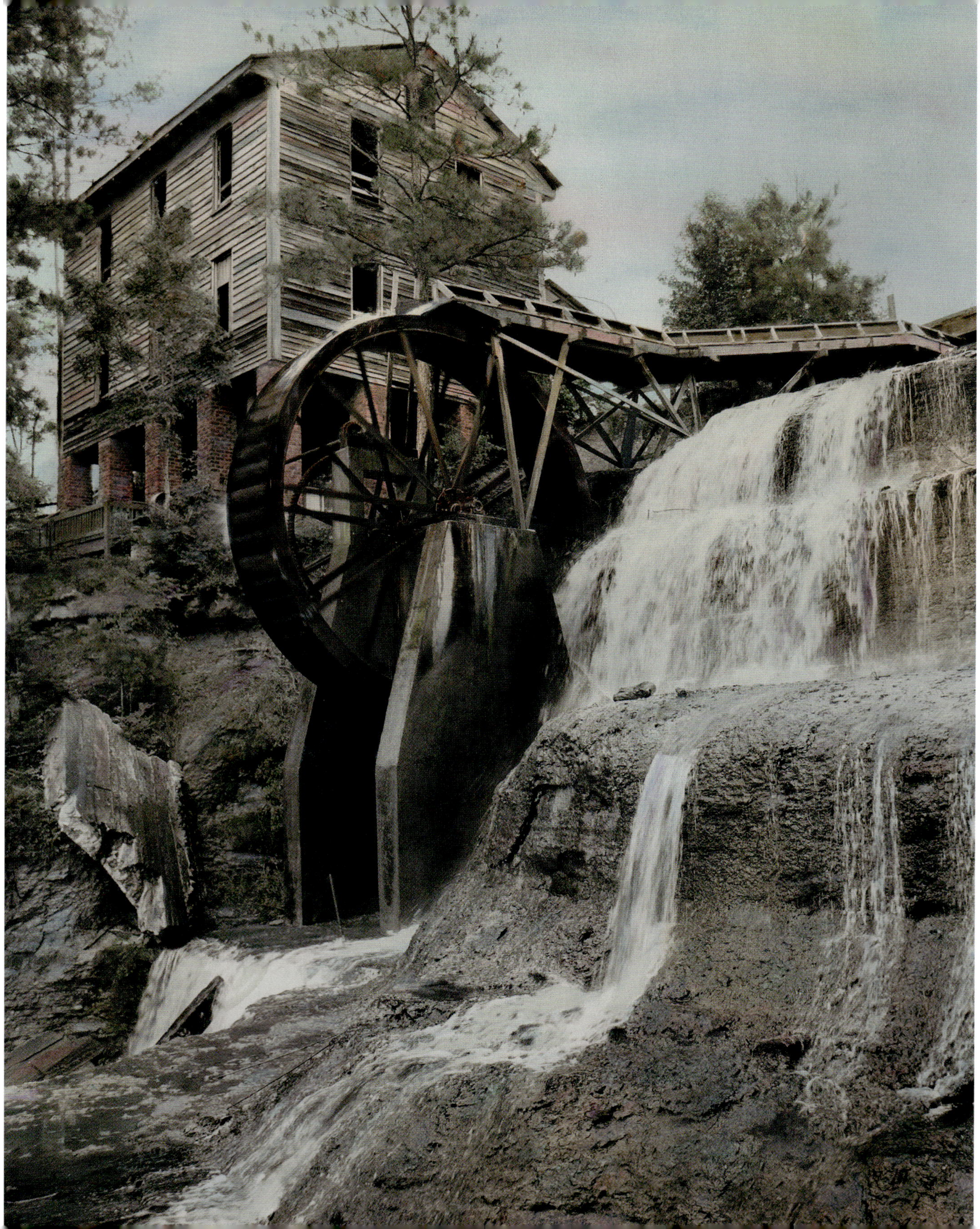

This vintage hand-tinted image offers a view of the mill from just above the Chunky River's edge. A series of flights of stairs and often-wet pathways provide access from the mill down to the river.

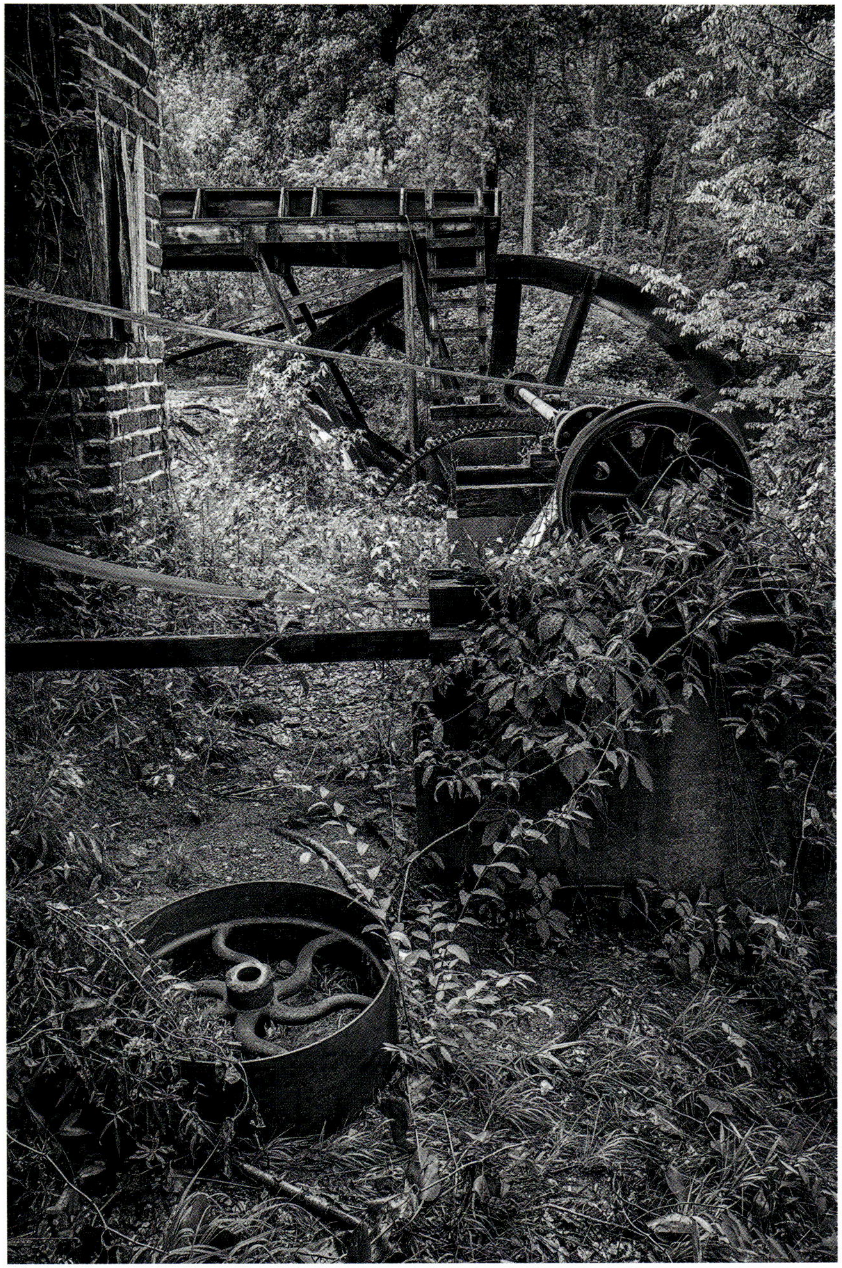

In this view from behind the mill, the waterwheel, main driveshaft, and belt linkage from the waterwheel to the mill can be seen, offering a better understanding of how the millwright envisioned the operation of the mill on the precipice of a steep cliff.

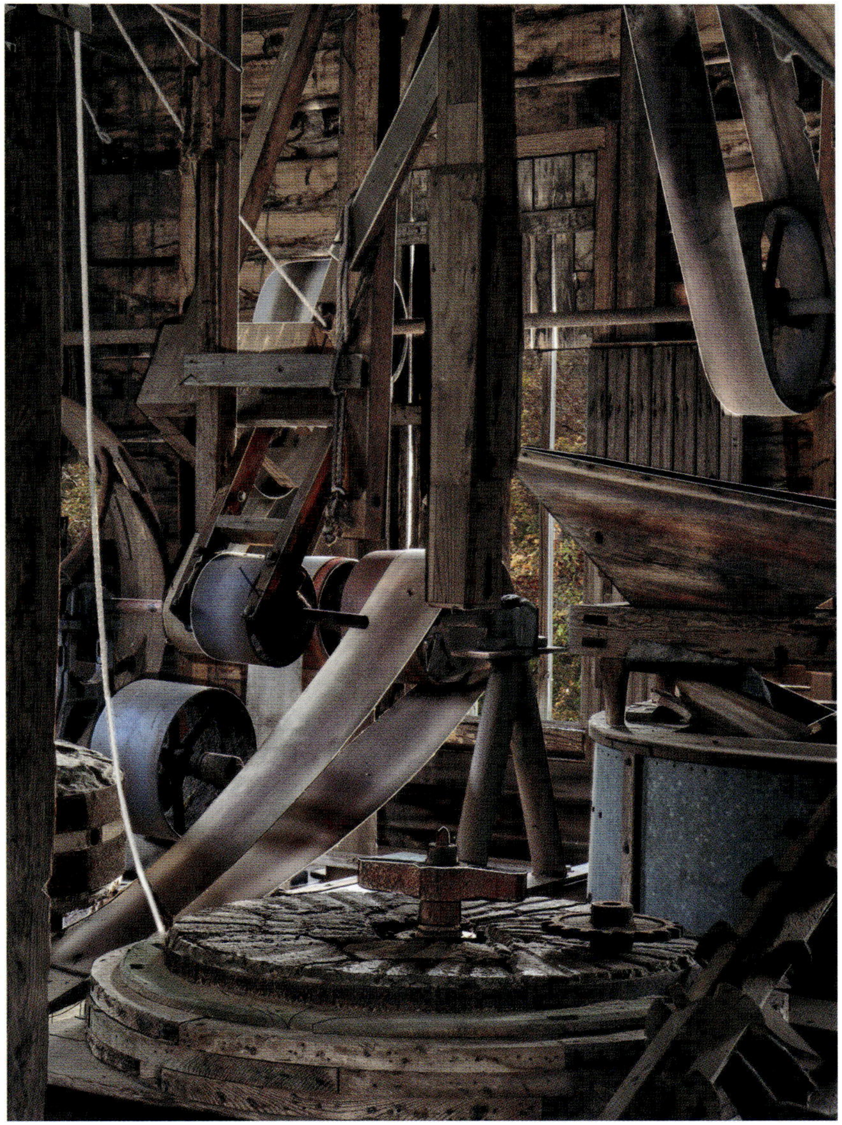

Inside the mill is a self-guided museum spread across three floors, with labeled machinery to promote an understanding of historic watermill operation. Family groups and young adults are often seen inside the mill studying and discussing the machinery and history of gristmill technology.

Sciple's Water Mill

DeKalb, Mississippi, 1790 / 1974

SCIPLE'S WATER MILL is one of the earliest, most authentic, and most remarkably well-preserved watermills in the South, and yet it is located away from the Appalachian Mountains, where most remaining watermills are found. The site includes the mill and a mill store, as well as a few other nearby structures. The mill began operation when northeast Mississippi was still essentially part of the Choctaw Nation.

The millstones used today are believed to be about 130 years old, although some accounts claim that they date all the way back to 1790, which is unlikely. The lower, stationary, or "base" stone weighs 2,100 pounds. The upper, rotating, or "runner" stone weighs in at 1,600 pounds. An 1880 turbine provides about fifty horsepower to grind corn or wheat. The mill can grind thirty-two bushels of corn per hour.

Until the 1950s, the mill also operated as a cotton gin and a sawmill. The sawmill could cut ten thousand board feet of lumber in a day. The cotton gin produced as much as twenty-five to thirty bales per day and was located on the second floor of the mill, which was destroyed during a 1973 tornado. Much of the mill was rebuilt after the storm, and the gristmill reopened in 1974.

The original mill had been built in the 1790s by a Dr. Hunnerly from South Carolina, but four different families owned the mill before it was purchased by the Sciple family sometime in the decade before the Civil War. The Sciples have never changed the price of their milling services—the toll is one-eighth of the corn or wheat that is ground. Farmers never pay money for the grinding of their grains at the Sciples' mill.

As was common for Southern watermills located on flatter terrain below the Appalachian and Ozark Mountains, Sciple's Water Mill was built straddling the Running Tiger Creek. The dam is located below the sawmill portion of the mill. This design provided for efficient use of the available head of water from the dammed-up creek.

The dam at Sciple's Water Mill holds back a relatively shallow millpond that at times can cover as much as twenty to thirty acres above the mill. During dry weather, the mill can drain much of the millpond with a day of grinding corn. Because storms with heavy rainfall are common across Mississippi, Running Tiger Creek can become a raging torrent at times—hence the name "Running Tiger"!

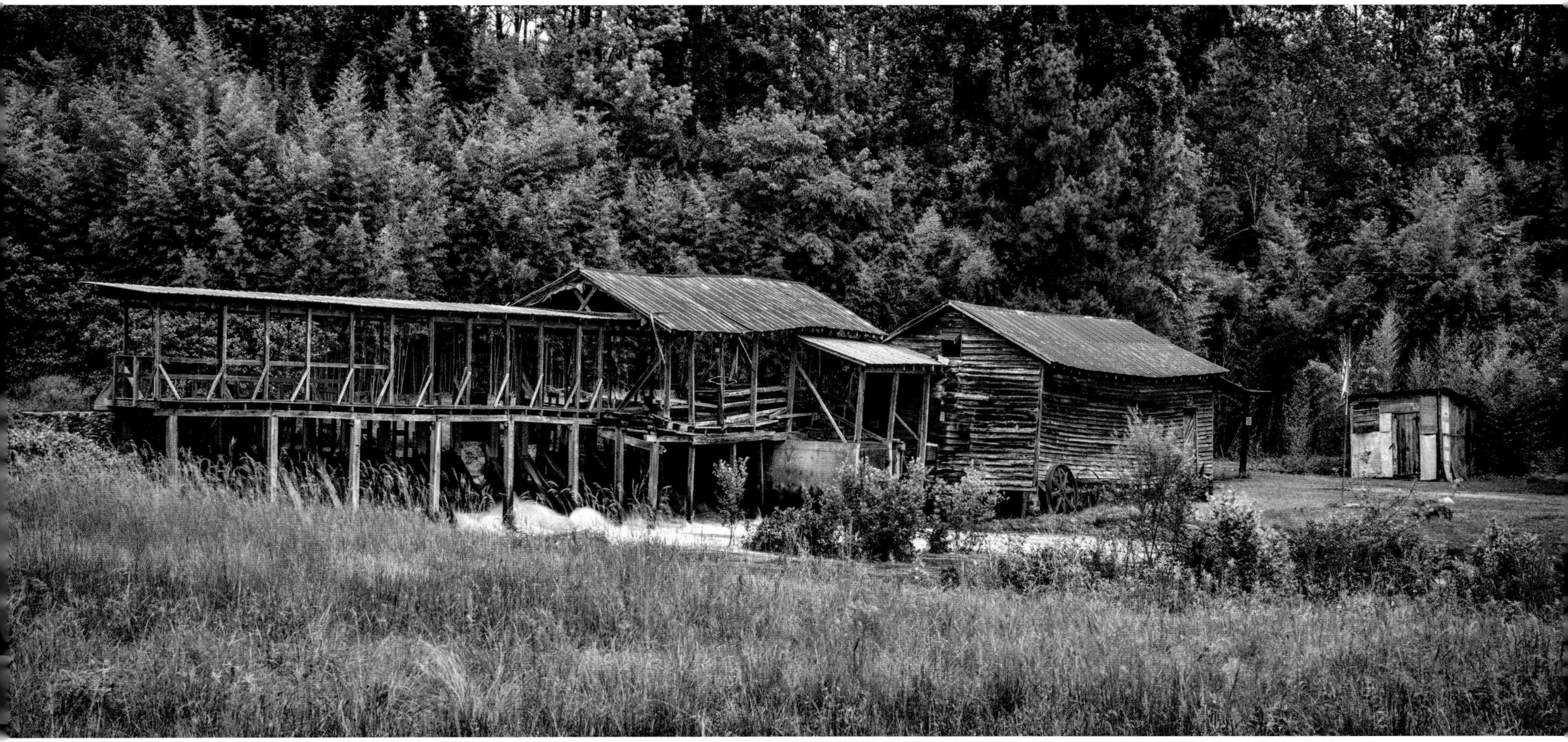

In this image of Sciple's Water Mill straddling Running Tiger Creek, water rushes over the millpond dam after a heavy rainstorm.

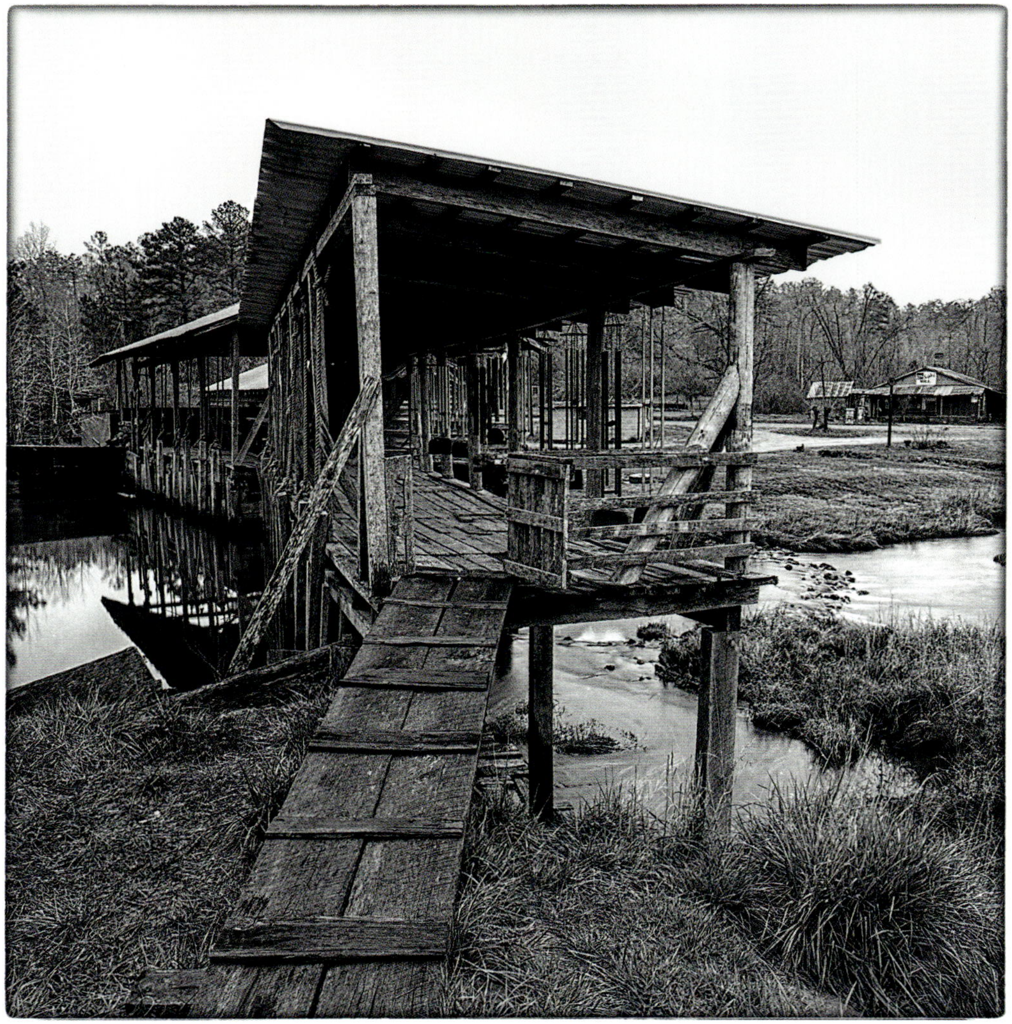

The Sciple's sawmill is still in place and looks as though it could be brought back into operation with a little cleaning up. Currently, this portion of the mill serves as a walking bridge over Running Tiger Creek. The mill store can be seen in the distance across the road.

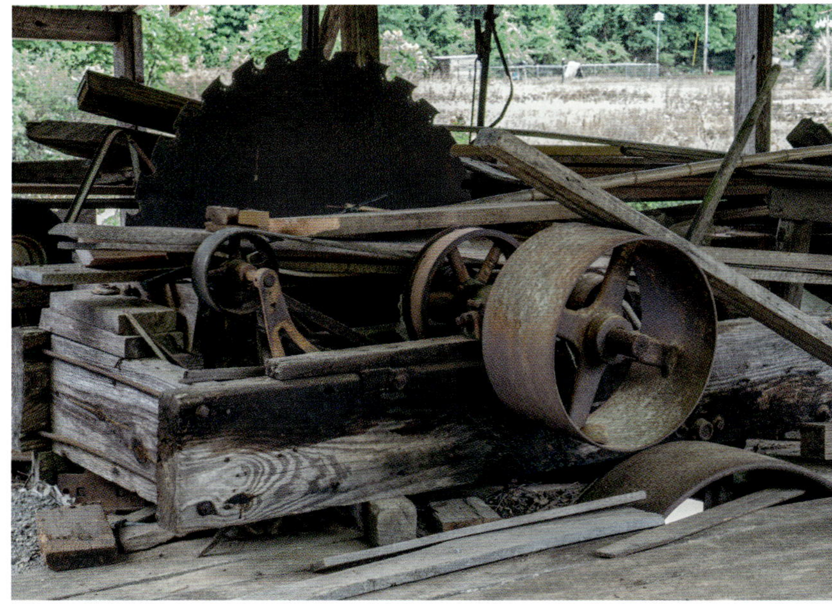

A large circular saw blade was used at Sciple's Water Mill until the sawmill closed in the 1950s.

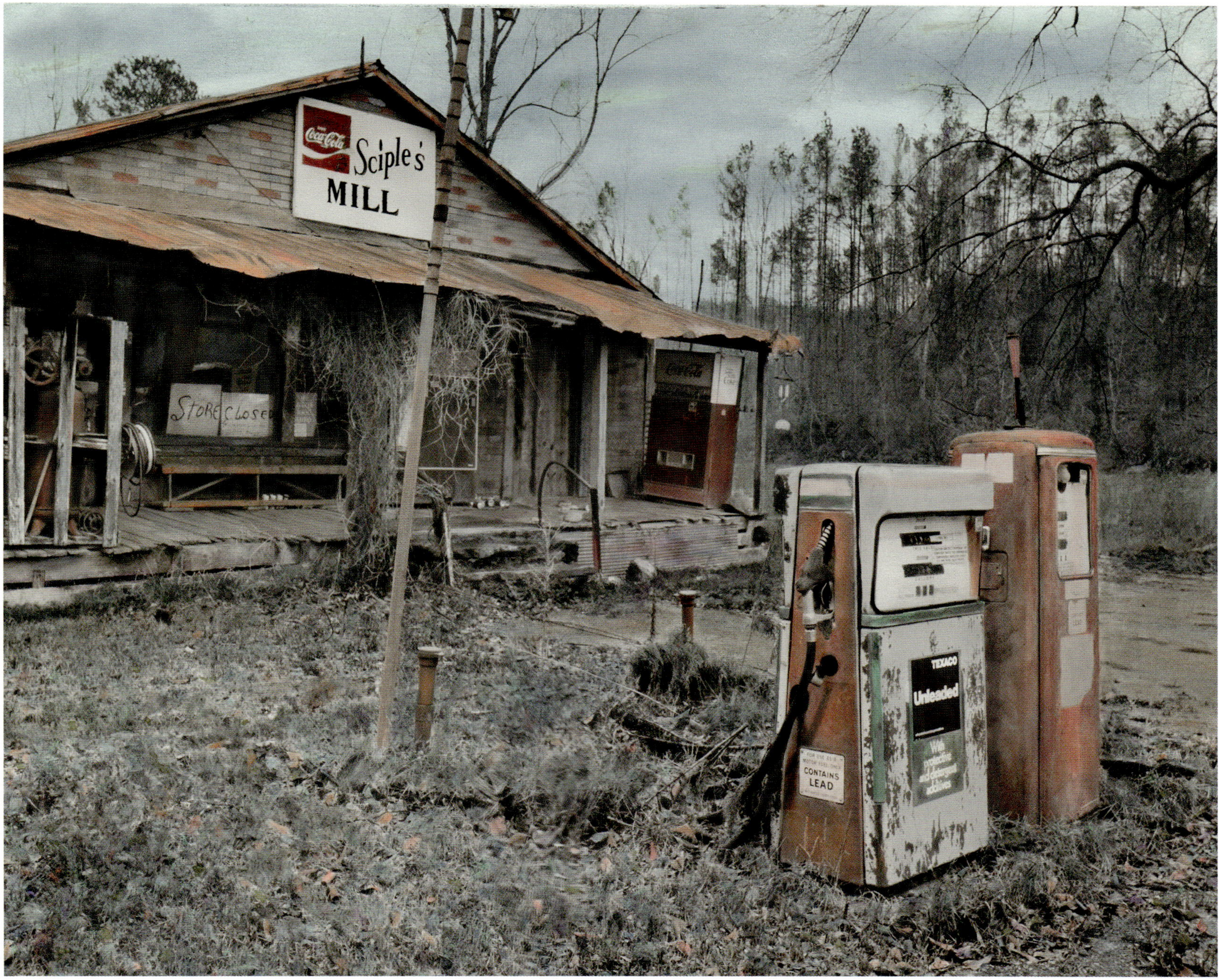

The Sciple's Mill store was closed and the door was boarded up when this photograph was taken decades ago. Old gas pumps and a soda machine were in place then and are still part of the scene today, some forty years later. Totally repurposed, the old store has taken on a new life as the "Watermill Opry," a popular Saturday night music hall!

Missouri: Ozark Mountain Springs

1800s and Early 1900s

As pioneers crossed the Mississippi River into the vast Missouri Territory beyond, they often found themselves settling in the Ozark Mountains. Their needs, hopes, and aspirations were similar to those of settlers in the Appalachians and the Deep South. They came to Missouri in search of homes for their families and a better way of life. Their basic needs included food, shelter, and clothing, and watermills could help provide those necessities.

What these settlers found in the Ozarks was a unique and amazing land. The mountains offered streams and rivers, much as in other mountainous and hilly areas of the South, but the Missouri landscape was also blessed with an abundance of natural springs. Estimates are that the mountains in Missouri are home to more than four thousand of these springs, and many produce millions of gallons of water each day. At some springs, settlers found water literally gushing from the ground or flowing from a mountainside and rushing away with incredible energy.

Resourceful individuals soon realized that these springs were well suited as sources of waterpower for mills. Unlike the flow of water in a creek, the flow of water from a spring is often consistent and predictable throughout the year. At certain springs, the flow was not only high in volume but also high in velocity.

As background, watermills in the Ozarks employed the familiar designs of undershot and overshot waterwheels, as well as water turbines after the latter technology arrived in the second half of the nineteenth century. At the largest springs, an overshot waterwheel, with its tall, high-maintenance dam and water delivery system, was not necessary. The rate of water flow from one of these springs was adequate to rotate a waterwheel with a minimal dam or almost no dam at all. Millers also soon discovered that watermills sited at springs were less likely to wash away in floods than mills sited along rivers and mountain streams, where the water flow could be unpredictable and even violent during storms.

Most Ozark watermills were gristmills. Sawmills were also built, but Missouri did not offer the vast, lush timber forests of the Appalachians or the Gulf Coastal Plain. Textile mills were also established in the Ozarks, but it was not a major cotton-producing area, and these mills were smaller and optimized more for woolen goods.

Watermills have a rich history in the Ozarks. Unfortunately, many of the region's old watermills have disappeared, but others remain, and some have been lovingly restored and preserved. Although a number of Missouri watermills were burned during the Civil War, others operated commercially into the twentieth century. At some sites, larger watermills replaced smaller mills in the years after the war, as the demand for milling services increased across the region.

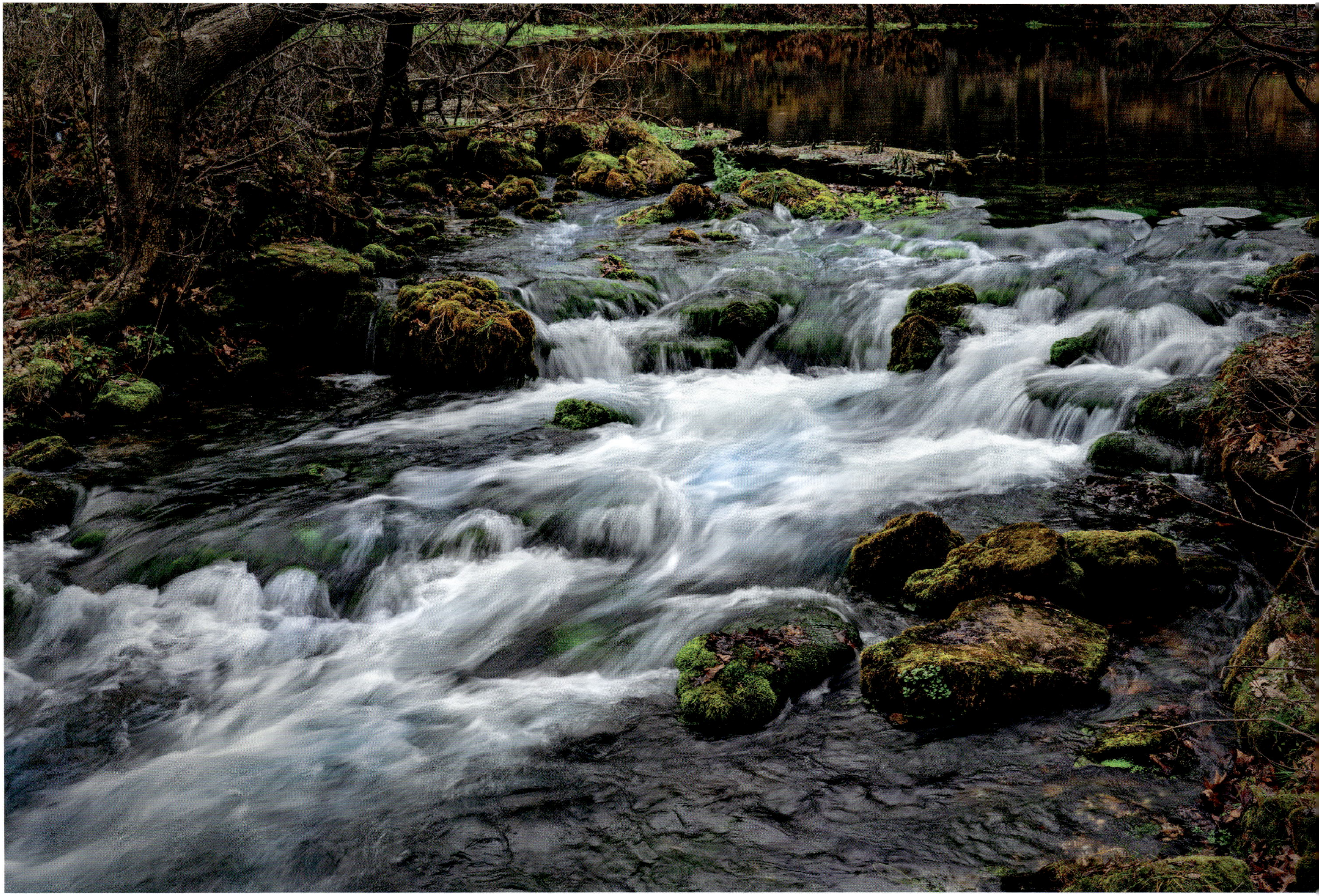

Settlers arriving in the Ozark Mountains found rushing spring water, as shown in this image. These springs provided an ideal source of waterpower for their mills.

Alley Spring Mill

Eminence, Missouri, 1894

THE WATERMILL at Alley Spring is one of the most recognizable in the country. It was selected as the engraving for the Missouri state quarter in the "America the Beautiful" coin series released in 2017. Alley Spring produces eighty-one million gallons of water every day from beneath an Ozark cliffside. The rich dolomite-blue water literally gushes from the ground and into a pool at the mill. The resulting stream forms the Jacks Fork River.

The first watermill was built here in 1868; the current structure dates to 1894. This watermill uses a turbine rather than a waterwheel because of the era of construction and because of the volume and swift flow of the water from the spring. Alley Spring Mill is a three-and-a-half-story wood-framed structure that served the community by grinding corn and wheat until 1918. Today the mill is located in a park with other historic structures and is open for tours to see the original roller mill machinery inside.

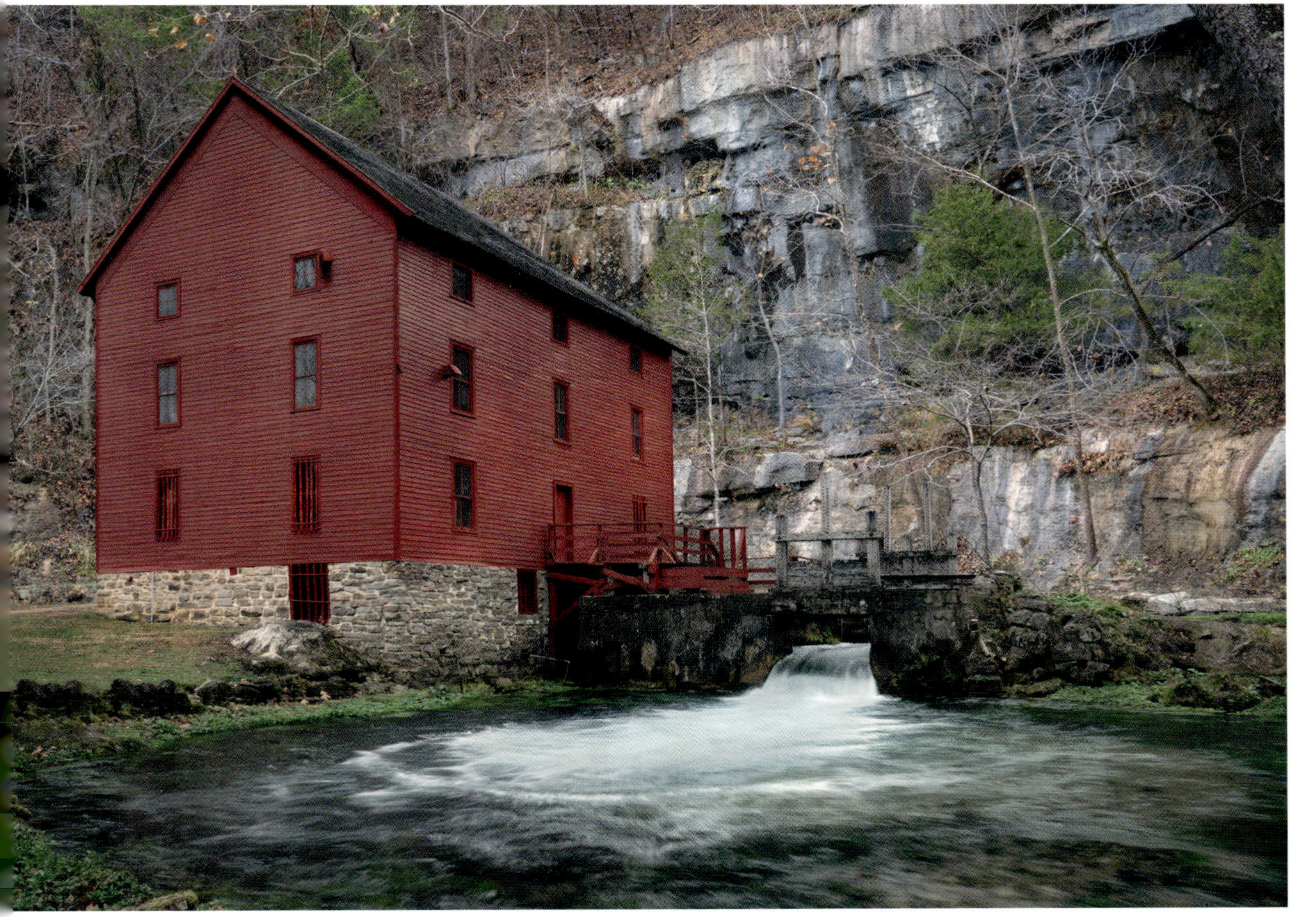

Alley Spring Mill is operated by the U.S. National Park Service. It is the focal point of a pleasant park that also offers other historic structures, walking trails, and picnic facilities.

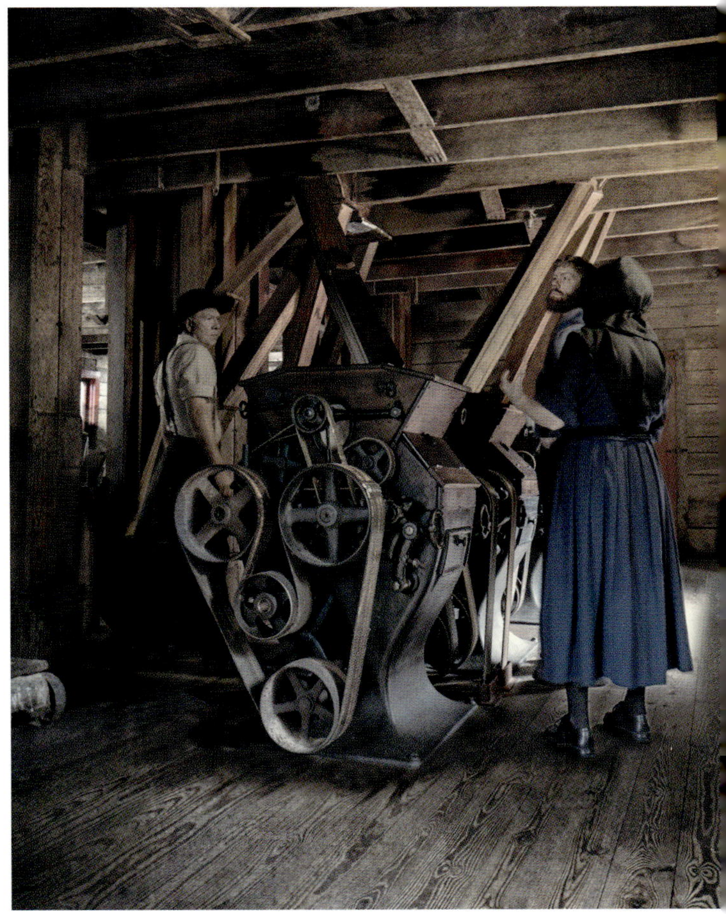

The steel roller machinery at Alley Spring Mill is shown here with the author and his son, Kevin Boyd (right), standing nearby for scale and perspective. These historic steel roller mills could grind fifty to sixty barrels of wheat or seventy to eighty barrels of corn in a day. Because this was an automatic mill, grain could be dumped into large bins in the basement and delivered to the roller machinery through elevator belt cups.

Reed Spring Mill

Centerville, Missouri, 1881 / 1973

REED SPRING MILL is one of the most idyllically picturesque watermill sites in the world. The mill would be even more famous if this area of the Missouri Ozarks were less remote and isolated.

The mill found at Reed Spring today is actually a beautifully weathered 1973 replica of the original mill. The original watermill was disassembled and sent to San Francisco for exhibition at the Golden Gate International Exposition in 1939. After the exposition, the mill was not returned to Missouri but instead went to Washington, D.C., where it is stored at the Smithsonian Institution.

The original mill at the site had been built in 1881 to grind corn and wheat. For a time, Reed Spring also powered a sawmill, as well as an electric generator that was the only source of electricity in Centerville until 1929. To generate electricity, the mill was switched from a waterwheel to a turbine that could supply the power needed to operate a dynamo. Wires were strung from the spring into town, a distance of about half a mile.

Reed Spring forms two millponds, both located behind and above the mill. A large millpond and spillway can be seen beyond and to the left of the mill in the accompanying image. This spring actually has several outlets, and today water flows not only over the waterwheel but also under the mill structure en route to the Black River. As at Alley Spring, the water is deep blue due to dolomite that has leached into it from the local bedrock.

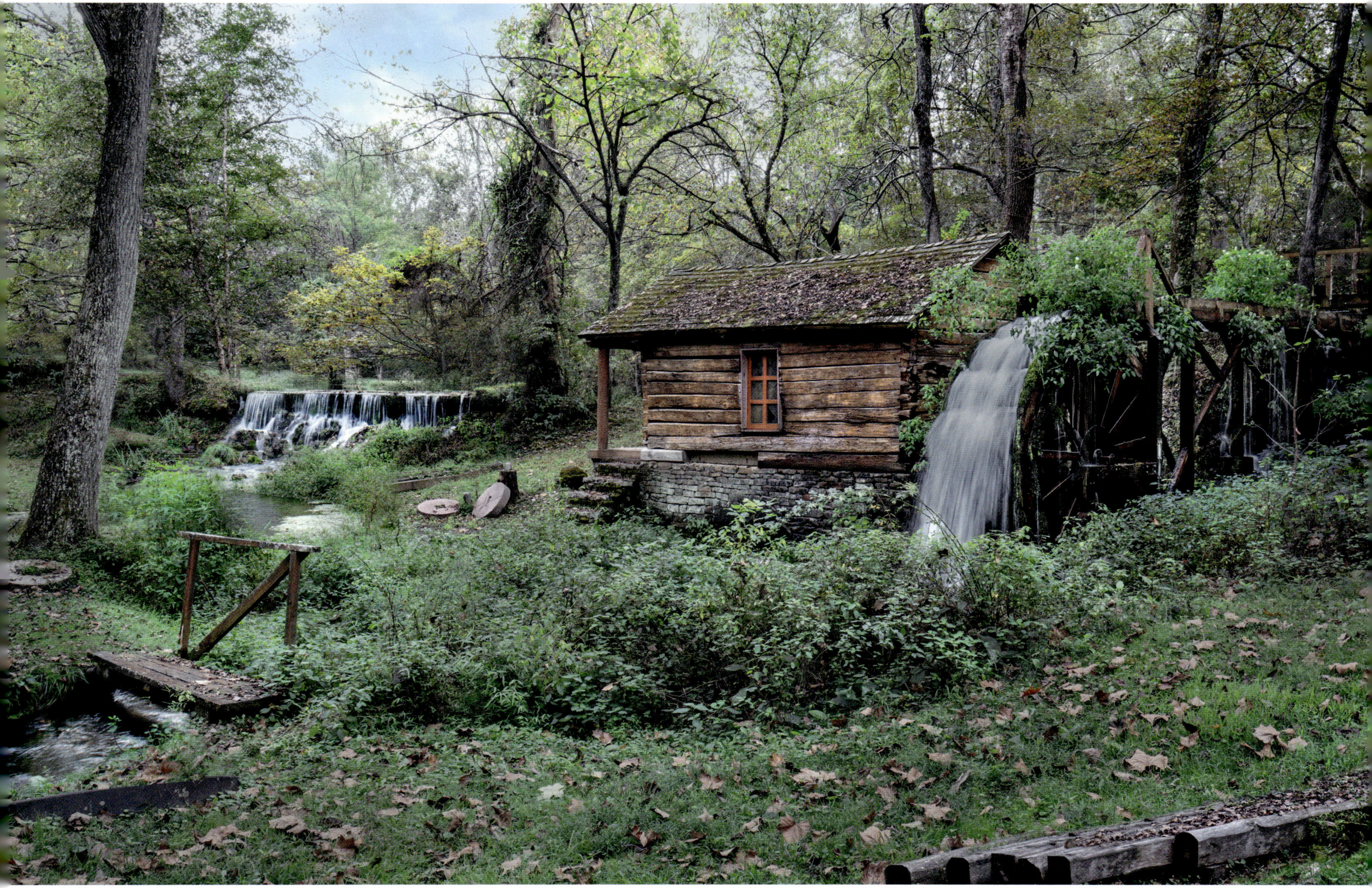

Reed Spring Mill sits along a narrow one-lane road in a lush valley in the northern Ozark Mountains.

Markham Springs Mill

A Few Miles West of Williamsville, Missouri, Late 1930s

THIS MILL was constructed to provide electric power to an adjacent estate home built by the Fuchs family. Ironically, the mill never operated as a generator because public electric service reached the area during this same period. The unusual-looking stone mill, which some refer to as a "wheelhouse," and the large metal waterwheel are still in reasonably good structural condition after all these years.

The site around the mill includes a large pond that is fed by six springs to produce more than five million gallons of water per day. Due to the presence of this rushing source of water, mills have operated here since before the 1850s. In the late 1800s, a large two-story gristmill with a fourteen-foot overshot waterwheel ground corn and wheat at a location near where Markham Springs Mill stands today. For a time, a sash sawmill used the power of the fast-moving water to cut more than fifteen hundred board feet of lumber per day.

Jefferson Markham purchased the property in 1901 and ran a gristmill here until 1907. After his mill ceased operations, the site stood idle until the present mill was constructed in the late thirties. In 1965, the National Park Service acquired the mill and the surrounding property, and it is now a public recreational area. Because of its isolated location and how rarely it is mentioned on the internet or in social media, few people, even in Missouri, know that this mill exists.

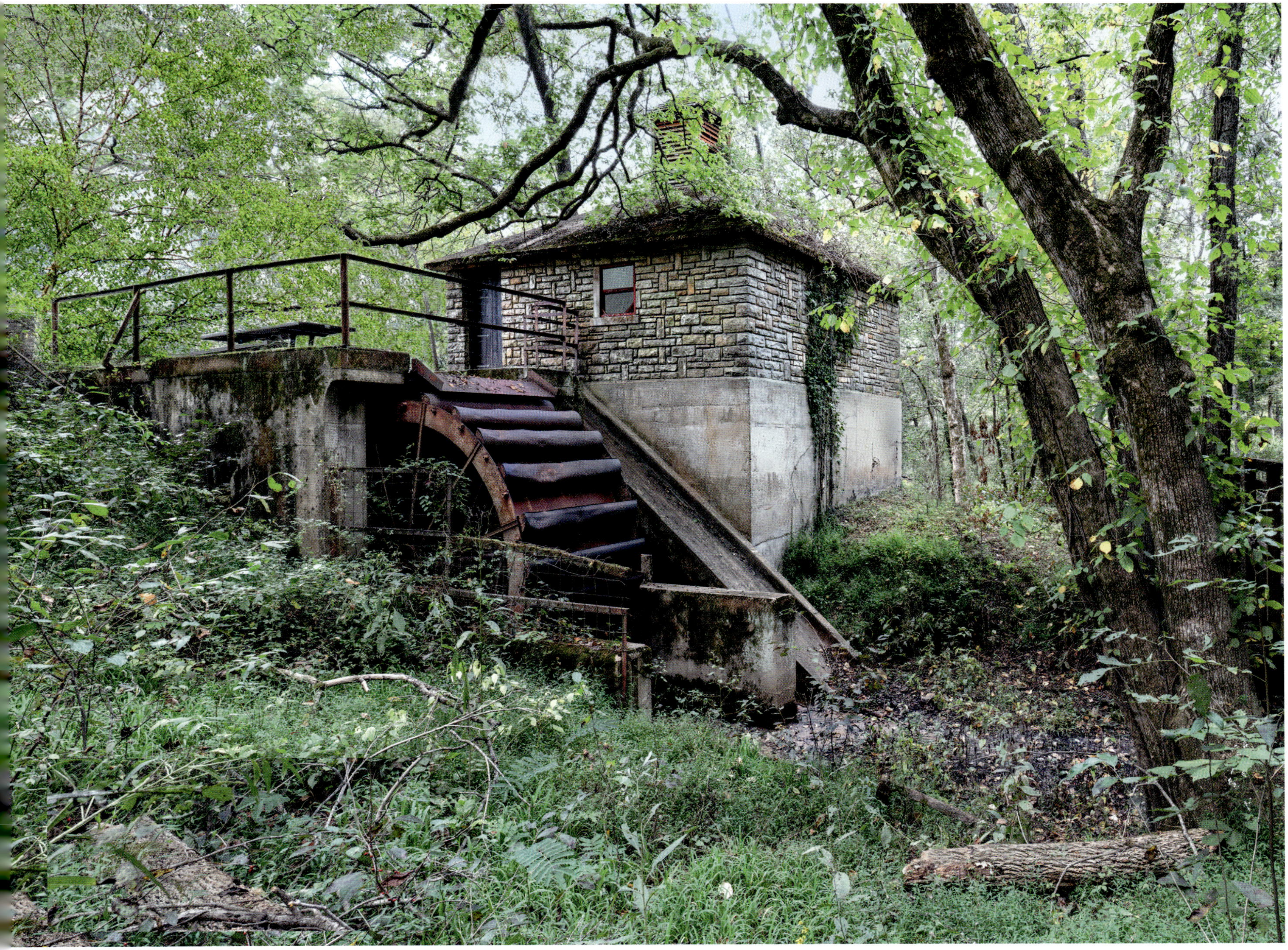

This lovely old mill at Markham Springs is almost unknown, even within Missouri.

Hodgson Water Mill

Tecumseh, Missouri, 1861 / 1897

LIKE ALLEY SPRING MILL, Hodgson Water Mill is a familiar image to many across the country, including supermarket shoppers. A drawing of this mill appears on grocery store packages everywhere as the logo for a full range of popular baking products, including flours and meals, noodles, yeasts, and various mixes for breads, cakes, muffins, cookies, pancakes, and waffles.

The historic mill started as a family business and grew and expanded until operations ended in 1976. Production of Hodgson Mill products subsequently relocated to larger facilities in Effingham, Illinois, and more recently Hodgson Mill became part of Hudson River Foods in New York.

The spring that powered the mill still produces twenty-four million gallons of rich, blue water a day, water that flows under the mill and in front of the building to form Bryant Creek. The first mill at the spring dated to 1861, and the current, familiar structure was built in 1897. Although a large white waterwheel is prominently displayed at the front of the mill, the machinery inside was turbine powered.

In addition to grinding grains, the mill provided electricity to a cotton gin, a sawmill, and a clothing mill. The busy mill also served as a general store and a post office.

This mill and Alley Spring Mill are the most visited and photographed watermills in Missouri and are among the most famous watermills in the United States. The exterior of Hodgson Mill appears to be cosmetically preserved and in excellent condition today, but the interior shows damage from a major 1982 flood in which the water level reached all the way up to the main milling floor.

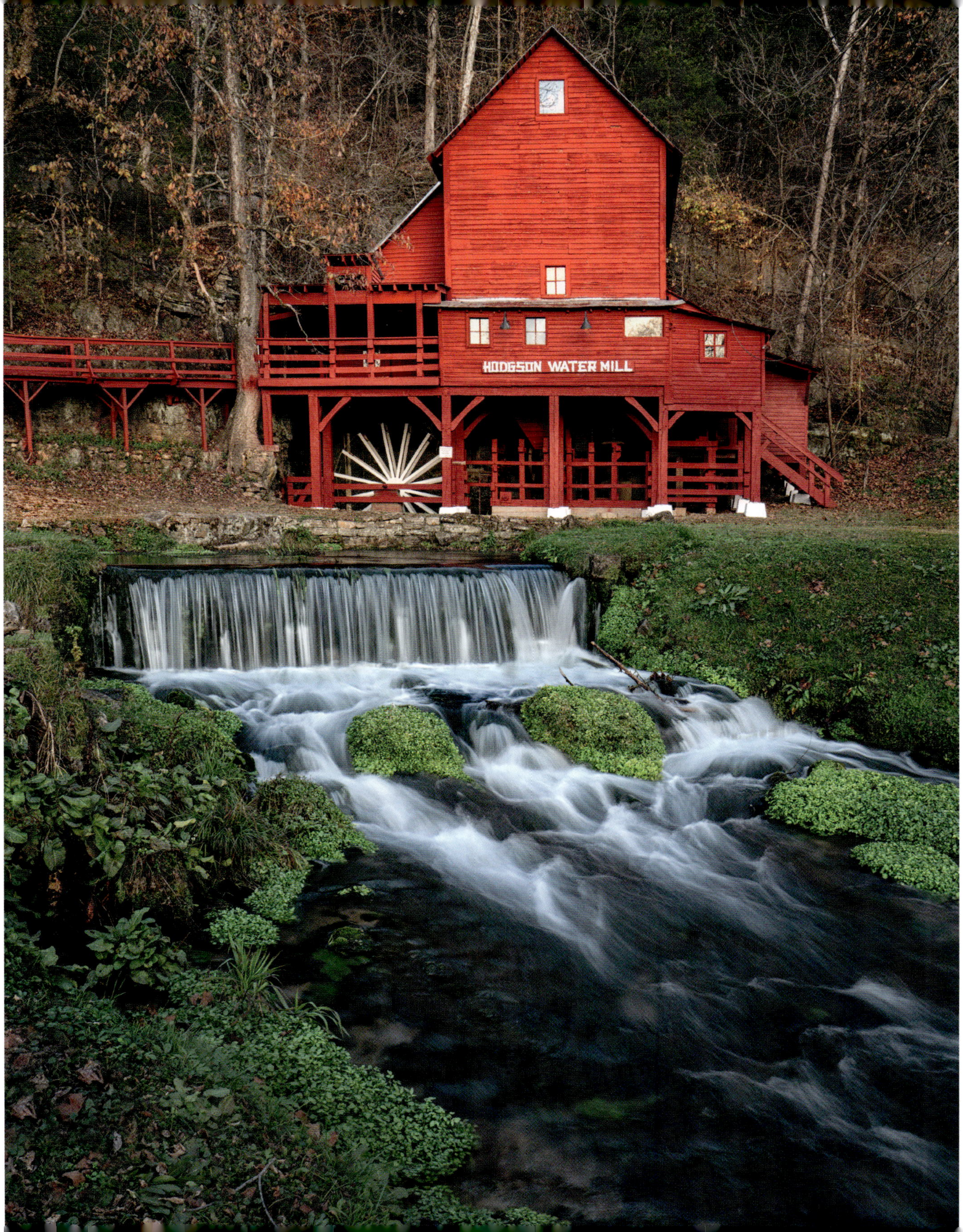

This image of Hodgson Water Mill probably looks familiar to many supermarket shoppers across the country.

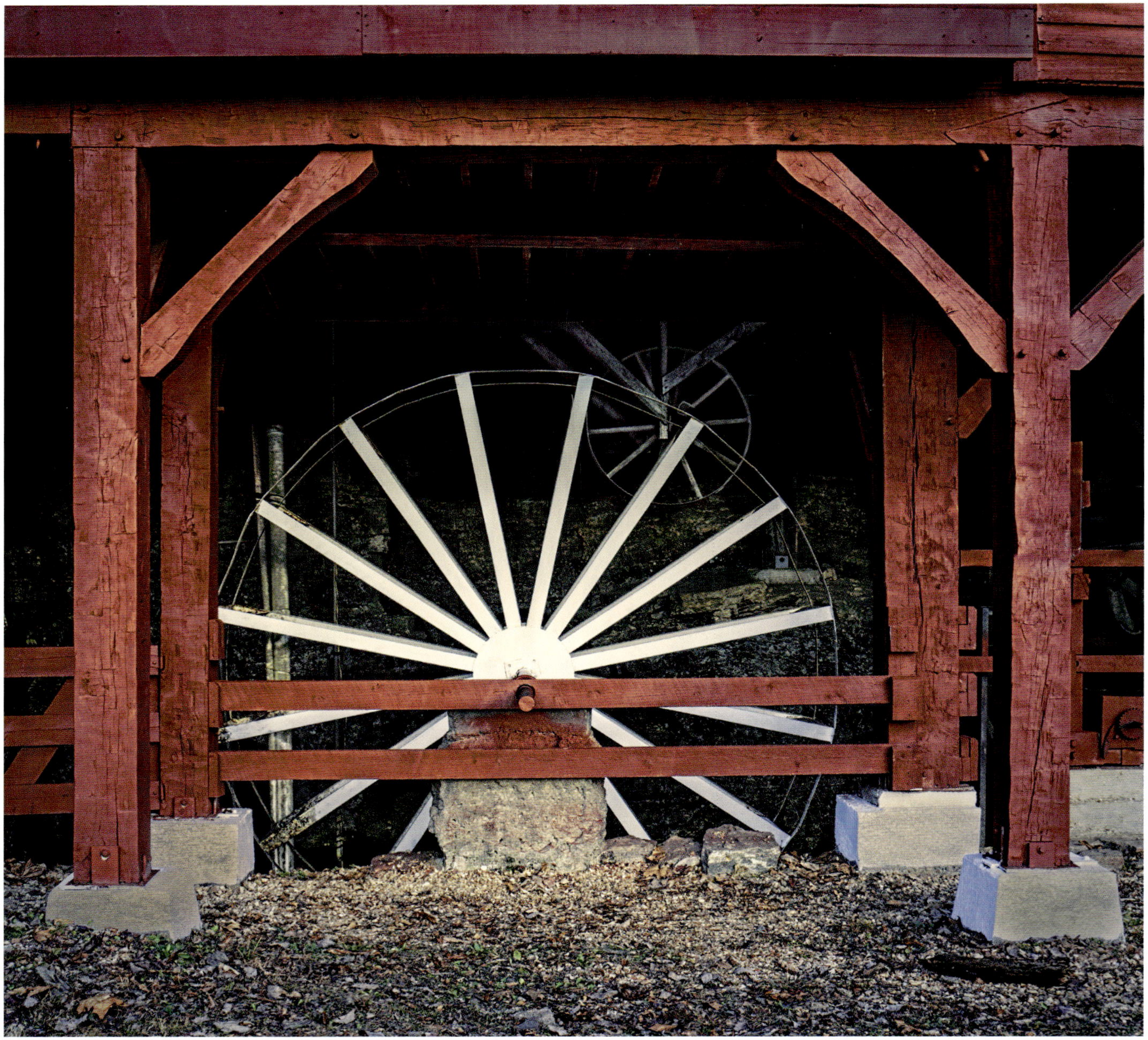

The large white waterwheel at Hodgson Water Mill never actually powered the mill.

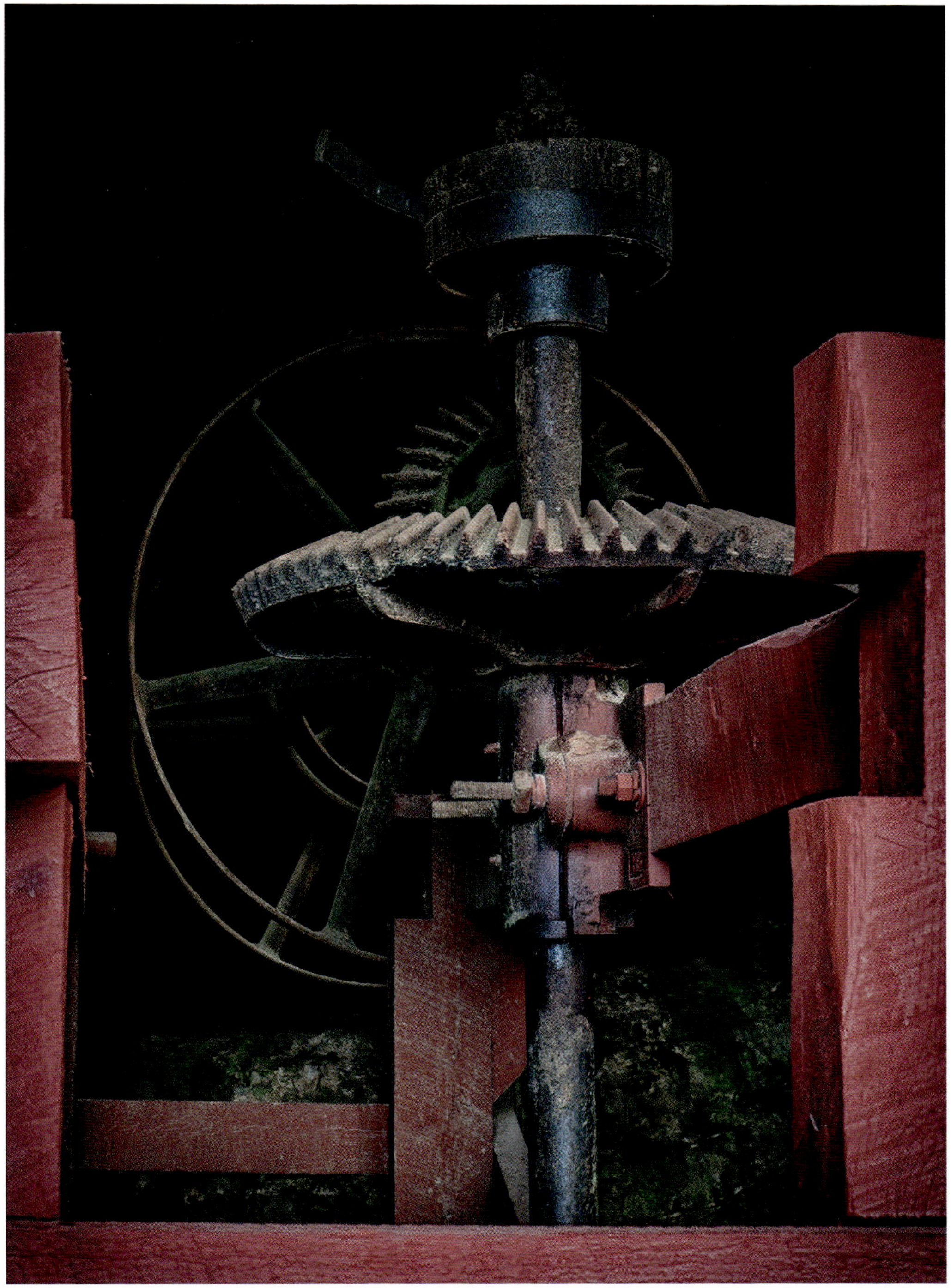

This image is from below Hodgson Water Mill. It shows the excellent design and condition of the old metal gearing.

Falling Spring Mill

Mark Twain National Forest, Missouri, 1927

LOCATED DEEP in the southern region of Mark Twain National Forest, Falling Spring Mill can be a little difficult to find. Visitors must navigate a mostly unmarked and undeveloped woodland roadway that winds for several miles through dense forest before arriving at the site of the mill and the spring.

This Ozark spring is set up a little differently than others. The water at the spring spills over a rock ledge as a waterfall and into a pond. The spring produces 500,000 gallons of water per day and once powered two watermills, but only one of the mills remains.

This watermill dates to about 1927. It features a large overshot waterwheel that was used to grind corn (mostly for animal feed), saw roofing shingles, cut firewood, and generate electricity for the area, which included several cabins and structures in addition to the two mills. Falling Spring Mill has not been restored, but it is in reasonably sound condition. The interior is open, and some original equipment and machinery can be examined inside.

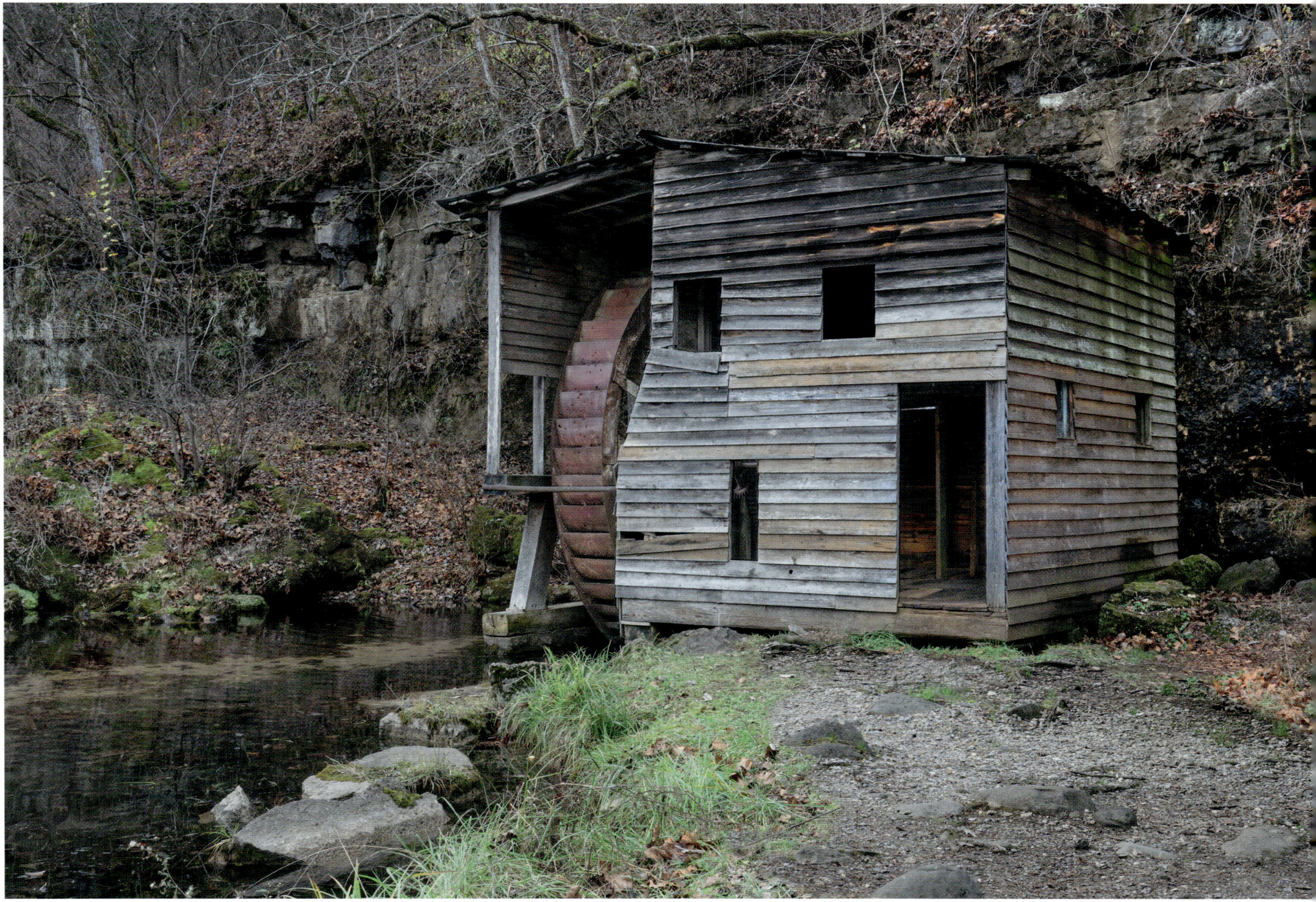

Falling Spring Mill is one of the most incredible destinations in the expansive Mark Twain National Forest.

Rockbridge Mill

North and West of West Plains, Missouri, 1868 / 1888

THIS SCENIC MILL is near the more famous Hodgson Water Mill. Today the site is known as the Rockbridge Rainbow Trout and Game Ranch and is beautifully maintained with a restaurant, lodging, and several other historic structures. The perpetual water flow at this location comes from four sources that feed into Morris Spring and Spring Creek, with about fifteen million gallons of water per day rolling over the mill impoundment and into a trout fishing basin below.

Back in 1841, a group of families left Kentucky with oxen- and horse-drawn wagons in search of a new life further west. Six months later, the group arrived in the area where Hodgson Water Mill is now located. Some of the families ultimately settled a little further north and established the Rockbridge community. They constructed homes and a mill, but everything was burned by the Union army during the Civil War.

In 1868, B. V. Morris rebuilt the mill at the present location. As was common across the South in this era, a small town soon sprang up around the mill, including a blacksmith shop, a store, a church, a school, a Masonic lodge, and a large building now called the "White House." A bank and a post office came later. The mill was enlarged in 1888, and a two-story general store was added. Sadly, the bank closed during the Great Depression, and the mill was shut down in the late 1940s. Then in 1954, the fortunes of the old mill changed for the better, and it was transformed into a private trout hatchery, and eventually into a fishing resort!

Over the years, the mill has survived several major floods, most recently in 2017; this last flood caused serious damage and required extensive restoration work.

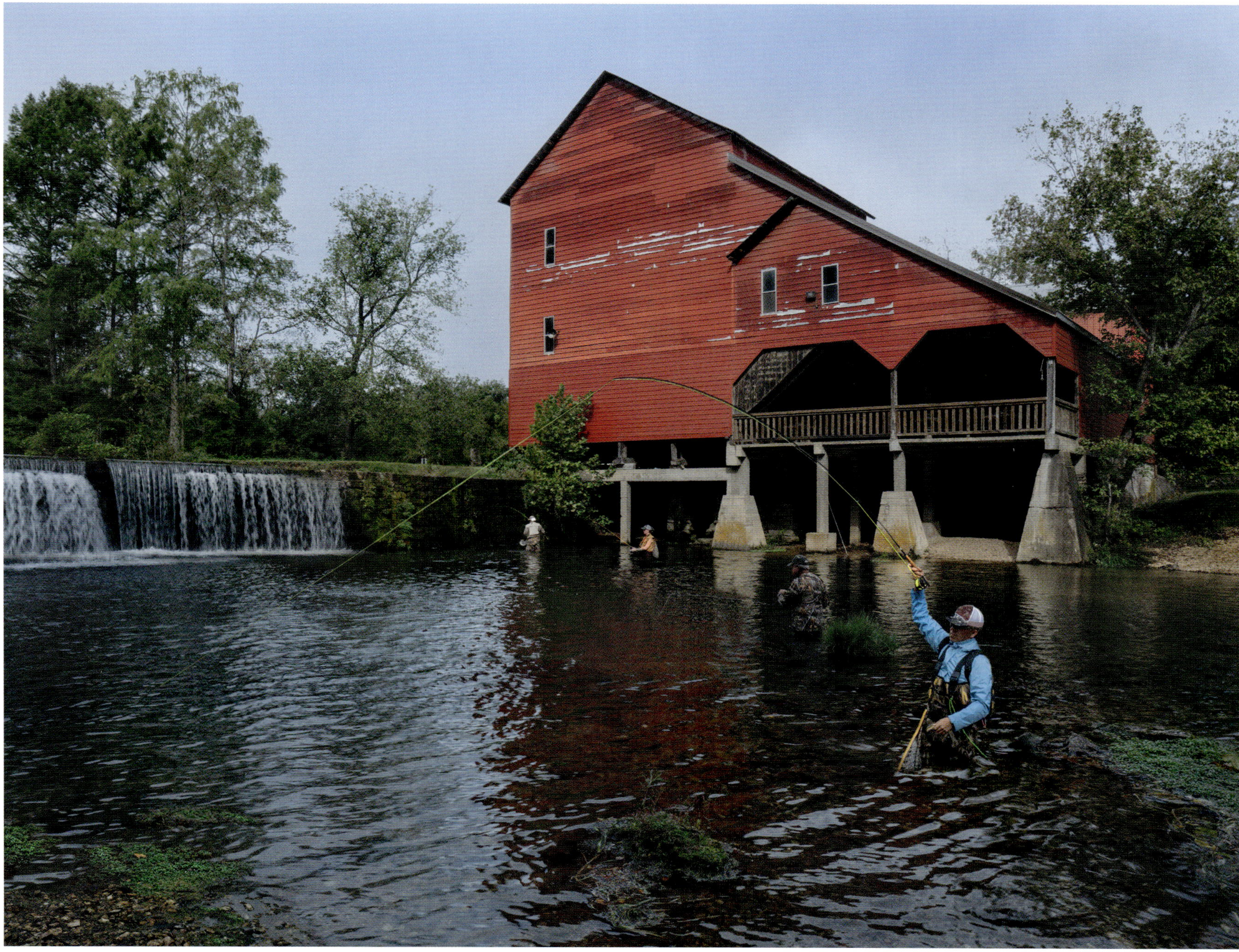

Thanks to the plentiful trout fishing and an excellent nearby restaurant, Rockbridge Mill has been transformed into a popular travel destination.

Topaz Mill

North and West of West Plains, Missouri, 1895

ALTHOUGH TOPAZ MILL is near Rockbridge Mill, getting there is like taking a trip back in time. The last one-and-a-half-mile stretch of the journey is over a one-lane gravel road that turns downhill into a hollow and winds through richly forested areas. But the mill and the overall experience are definitely worth the trip!

Topaz Spring was first discovered by Europeans in about 1818 or 1819. About ten million gallons of water flow from the spring daily into a pond that has been elevated by twelve feet to power the mill's turbine. The name of the mill and the spring may come from the pale blue color of the water.

The first watermill on the spring dated to about 1840, while the current three-story mill was built in 1895. The three floors of the structure were necessary for the operation of automatic milling machinery. The mill features all the elaborate machinery from the late 1890s, including steel roller mills, a bolter, a smutter, and a maze of belts and chutes. The mill's operations are said to have transported wheat up and down nine times from when it arrived until it was finally refined into flour! The mill still operates today, but only to grind corn on the main level. The upper floors serve as an educational milling museum.

In the late 1800s, the mill and the area around it became a community gathering place. It was like a small town, with a blacksmith shop, a tomato-canning factory, a service garage, a general store, a barbershop, and a post office. As occurred at Rockbridge, most of these businesses had closed by the 1940s. The grinding of flour ended in the 1920s, and corn milling concluded in the 1930s. Post office service was discontinued in 1943, and the general store closed two years later.

The O'Neal family purchased the site in 1957, and a third generation of the family, Joe Bob and Betsy O'Neal, care for and operate the mill today, giving tours of the mill and of the adjacent store that has become an incredible museum of vintage items from the area.

As is evident in the accompanying image of the structure's exterior, Topaz Mill is painted red. Many old watermills, barns, covered bridges, and other wooden structures are painted red. Why was red so popular? Understand that this was not to make them bright and beautiful. Through the early twentieth century, red paint was often the cheapest color. Red was also thought to last longer. Over time, the color just fades to a softer look, but this color is historic and authentic.

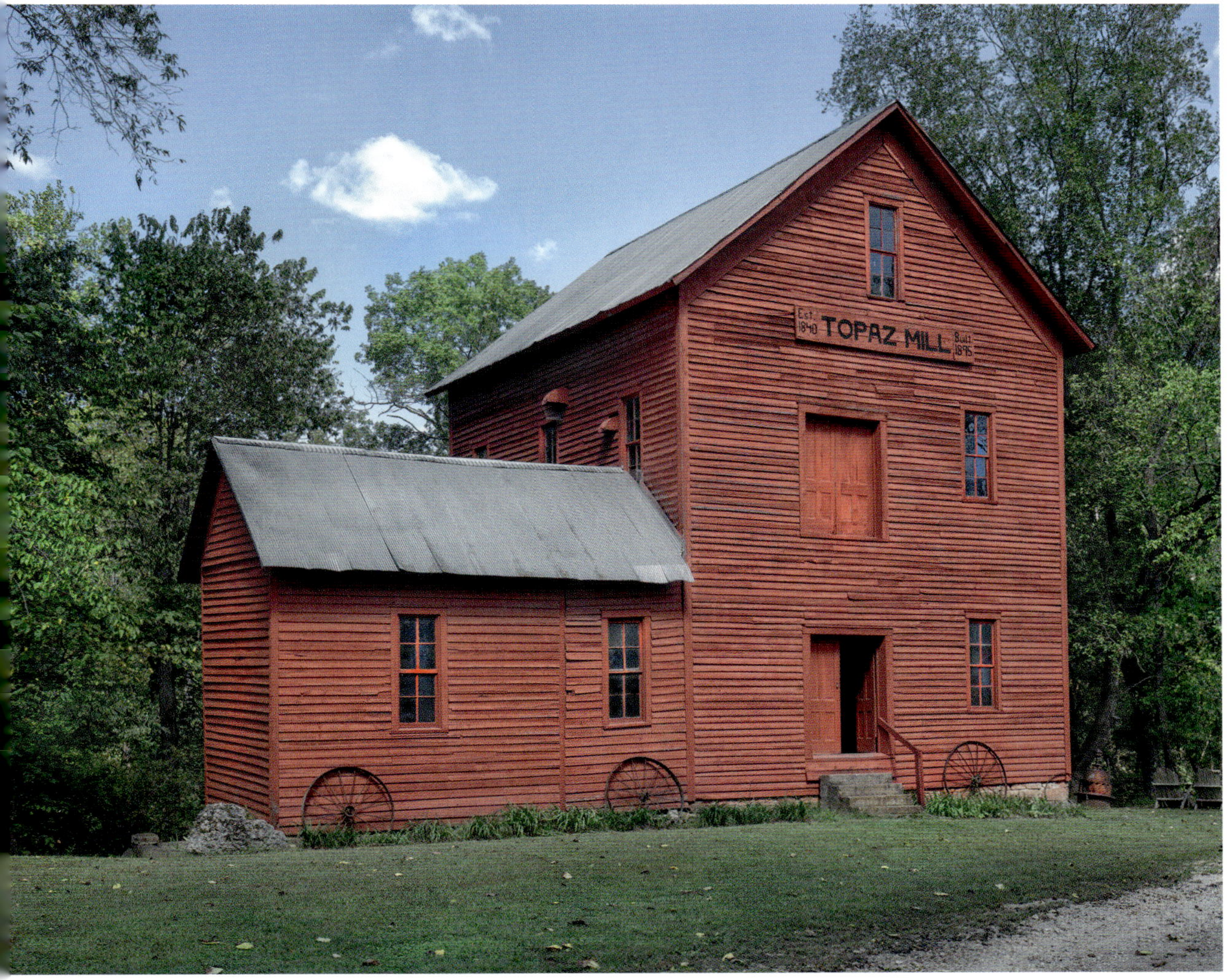

According to Joe Bob O'Neal, who lives onsite at Topaz Mill, red paint was often used on mills because the pigment would last longer and protected wood better than other available colors.

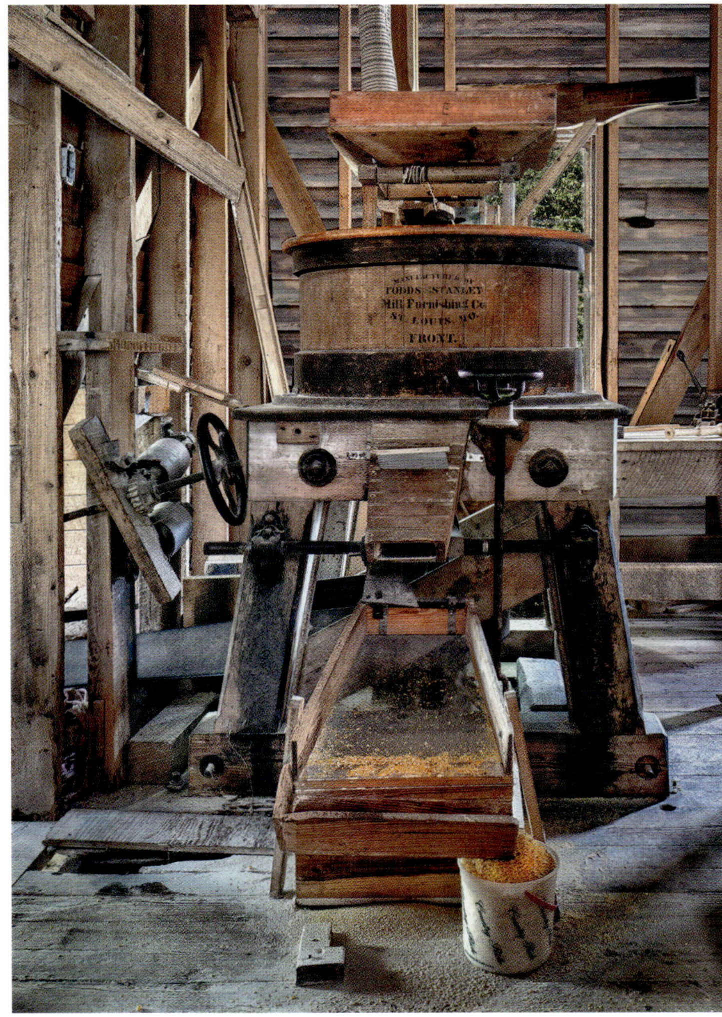

Topaz Mill was shaking and grinding corn when this photograph was taken. To grind corn, the turbine could operate at only half power; for wheat, the full power of the turbine was required.

The Watermills at Allison Woods Estate

Statesville, North Carolina, Circa 1928 and 1933

ALLISON WOODS has a 250-year-plus history with many generations of the Allison family. Back in the 1700s, the king of England granted the Allison family eighteen hundred acres of land in what would become the Statesville area of North Carolina. At the time, the family was living in the Philadelphia area. There were five brothers, and one of the brothers traveled with his wife and child to see the property. The trip took two months.

Upon arrival, the brother was impressed with the beauty of the countryside. He soon set up a leather tannery business based on the tanning process he learned from Indigenous Americans living in the area. His pioneer tannery business became successful. Wagonloads of leather were sent to Charleston and Savannah for shipment to Europe. At one point, the tannery processed two thousand handsome saddles for the British army.

Over the years and through succeeding generations, the Allison family built up and developed its land with money made from the tannery, and the property gradually evolved into an estate. Enslaved laborers did much of the work on the estate. Some years later, the Allisons left the estate and moved into Statesville, and the neglected property fell into disrepair.

Between 1926 and 1936, William Allison began work to redevelop the family estate as a recreational area for his family and their friends. As part of this development, he had two watermills built on lakes at Allison Woods. The first watermill was built in 1928 and was called the "Upper Mill." In 1933, a second and more elaborate mill was constructed on a lake further down the property, and it was known as the "Lower Mill." Neither of these mills was a gristmill, although a gristmill had operated on the estate during an earlier generation.

The Allison Woods Upper Mill was built to generate electricity and pump water. This small mill featured numerous windows, a chimney, an interior sitting bench, and a water fountain. The brick and stone structure is almost dwarfed by its large waterwheel. The Upper Mill became part of a complex of structures associated with an impounded "upper" lake at the estate.

The three-story Lower Mill was constructed of stone recovered from blasting at the site to prepare for a lake impoundment just above this mill. The nine-acre "lower" lake was located over what had been a quarry. This structure served as a sawmill and also generated electricity and pumped water.

Interestingly, the Lower Mill featured distinctive architecture representative of the bungalow/American Craftsman style. This mill was designed with 229 different types of woods, including poison ivy, and was conceived to feature "all the appointments of a fine home." For example, the door for the mill came from a mansion in Philadelphia.

As the years passed, the estate again fell into disrepair and was essentially overgrown with vegetation. Once more, a generation of Allisons became interested in sharing the property with area residents, and especially with children and environmental and outdoor groups. By the 1980s, work began to clear the site as an environmental education center. Today the property is known as Allison Woods Outdoor Learning Center and has official recognition as a North Carolina Environmental Education Center. More than six thousand students visit and learn on the beautiful estate each year.

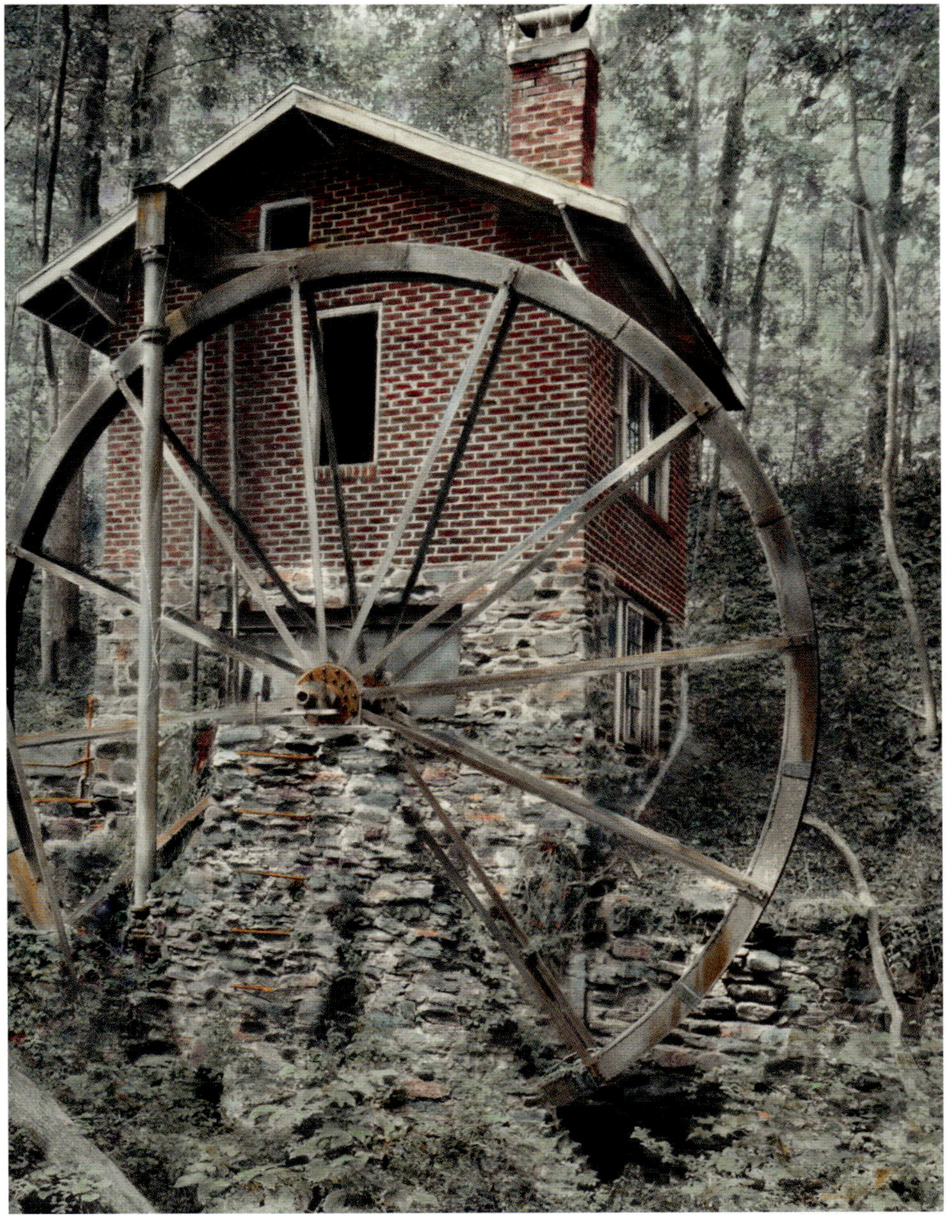

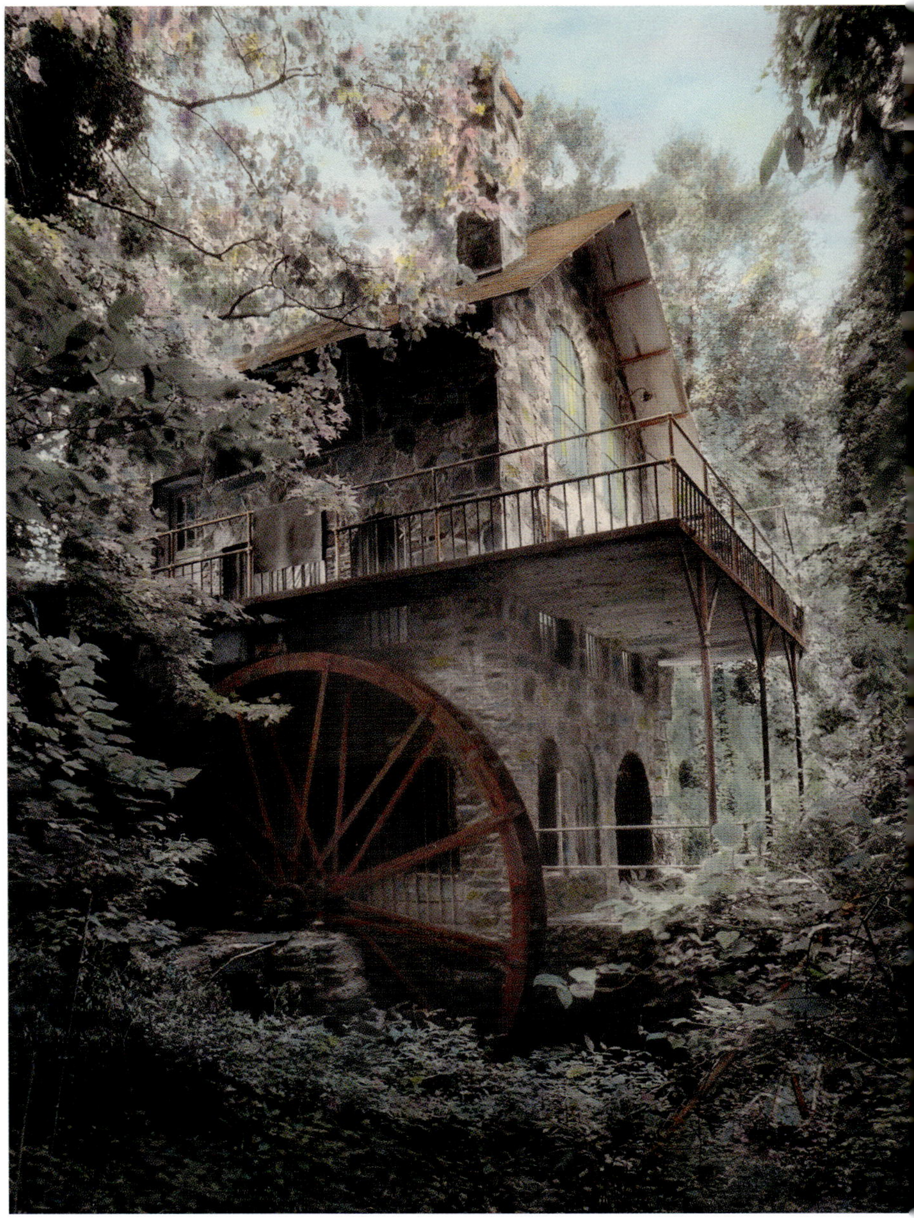

The Allison Woods Upper Mill (left) and Lower Mill (right) are shown as they looked during site reclamation work in the 1980s. At this time, the mills were almost overcome by the lush vegetation of central North Carolina. As at other sites across the South, unattended watermills in this region can be completely consumed by forests in just a few seasons due to abundant rainfall and long growing seasons.

Murray's Mill

Catawba, North Carolina, 1913 / 1980s

SET INTO A SCENIC COVE along Balls Creek in gentle hills east of the Appalachians, Murray's Mill is the principal attraction of the Murray Mill Historical District, which includes more than twenty structures dating to as far back as the 1880s. The Wike and McNeil Mill had been built even earlier just upstream of the present mill location.

William Murray built his mill here in 1883 and powered it with a turbine. In 1913, soon after William's son, John Murray, took over, he replaced his father's mill with the current mill building and exchanged his father's turbine for a twenty-two-foot overshot waterwheel. In 1938, John's son, Lloyd, continued in the family business. He raised the arched dam by six feet and brought in a twenty-eight-foot waterwheel for more power. That wheel drove the milling machinery until commercial operation ended in 1967, when high taxes and tighter health regulations forced the remaining mills in the area to close.

The Murray & Minges General Store sits just a few feet up and across the road from the mill. Curiously, the store originally sat where the mill is sited today, but in 1911, it was rolled on logs to its current location. This move made way for John Murray's two-and-a-half-story mill, with its tall brick basement.

After the mill sat idle through the 1970s, the Catawba County Historical Association began restoration in 1980, and the mill reopened by the late 1980s. Today it is operational to grind corn and wheat. Through the years, French buhr stones were used for corn and Eureka roller mills processed wheat.

The mill's general store stocks an interesting inventory of mill and gift items and is operated by a well-informed staff who can answer questions and share information.

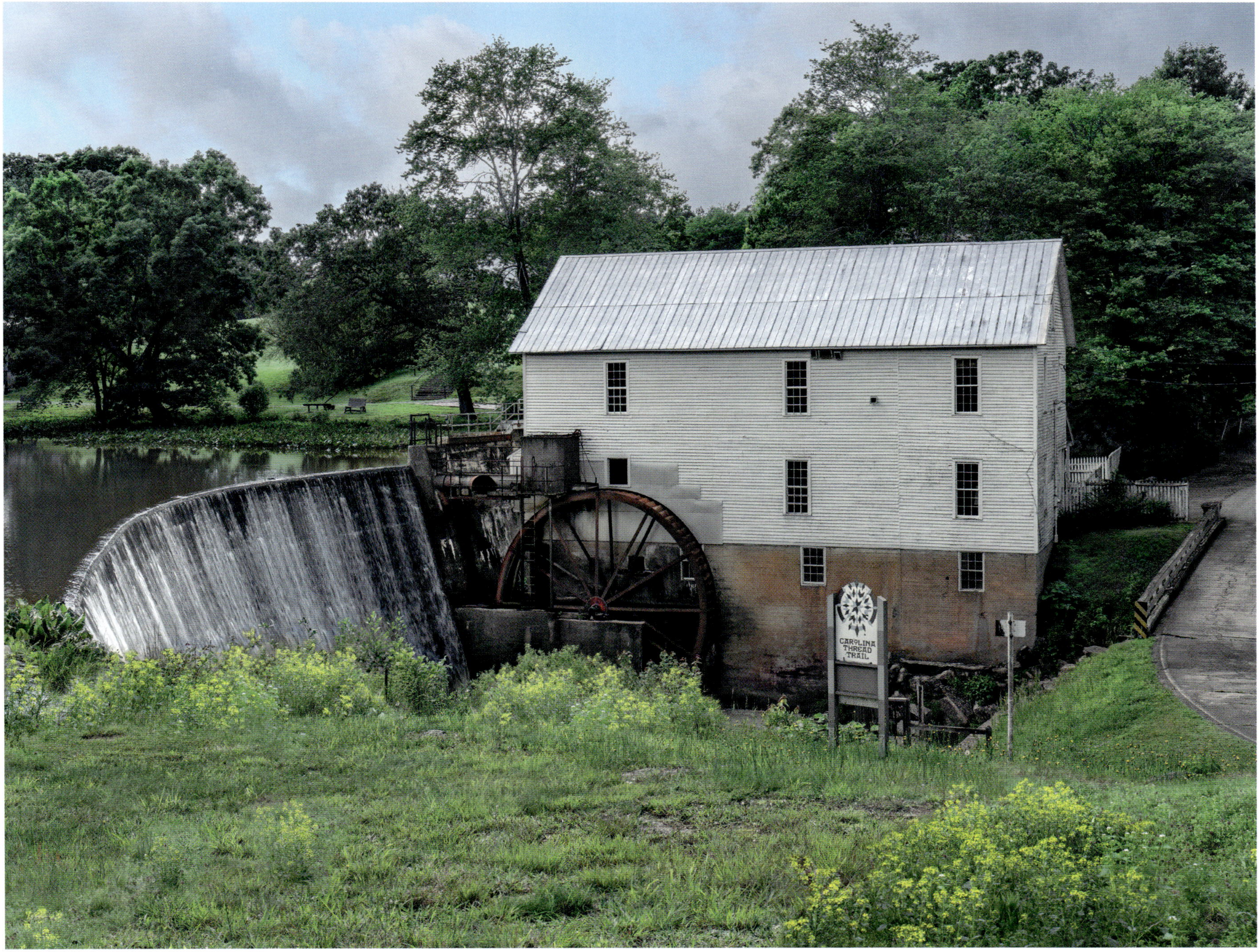

The tall dam at Murray's Mill has an arched design. An arched dam is sometimes used in canyons and along streams with steep banks and walls. As water pressure pushes against the dam, the concrete structure flexes and in turn presses into the stream walls for added strength.

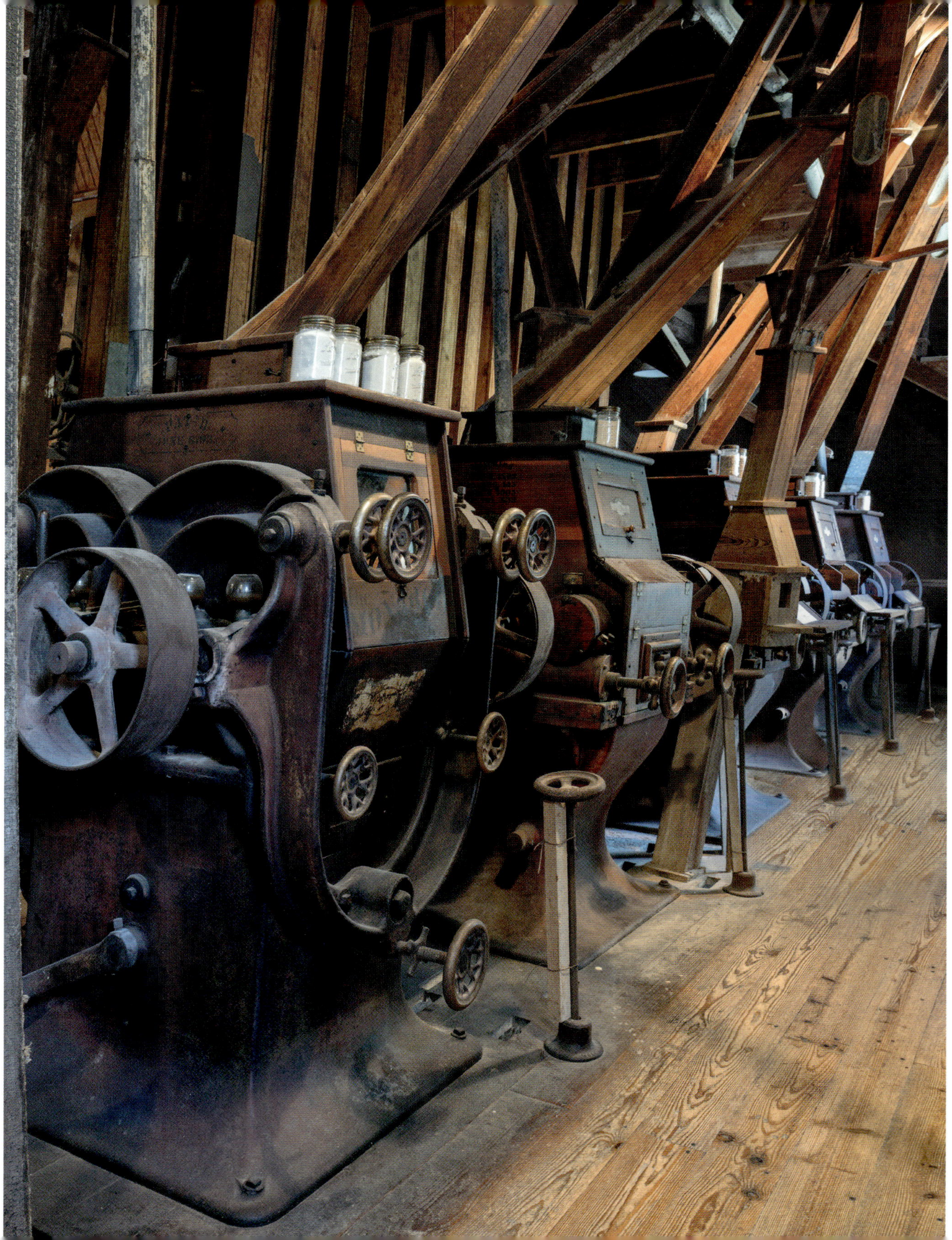

This row of Eureka roller mills was used to process wheat. The square wooden shoots above the mills automated the handling and delivery of the grain to the machines without heavy lifting or manual feeding from baskets, sacks, or barrels.

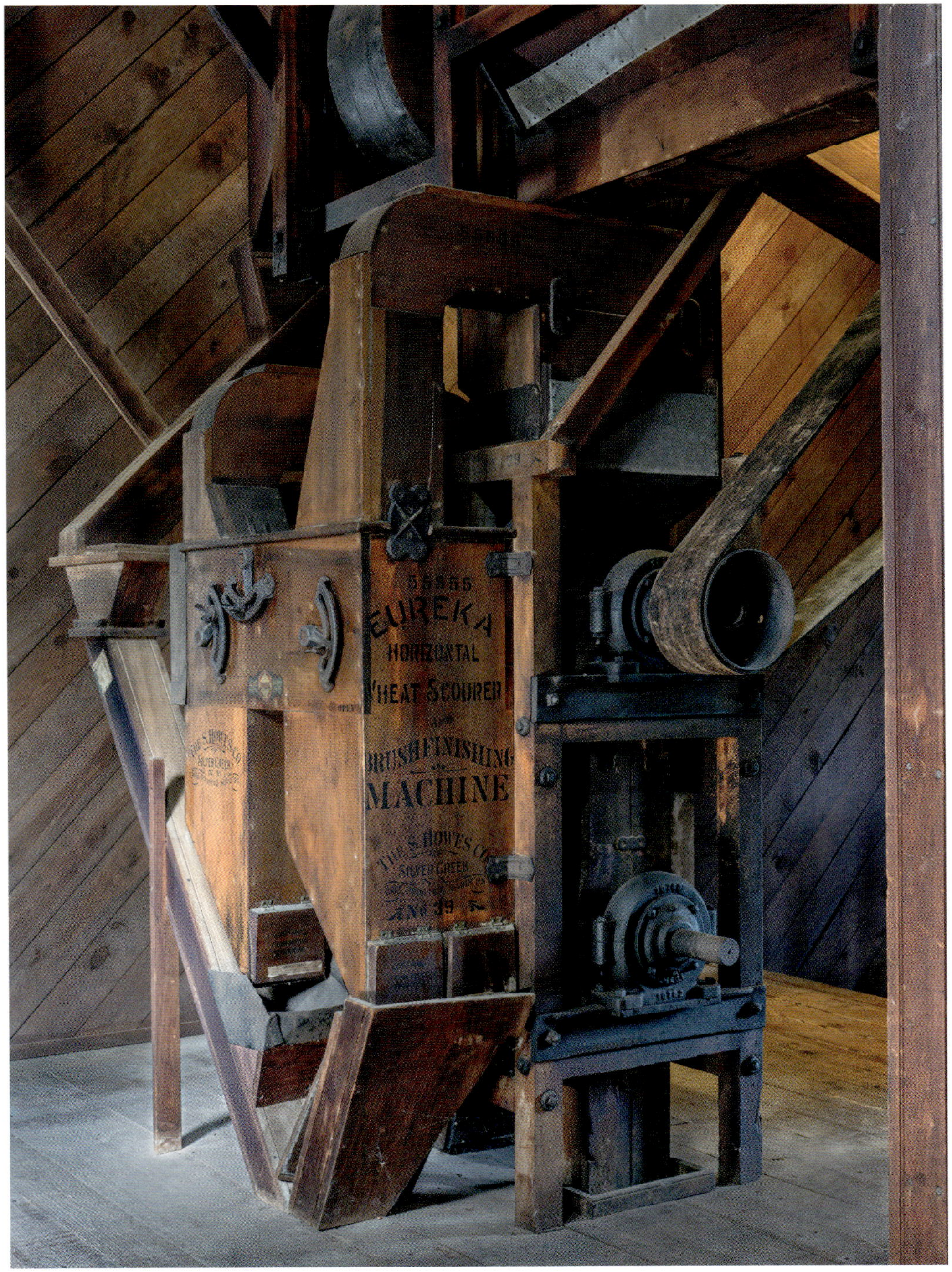

This beautiful wooden machine was used to clean seed before grinding. Screens could be exchanged in the machine to optimize operation for wheat, corn, rye, barley, grasses, and other grains. Notice that there are no safety shields or other protections for workers, as would be installed on similar equipment today.

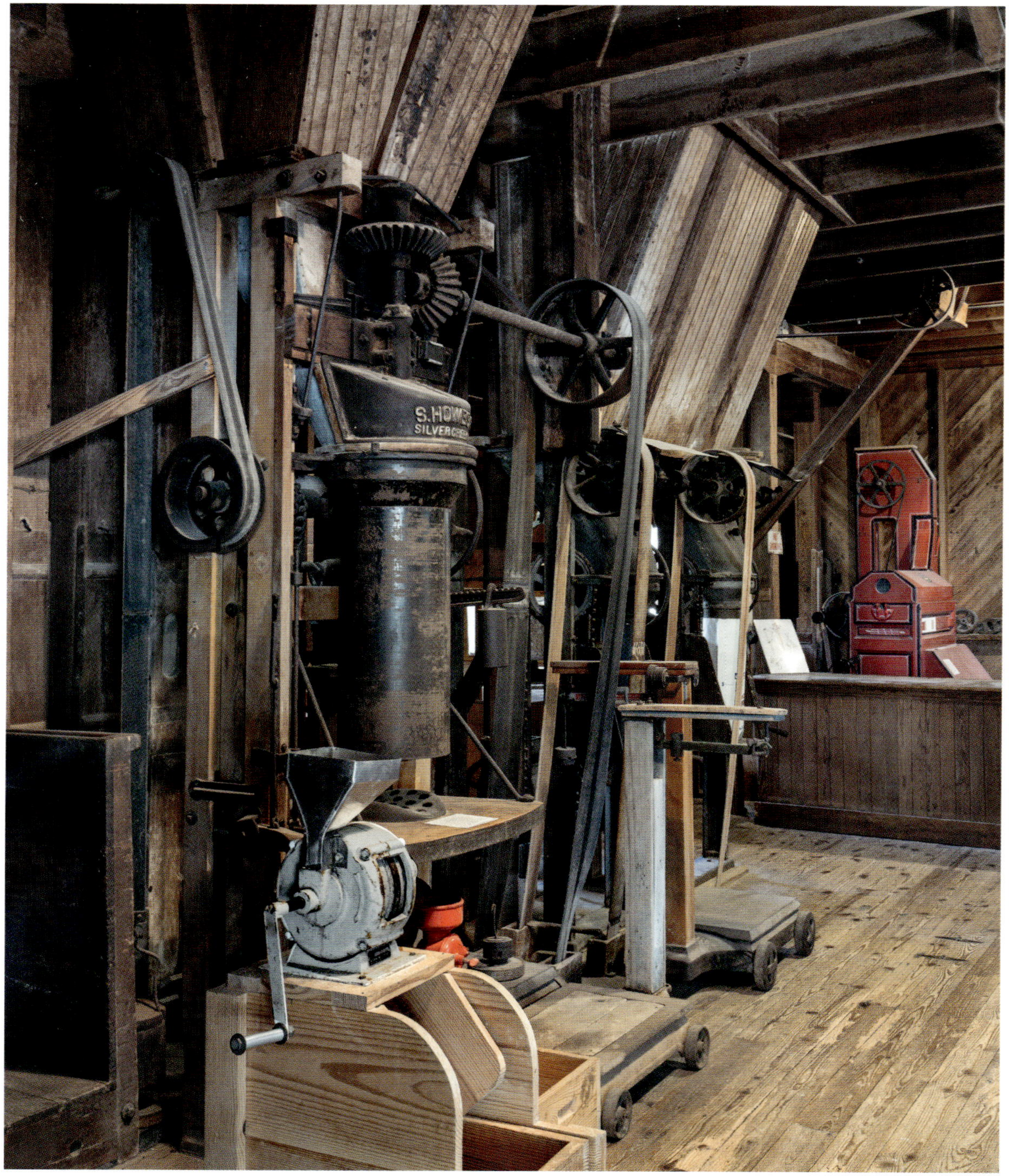

These grain-packing machines (at left) and the customer service counter in the background are just inside the front door to the mill. During operation, this area greeted farmers, friends, and customers with orders for the miller.

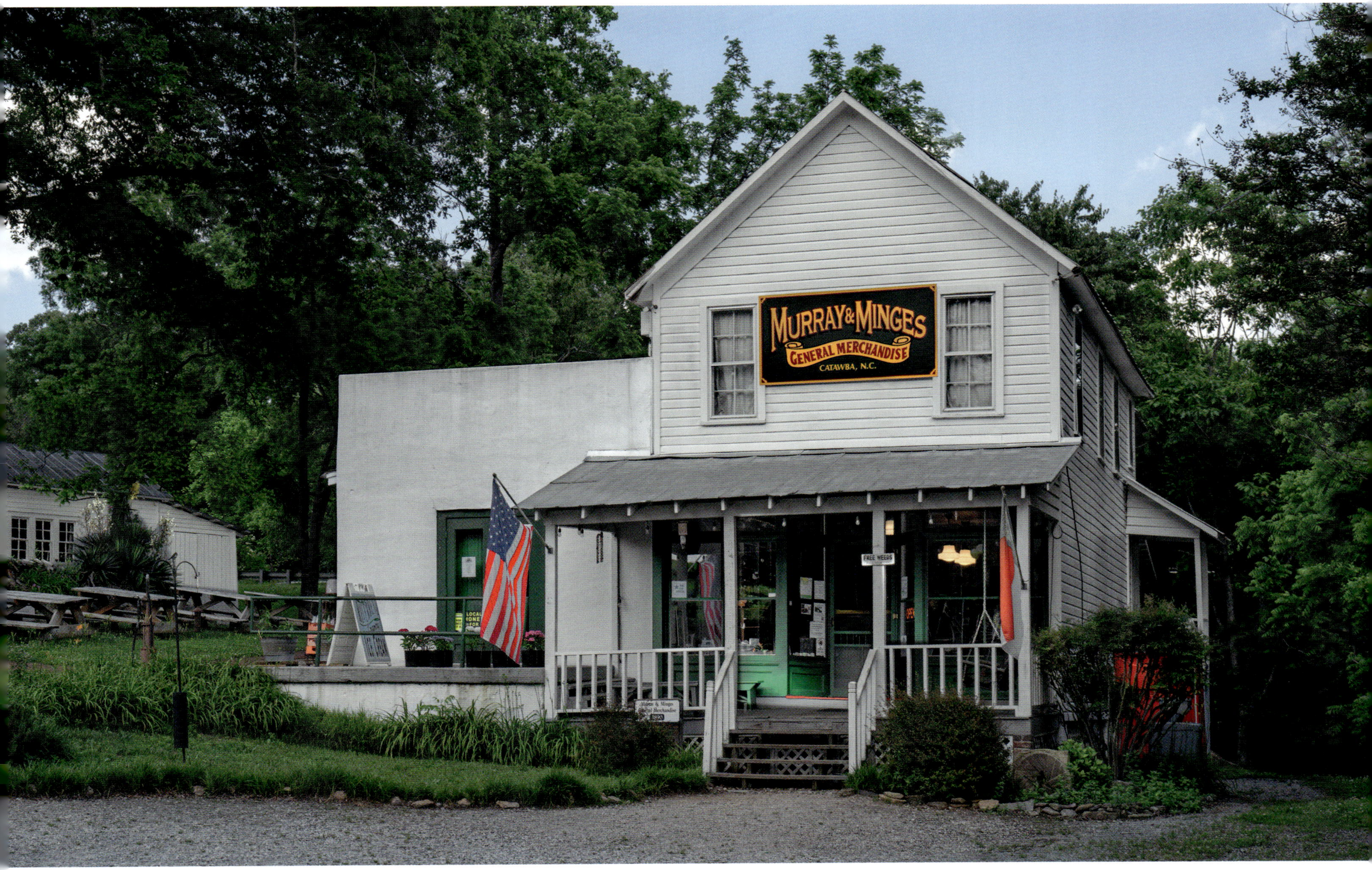

The old Murray & Minges General Store sits just across a closed road from Murray's Mill. The store serves as a visitor information center and as a lovely, well-stocked gift shop.

Francis Grist Mill

Waynesville, North Carolina, 1887 / Early 2000s

THE RESTORATION of the Francis Grist Mill is a model for how the preservation process to save an old watermill can be organized in a local community. By the early 2000s, this 1887 structure was in peril and close to collapse. Members of the community formed the Francis Mill Preservation Society to save the mill. Work began in 2004, and in 2007, a big new waterwheel turned for the first time in decades. Significantly for this site, the long, tall wooden flume that had delivered water to the wheel for years had finally fallen in the 1980s and thus had to be rebuilt before the waterwheel could be operated.

So how did the restoration work on the mill proceed? The preservation society, which included members of the Francis family, moved quickly and planned its work through a series of annual workshops. Each workshop was carefully scheduled to address specific projects toward a total renewal. The most critical structural work was tackled first.

The original twenty-four-foot yellow poplar waterwheel had been replaced with a steel wheel from Fitz Waterwheel Company in 1914. Almost a hundred years later, the wheel needed to be replaced again, and the flume and five support pillars had to be totally reconstructed. At one point, the society asked folks in the community to donate $100 each toward the work, with $100 said to be the approximate cost per bucket for the wheel. By proceeding with the annual workshop strategy, the society was able to fully restore the mill and make it operational again!

Annual music festivals and other events are held at the mill as fundraisers and to share the restored landmark with the community. The society also recognizes funding for the restoration work from the Society for the Preservation of Old Mills (SPOOM), the Society for Industrial Archaeology, the Terence L. Mills Memorial Preservation Fund for North and South Carolina, the Haywood County Community Foundation, the Janirve Foundation, the Steele-Reese Foundation, the James McClure Clarke Foundation, the Bethel Rural Community Foundation, and the Blue Ridge National Heritage Area, as well as from many individuals, local businesses, friends, and neighbors.

In September 2024 Hurricane Helene destroyed the millpond at the Francis Grist Mill, filling the pond with mud, silt, and debris. Restoration was completed in early 2025, and two hundred red-twig dogwoods have been planted in an attempt to stabilize the creek bank and hillside.

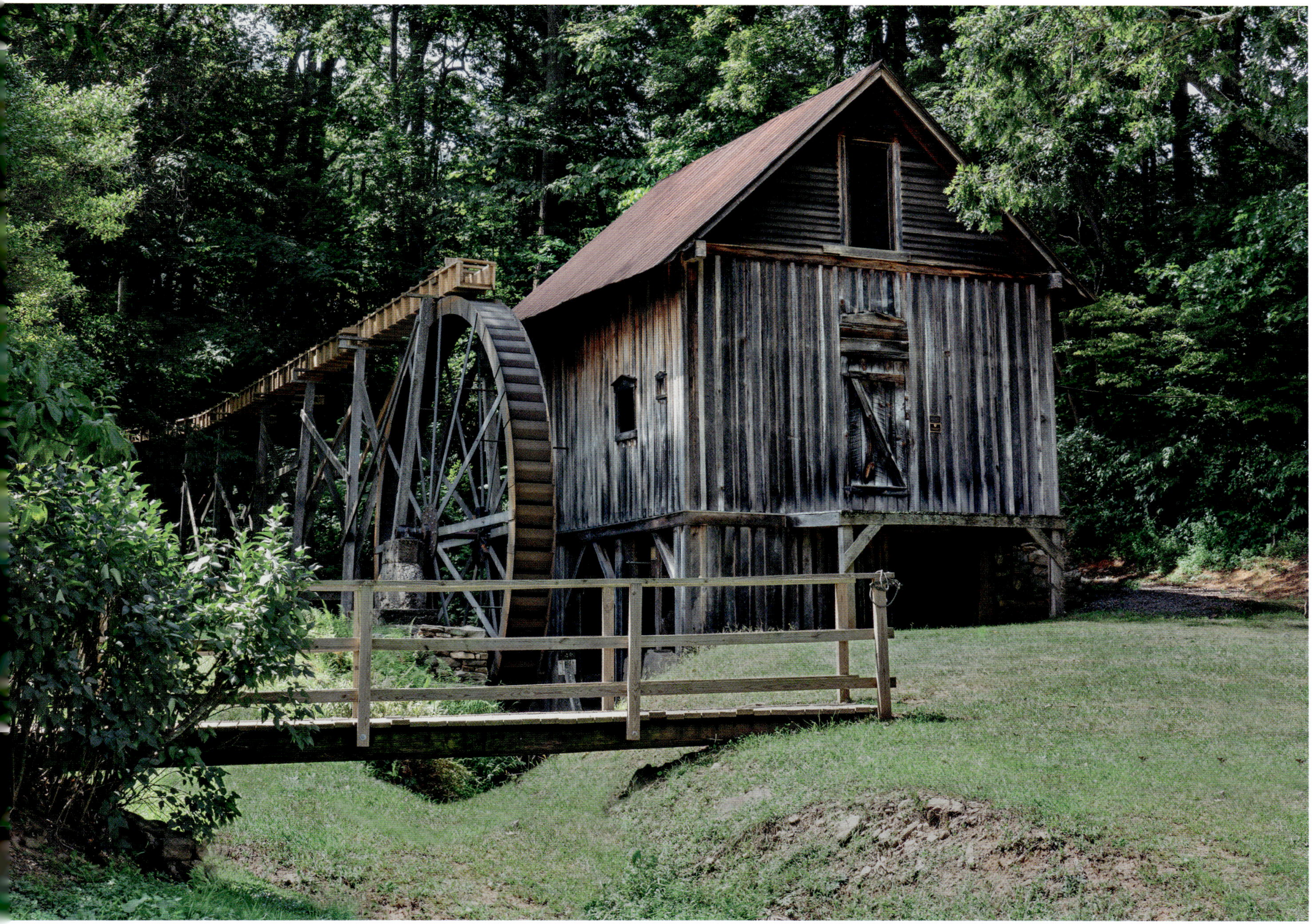

Back in the 1880s, William Francis moved from Tennessee to western North Carolina hoping to become successful. He struck up a courtship with Polly Almon, the daughter of what was supposedly the richest family in the western part of the state. They married, the Almons gave them one square mile of land as a wedding present, and the young couple built the Francis Grist Mill, which their descendants own to this day.

Mingus Mill

Cherokee, North Carolina, 1886 / 1930s

THE BUILDERS AT Mingus Mill chose an ideal location for providing water to a large gristmill without the need for a dam and millpond. Water is channeled directly from the flow of springs and Mingus Creek up the hill from the mill site. From the creek, water is directed through a long raceway to the mill. A tall, substantial wooden flume then delivers the water over the final stretch of its journey. The water gains speed as it flows down the hillside.

The existing mill, built in 1886, is not the first watermill at the site. An earlier mill had been located here since about 1800. That mill was powered by a large wooden overshot waterwheel. Mingus Mill uses an early iron turbine that is located below the mill floor.

Dr. John Mingus paid Sion Early $600 to build the present mill. In the 1880s, this area was at the center of an extensive logging operation, so all the wood used to build the mill was locally sourced. The mill is constructed primarily of tulip trees, known as yellow poplar trees in other areas. These trees can be some of the largest in the South, and their wood is known to exhibit a long lifespan. The wood used for Mingus Mill has stood the test of time, as the exterior siding and the support beams have never been painted and are still mostly in place today as the mill approaches 150 years in age.

Water flow from the flume to the mill is regulated by a clever spillway design on the side of the flume. If the water flow is too high, part of the water in the flume can be released onto the ground before it reaches the mill.

At the lower end of the tall flume, the water drops twenty-two feet into a penstock that rotates the turbine. As mentioned already, unlike most watermills of this era in the South, Mingus Mill never used a waterwheel.

Inside, the mill has two sets of millstones—one set for grinding wheat and the other for grinding corn. The millstones and their housings are set on an elevated platform with reinforced support beams that isolate and help protect the building from vibrations and shaking as the millstones and the drive mechanisms operate. Until 1930, when the property was sold to the National Park Service, Mingus Mill served as many as two hundred families in the area.

Since then, the mill and the grounds have been part of Great Smoky Mountains National Park. The park service restored the mill as a public works project during the Great Depression. The mill closes occasionally for preservation and restoration work, but it typically is open otherwise and offers grain-grinding demonstrations on most days, outside of the winter months.

Recently, extensive research and study has been undertaken to better understand a nearby cemetery for enslaved people and other Black residents of the area around Mingus Mill. Because the Mingus family was wealthy and had held people in slavery, it is likely that some of the people buried in the cemetery were involved in the construction of the first mill on this site. Local Black laborers were probably employed for the construction of the 1886 mill, especially for the heavy lifting and digging at the site. These laborers and their families may also account for some of those buried in the cemetery.

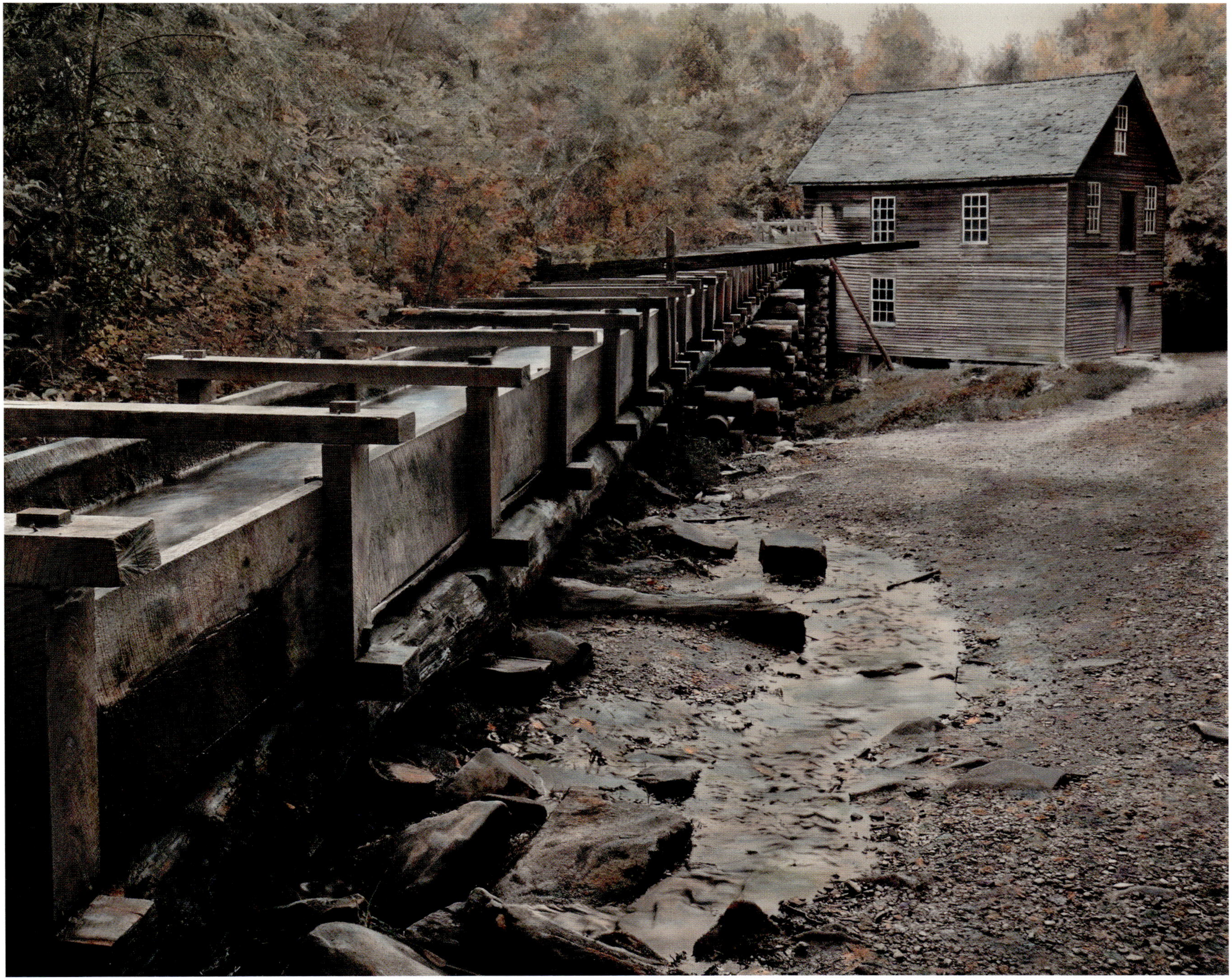

This is a vintage hand-tinted image of Mingus Mill by the author. The mill is a popular stop for many visitors to Great Smoky Mountains National Park.

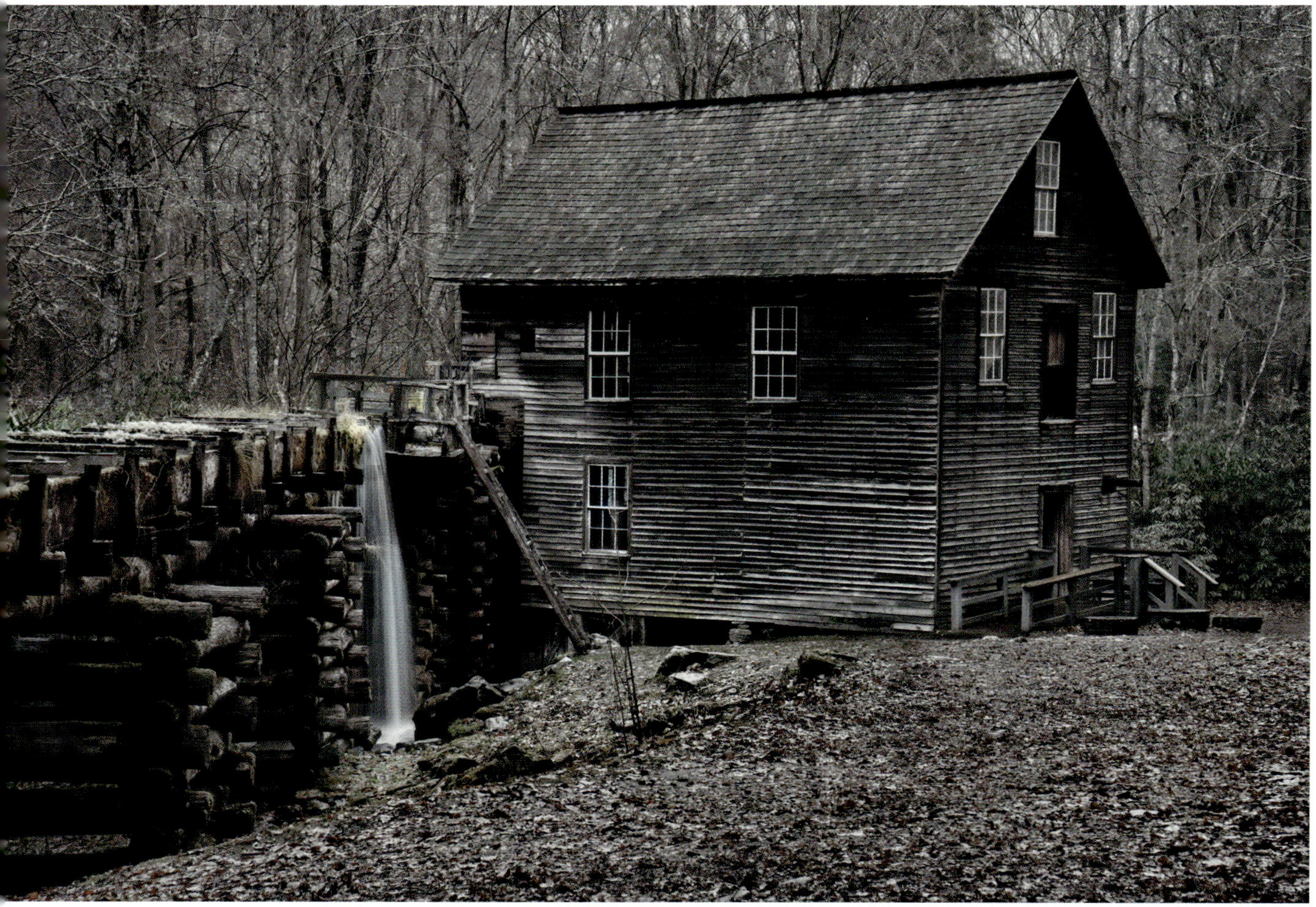

Mingus Mill is equipped with a flume "spill gate" that can be used to control the water flow to the penstock and turbine. If the flow of water through the flume is too high for turbine operation, the miller can climb the wooden ladder shown in the image and release water from the flume before the water reaches the mill.

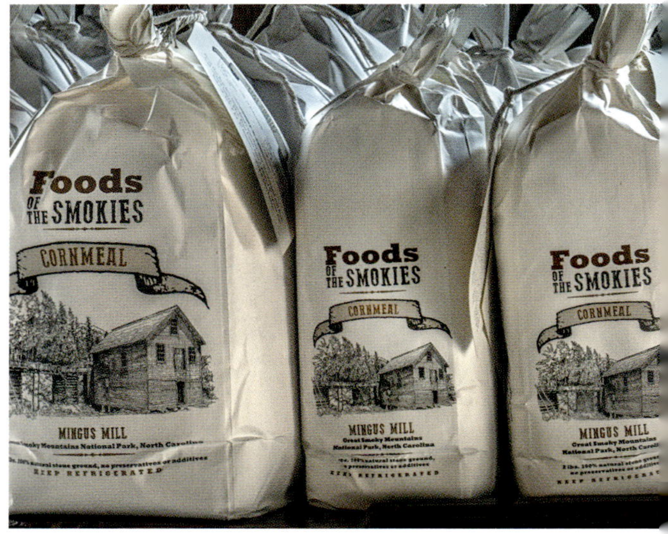

Some mills grind corn and/or wheat as demonstrations but do not sell the products. To sell cornmeal, grits, flour, and other grains as food for consumption, mills must meet local health and cleanliness standards, as at Mingus Mill.

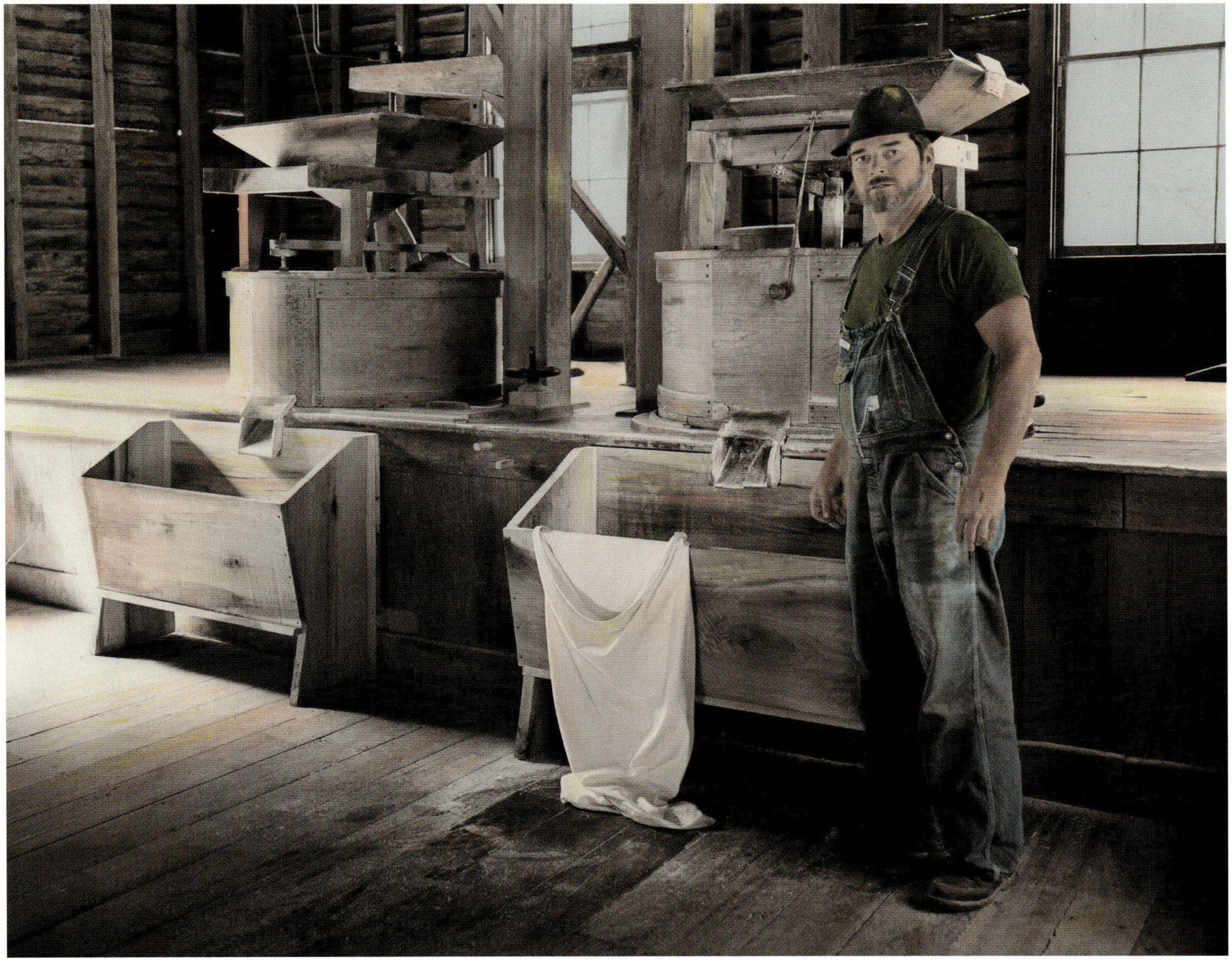

This 1986 oil-tinted photograph of the inside of Mingus Mill shows the miller, Jimmy O'Neal, with the two sets of millstones and the elevated platform that effectively isolates the building from destructive vibrations as the millstones and their drive mechanisms operate. O'Neal passed away in 2001.

Saunooke's Mill

Cherokee, North Carolina, 1975

SITED ALONG THE BANKS of the Oconaluftee River with the foothills of the Great Smoky Mountains in the background, Saunooke's Mill, constructed in the 1970s, certainly makes a remarkable impression on anyone with a fascination for watermills and waterwheels. The millstones have been grinding grains with a large waterwheel for fifty years, proving that in the right market and with a commitment to hard work, watermills could and can be successful in the late twentieth and early twenty-first centuries.

Located on the reservation of the Cherokee Nation, this mill has always been owned and operated by Indigenous Americans. The land has been in the Saunooke family for many generations, including former Principal Chief Osley Saunooke and former council member Bertha Saunooke. Their son Charles and his wife, Carol Saunooke, searched the mountains for millstones and a waterwheel for their mill. They got Saunooke's Mill up and running and added a general store and gift shop for their all-natural products.

In 2019, the mill passed to the current owner and operator, John Parker, who is also a member of the Eastern Band of Cherokee Indians. Thousands shop at his mill every year, with many making regular trips for the stone-ground grits, cornmeal, and other items. The mill's products are also available online, and each purchase, as at other operational mills, helps to assure the continued survival of this landmark.

The fact that this mill was established by Indigenous Americans is especially significant. At most Southern mill sites, the land was taken from Indigenous Americans before a mill was built.

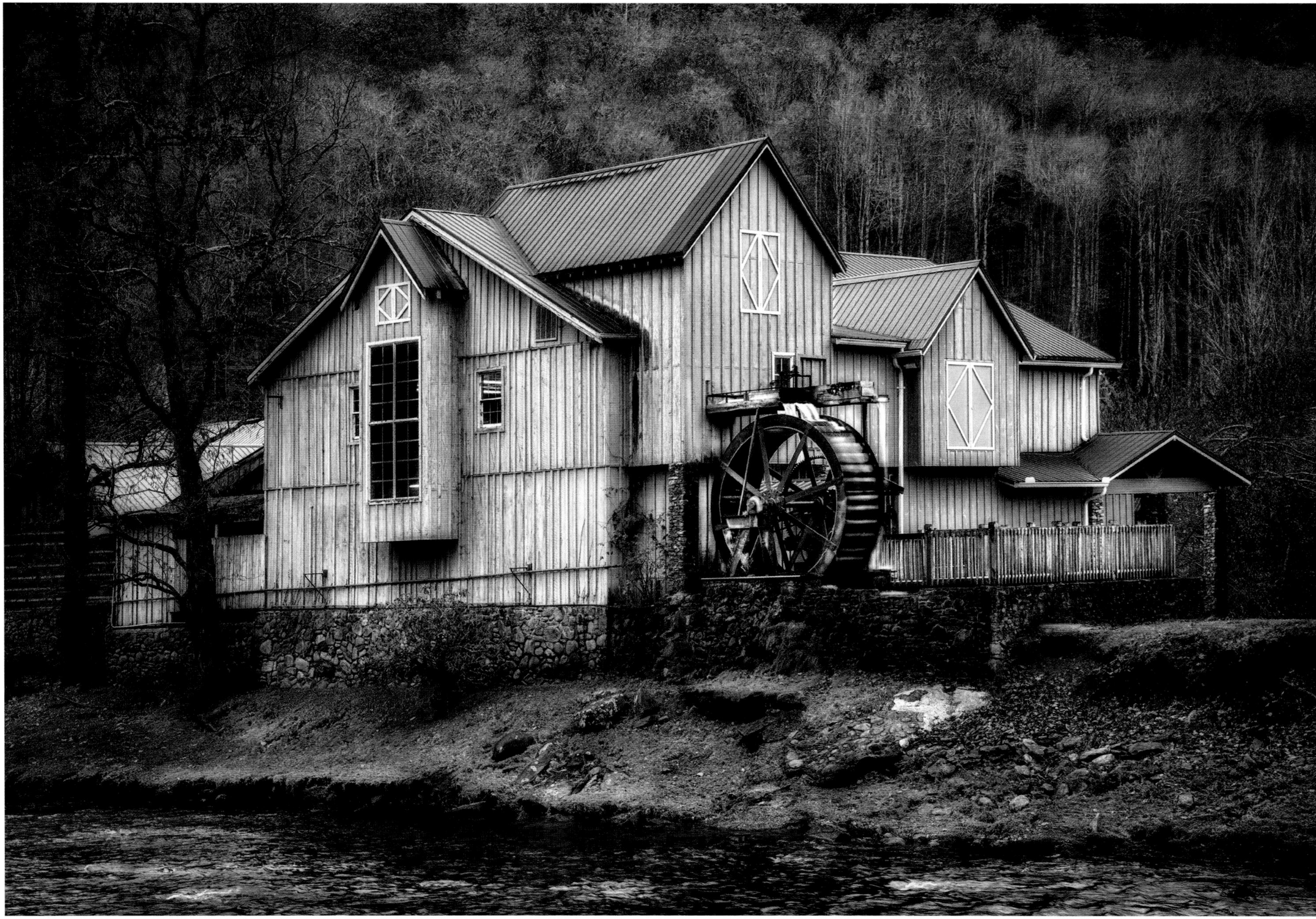

Saunooke's Mill and store sit alongside Oconaluftee River at the east entrance to Great Smoky Mountains National Park. The waterwheel still provides the power for turning the millstones.

The Old Mill of Guilford / Bailes Old Mill

Oak Ridge, North Carolina, 1818 / Several Restorations

THE OLD MILL OF GUILFORD, sometimes known as Bailes Old Mill, is one of the most historic and oldest operating gristmills in the United States. The mill has been a popular tourist attraction for years, and artists and photographers have traveled from all around the world to see it for themselves. Its red-and-white exterior and giant twenty-four-by-four-foot waterwheel, along with the classic, picturesque setting, provide a scene that thousands have enjoyed.

The first mill here along Beaver Creek began operating in about 1767; this was a horizontal tub mill. Although the area was still a frontier at the time, Indigenous Americans never threatened the mill. British troops under General Cornwallis seized the mill in 1781 to grind grain for the British army before the Battle of Guilford Courthouse. The current structure and the first waterwheel date to about 1818, or to 1822 by some accounts. (So much has been written about this mill that conflicting dates and information can be found in the literature.) During the Civil War, the area around the mill remained solidly in Confederate control, so there was never a concern that the mill might be burned.

The mill is constructed of heavy timber and mounted on a substantial fieldstone foundation. Sometime after 1913, the mill was converted to a roller mill, and the waterwheel was replaced by a turbine. When Clarence Bailes purchased the mill in 1954, he removed the turbine and installed the current waterwheel; he also removed the roller mill equipment and made upgrades to the building and property.

For years, an elevated wooden flume delivered water to the wheel from a ten-acre millpond above the mill site. In 1932, Highway 68 cut through the property, and a twenty-six-inch-diameter steel pipe was installed under the road to replace the flume.

In recent decades, upgrades and improvements have been ongoing at the mill. The site actually closed for about a year in the mid-1970s. The mill was reconfigured to allow electricity from the waterwheel (and later from the local electric grid) to power the equipment in the mill. In the spring of 2024, Guilford's water pipe and electric generator were repaired to restore the electrical systems and effectively power the mill once more.

The mill offers an extensive range of popular and famous items in the mill gift shop. Favorite products include cornmeal, grits, and flour, along with mixes for gingerbread, pancakes, cookies, biscuits, muffins, scones, and hushpuppies. A Guilford specialty is the mill's sweet potato muffin mix!

It is said that each miller over the years has left a unique stamp on the history of the mill. For example, Henry Ledbetter kept as many as forty cats on the site when he ran the mill!

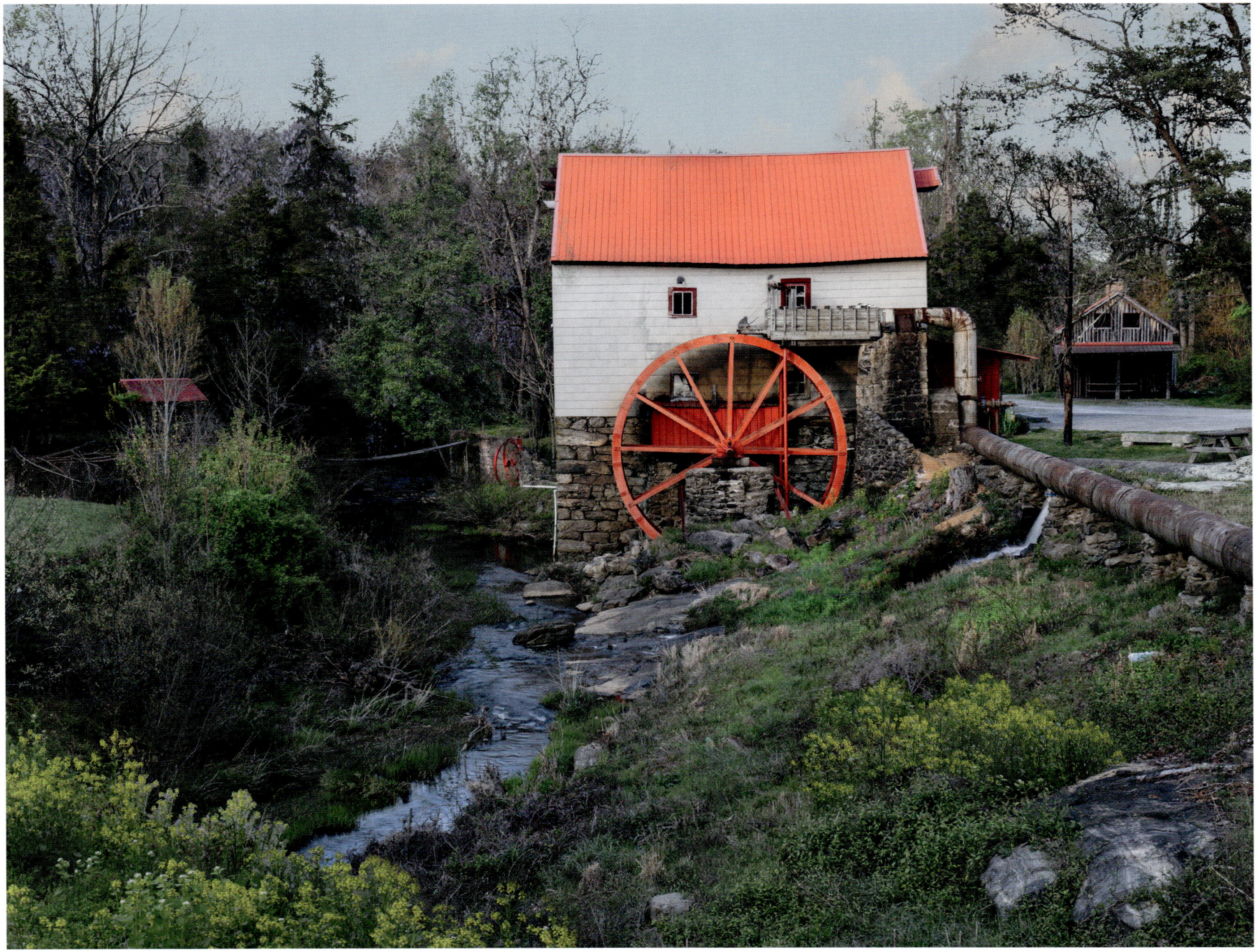

The famously picturesque beauty of the Old Mill of Guilford is obvious. This early spring image shows the view from Highway 68, with Beaver Creek in the foreground.

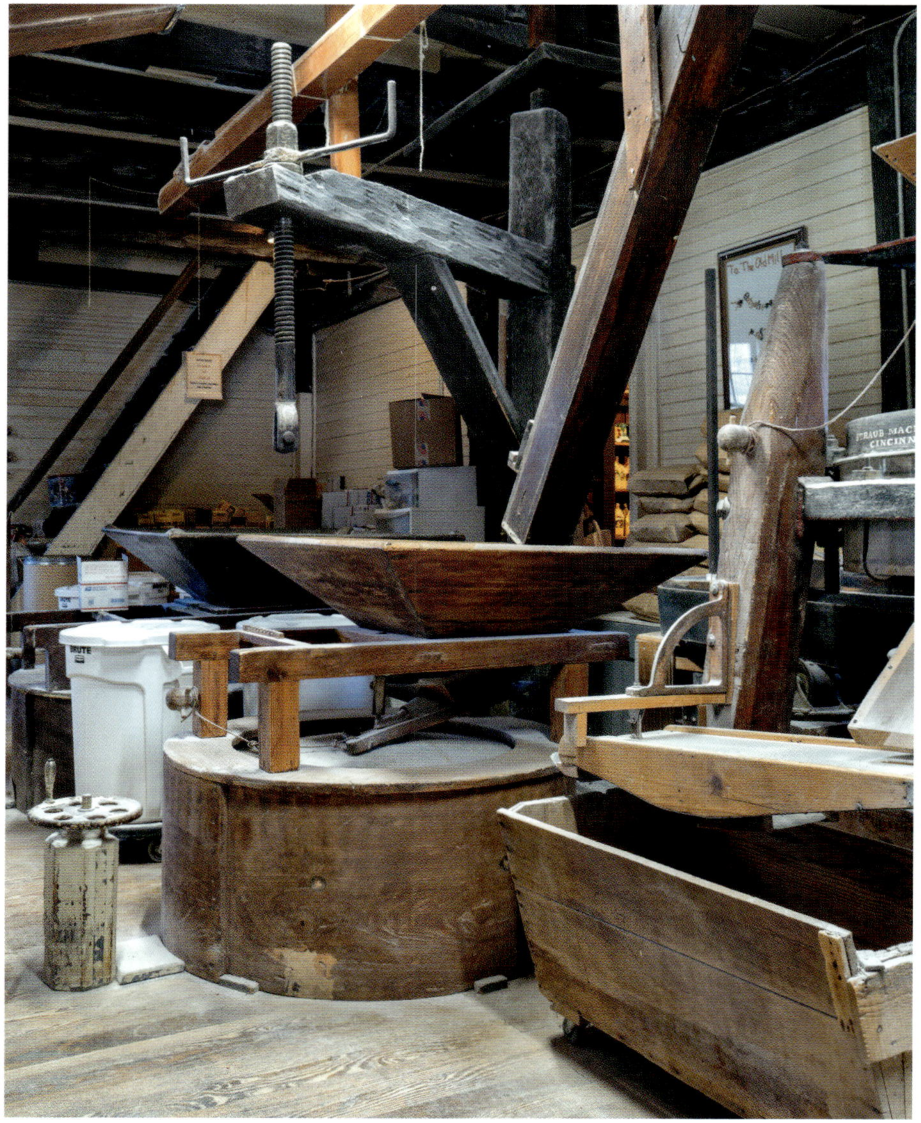

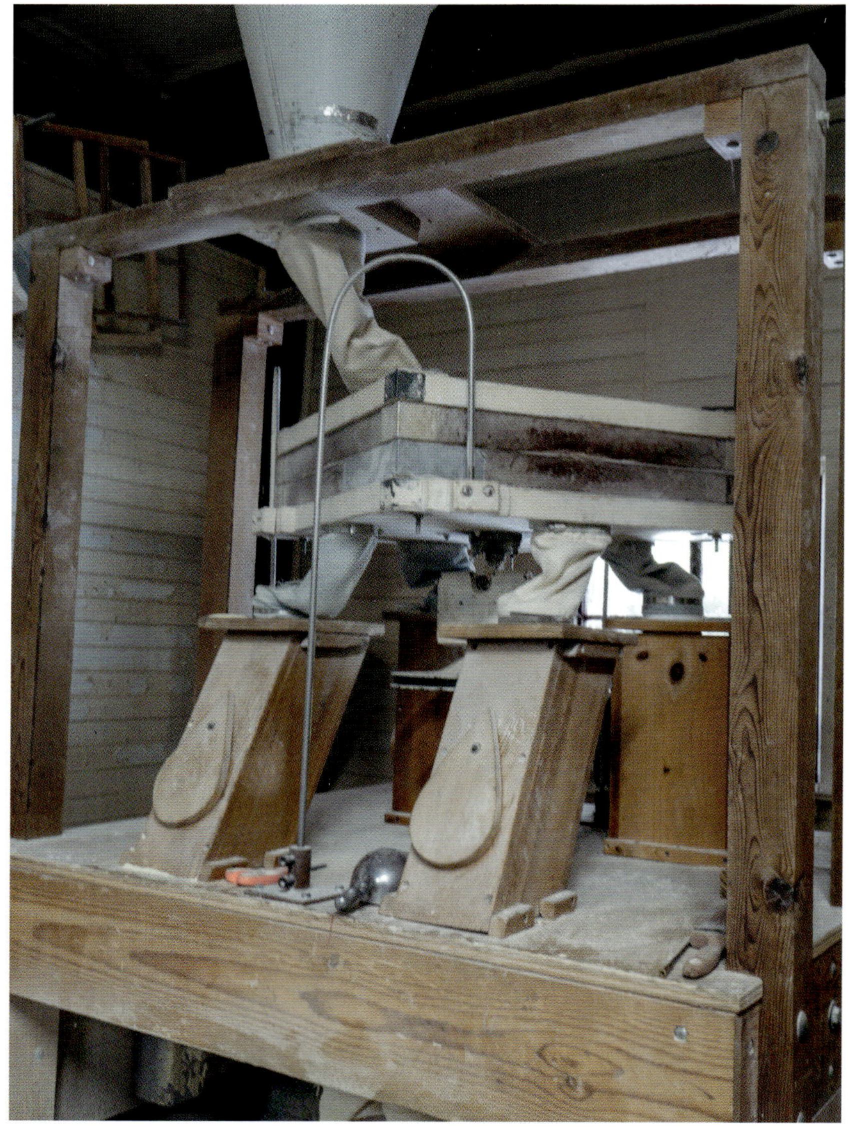

Although visitors are always welcome, this is a busy working gristmill. In the image at left, a circular millstone housing or “vat” can be seen below a hopper and a grain-feeding chute; a large wooden grain bin is located to the right of the vat. The image at right shows a flour sifter.

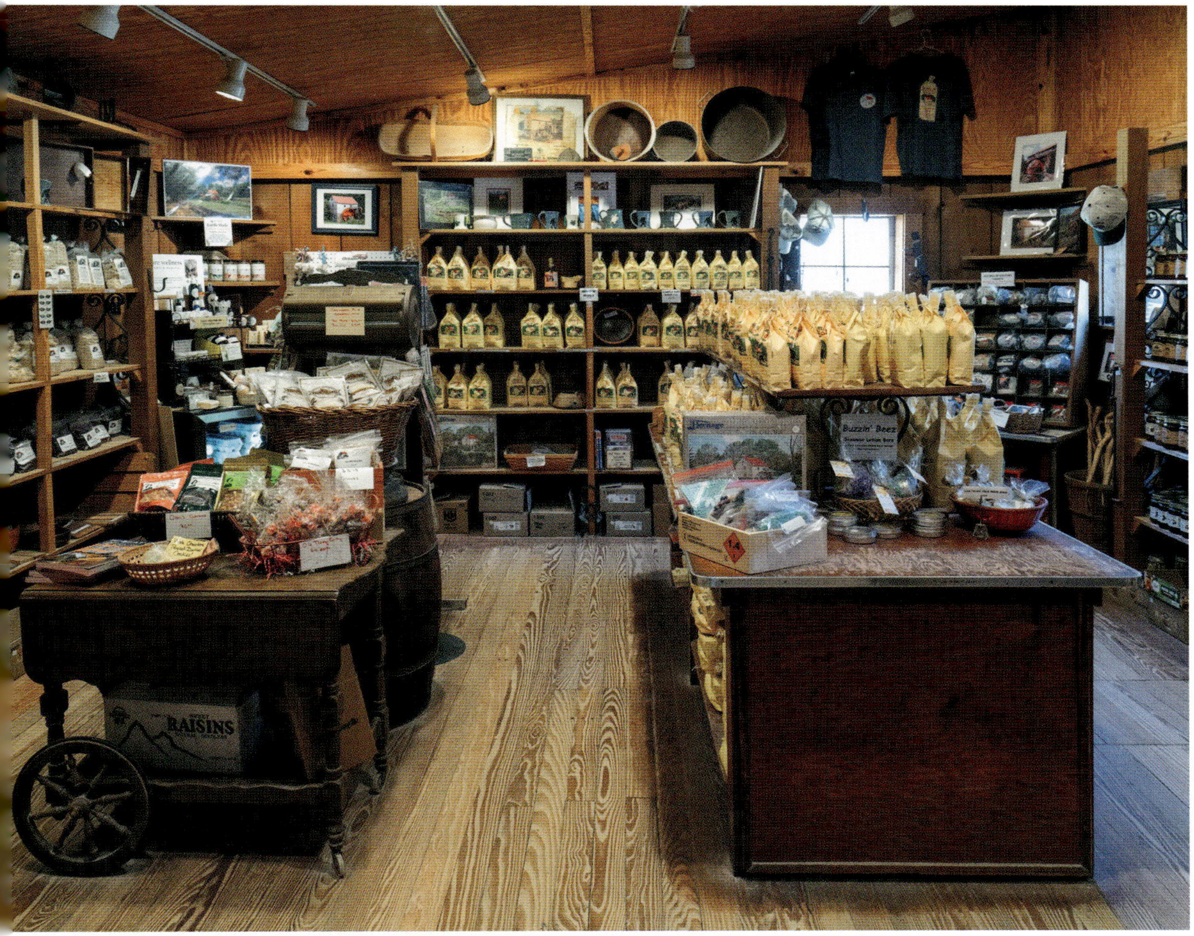

The mill store offers a wide range of meals, flours, and mixes, along with jams, jellies, and various gift items.

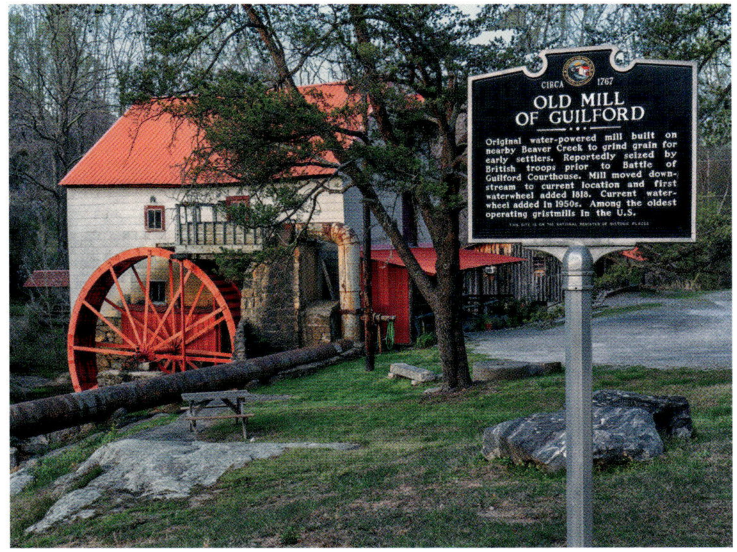

Several signs and markers around the site highlight the long and significant history of the mill. When the mill is open and operating, the parking area behind the sign and to the right of the mill is typically busy with visitors and shoppers.

Yates Mill

Wake County, North Carolina, Circa 1756 / 1990s

FOR ALMOST TWO CENTURIES, Yates Mill served the Wake County community by grinding corn and wheat, sawing lumber, and carding wool. Historical records show that some seventy watermills have served Wake County around Raleigh. Today, only the restored Yates Mill is in operational condition.

Phares and Roxanna Yates purchased the site during the Civil War. The previous miller was said to have been involved in a murder! The Yates family ran the mill until 1948, when it was sold to a local businessman, A. E. Finley, who constructed a retreat at the site for family members and employees.

The mill closed in 1953, and the structure subsequently experienced a decade of deterioration, until North Carolina State University purchased the property and a thousand adjoining acres in 1963. The university made efforts to stabilize and preserve the structure but had somewhat limited resources.

In 1988, Yates Mill Associates assumed the lead with a plan to bring the old mill back to life. Restoration was ongoing throughout the 1990s, thanks to a joint public and private community effort. Service restarted in 2001, and Historic Yates Mill County Park, with walking trails and boardwalks around the mill, opened in 2006 and has become popular with local residents and tourists.

In recent years, restoration of the watermill has been a costly, time-consuming, and ongoing proposition. A twelve-foot wooden waterwheel was installed in 1994. Waterwheels of this type generally have a working life of about ten to twelve years, but with care they can survive for as long as twenty years. In 2025, a new wooden wheel was installed at a cost of about $400,000. The wheel was fabricated by Ben Hassett of B.E. Hassett-Millwrights, a company that specializes in the restoration of early agricultural and industrial structures.

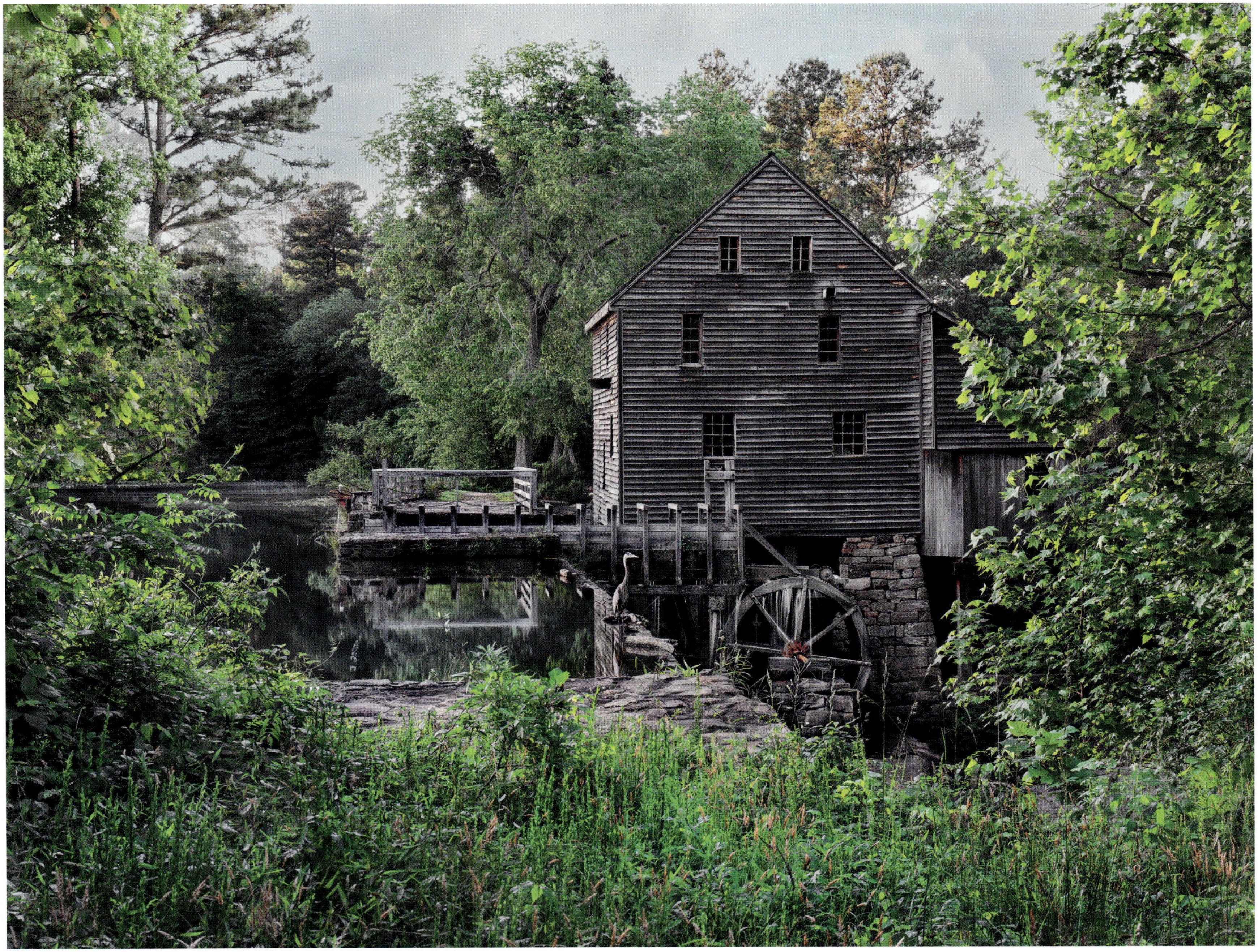

Great blue herons, such as the one standing on the Yates Mill dam in this photo, are common at Southern watermills where the fishing is good. These beautiful wading birds stand motionless until they spot prey.

Anderson Mill

Moore, South Carolina, Late 1700s / Early 1900s / Restoration Ongoing

THE AREA AROUND present-day Anderson Mill was for centuries occupied by the Cherokee, who valued it primarily as a good source of clean water and as a hunting ground. A treaty with the Cherokee Nation in 1753 opened up the area for British settlers, and nearby Spartanburg was initially established as a frontier trading post.

During this period, the earliest settlers continued to suffer violence from dangerous bandits in the area as well as attacks by the Cherokee, who obviously resented the occupation of their lands. A battle between the British and the Cherokee in 1761, along with other efforts to eliminate lawlessness, eventually made the area more secure for Europeans.

Throughout the 1750s and 1760s, Scots-Irish settlers from Pennsylvania and others groups from the mid- and south Atlantic coastal areas entered the region in waves of migration. Many settled along the bottomlands of the Tyger, Enoree, and Pacolet Rivers.

As early as 1760, settlers recognized the value and importance of the North Tyger River and understood that it was well suited to establishing watermills. One location along the river was selected as the site for what is known today as Anderson Mill. This is an area where the river runs over a rocky outcrop of native granite, and the elevation drops significantly over a rocky shoal that was called Pleasant Falls at that time.

Although the precise date of origin of the first North Tyger River watermill has been impossible to pinpoint, historical information indicates that a mill stood at the current Anderson Mill location by the mid-1780s, with some information pointing to a founding date as early as the 1760s. Regardless, timbers and some elements of the foundation seem to confirm the 1780s as the origin date, and historical records show that Spartanburg County was formally established on the grounds of the mill site in 1785.

During the decades that followed, the mill was known by a succession of names, including Nicholl's Fort, Nicholl's Mill, and Tanner's Mill. Its current name comes from James "Tyger Jim" Anderson, who acquired the mill in 1831.

Historically, the Tyger River has been subject to flooding, and much of the early mill was heavily damaged by the great flood of 1903. By 1905, the structure and much of the equipment had been restored, and the mill continued to serve the community until its closing in 1975. The mill was listed in the National Register of Historic Places in 1978 but stood idle for years.

In 2005, the Tyger River Foundation began studying the possibility of restoring the mill. In 2010, the Riveredge Homeowners Association donated the mill site to the foundation.

The foundation has made great progress in restoring the site to an operating mill, with the intention of periodically opening the mill to the public for educational demonstrations. An endowment is being established to assure that the mill is maintained in operating order for future generations to enjoy.

To date, the mill's restoration has been a long and challenging project made possible through the dedication, commitment, and persistence of Tyger River Foundation members and private foundations.

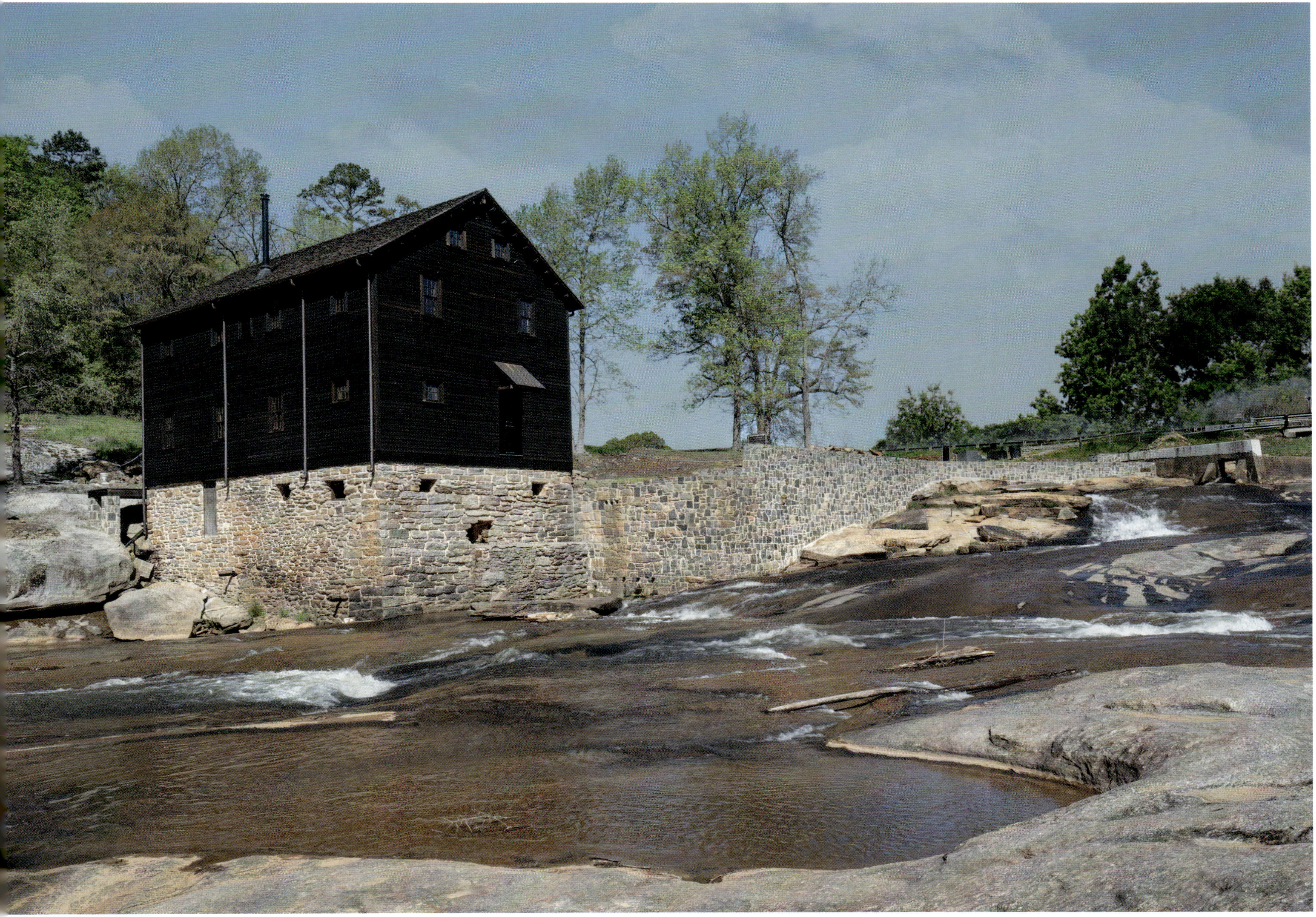

Anderson Mill is the oldest mill standing on its original foundation in South Carolina. The Tyger River Foundation is currently well along in its historic restoration of the property as an operating mill. As of 2024, when this image was made, the waterwheel had yet to be installed.

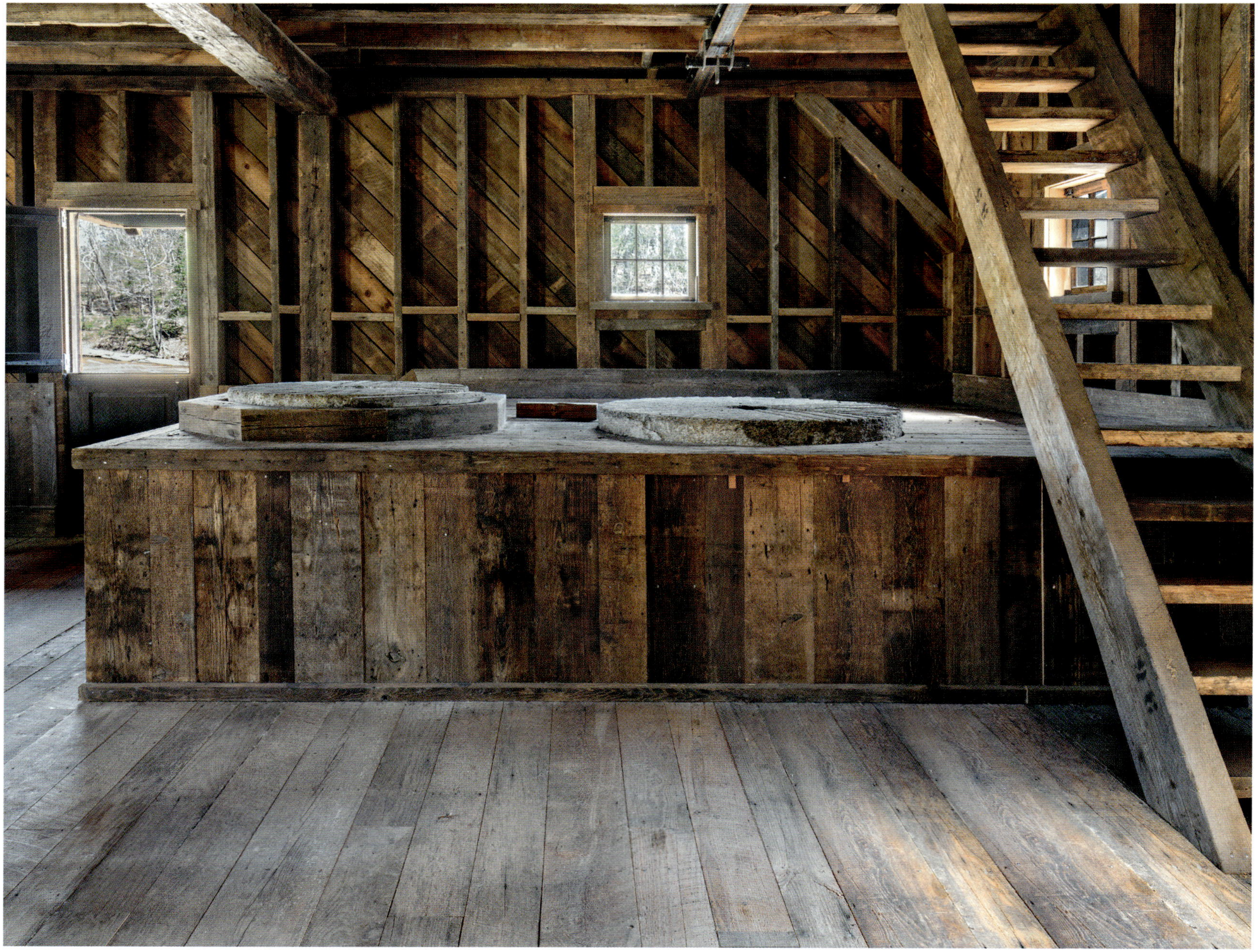

This interior view of Anderson Mill shows the status of restoration inside the mill in spring 2024. The elevated millstone platform was complete, and the base (stationary) stones had been installed. The "runner" (upper and rotating) stones and housing had not yet been added.

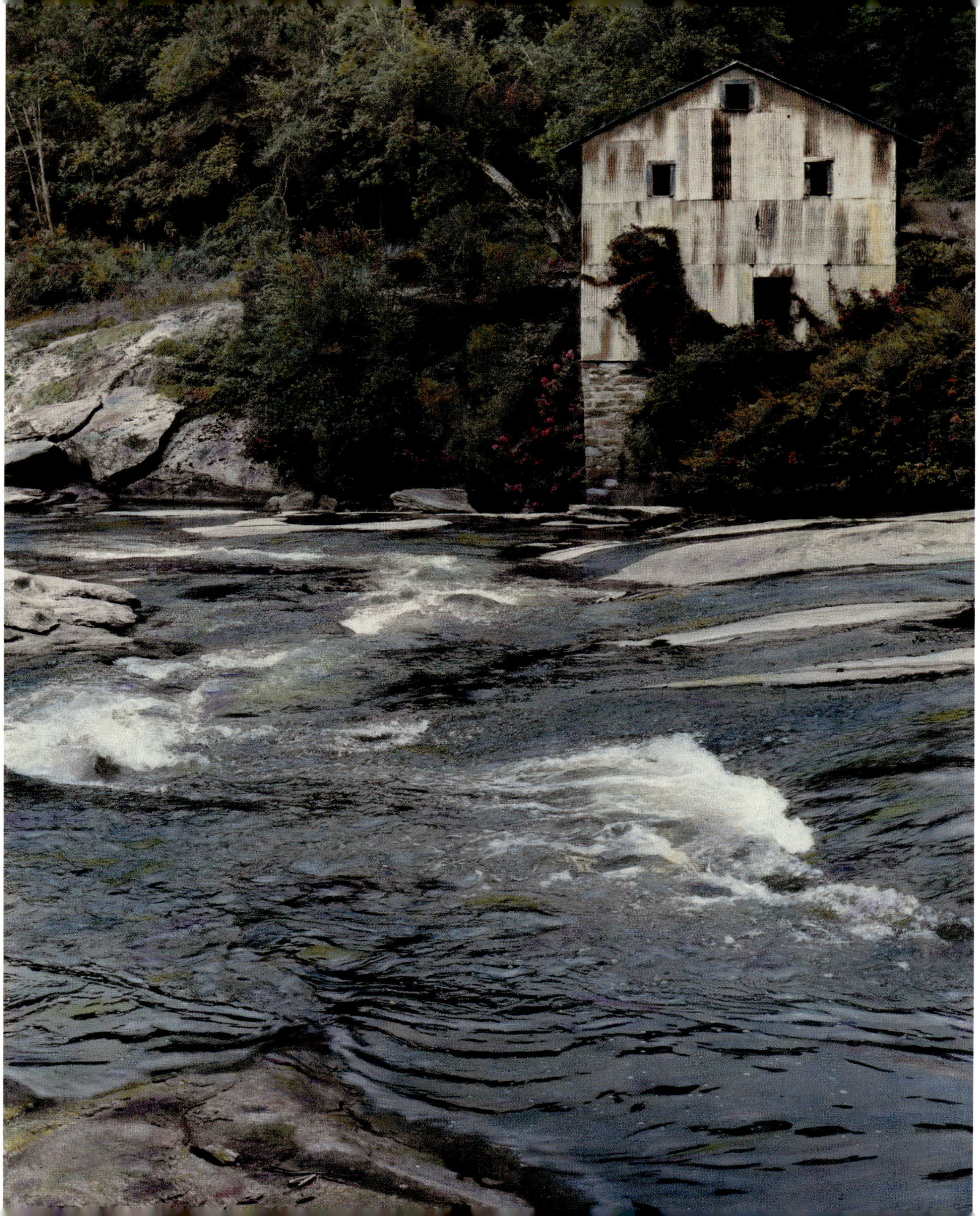

This early 1980s hand-tinted image of Anderson Mill with corrugated siding shows the mill with the waterwheel and raceway completely covered in wisteria. The site was in peril at this time.

Gilreath's Mill

Greer, South Carolina, 1839

ACCORDING to SPOOM records, there were 720 gristmills operating across South Carolina in 1882. No more than 25 of these mills remain today, and only a few are still operational. Gilreath's Mill, one of the oldest surviving sites, is not operational, but it remains sound structurally. It sits abandoned, just a few feet from a busy roadway.

The large waterwheel, three feet wide by twenty-four feet in diameter, is still in place. Interestingly, the wheel is located across the creek from the mill. This placement is said to have facilitated using water from both Clear Creek and nearby Shoal Creek, but the arrangement seems odd today without any remaining linkage or connection across the water. When the mill was operational, a ring gear around the perimeter of the wheel was coupled with a long geared shaft to transfer power across the creek to the mill.

The two-and-a-half-story wood-framed mill is constructed on a stone foundation. Two small sheds are attached to the main structure. Inside, most of the milling equipment is still in place. The doors and windows are missing, but as of 2019, there were no signs of the vandalism or graffiti so common at many historic sites today.

The mill operated for at least 111 years before closing in 1950. Local tradition and some sources date the mill to 1814, but more substantiated records show that it began operating in 1839. Historically, a second waterwheel and structure were located just beyond this site to run a sawmill. The earthen dam that held back the millpond for both mills is long gone.

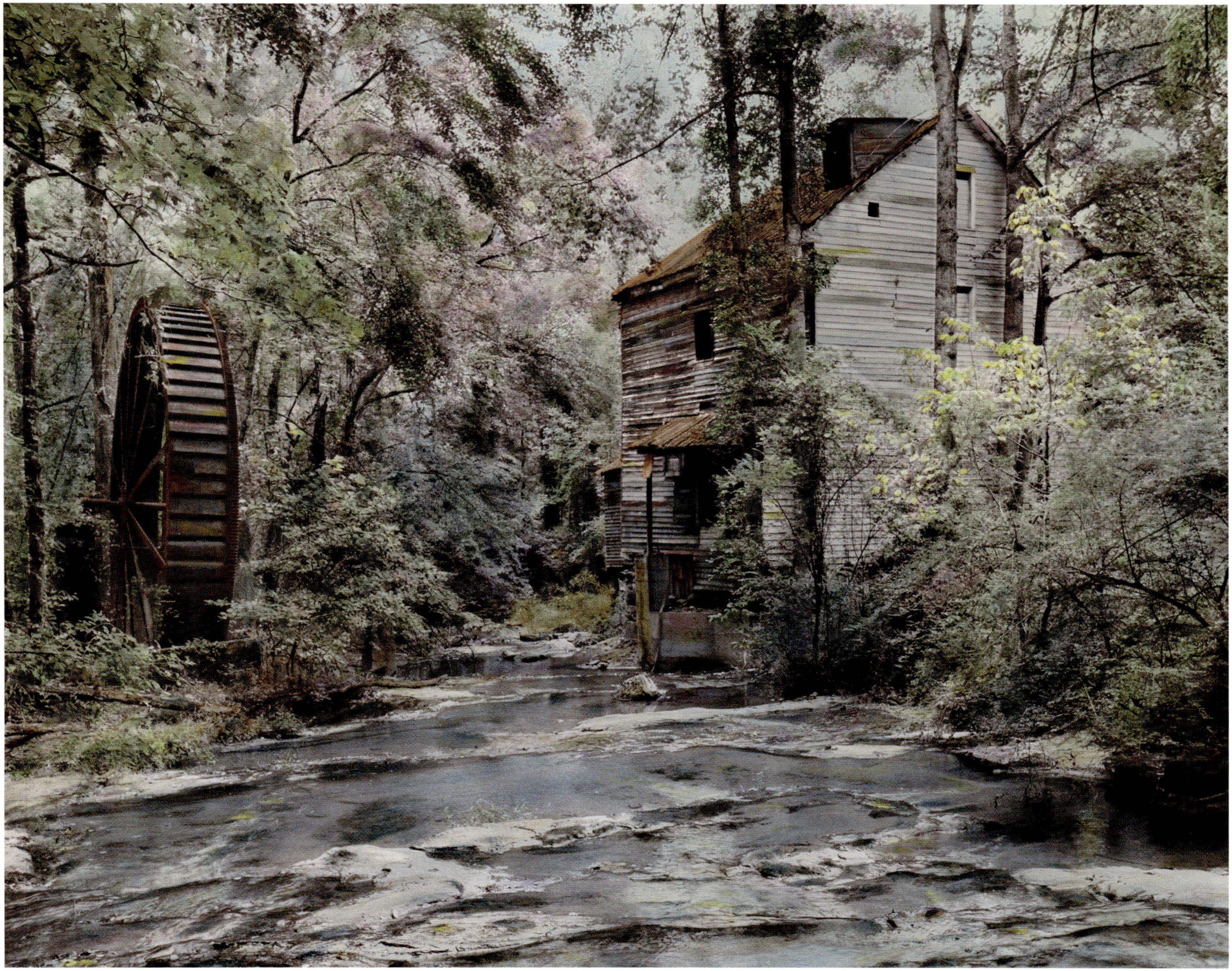

This hand-tinted image shows Gilreath's Mill with its twenty-four-foot waterwheel located on the opposite side of Clear Creek from the mill. The linkage between the wheel and the mill has been gone for at least fifty years.

Hagood Mill

Pickens, South Carolina, 1845 / Mid- to Late Twentieth Century

HAGOOD MILL is a fully restored and operational gristmill dating to 1845 that is on the site of an older watermill that had operated as early as the 1790s. James Hagood constructed the newer mill, and several generations of his family operated it until it was retired from commercial service in 1966. The mill was set up to produce cornmeal, grits, flour, and animal feeds. In later years the mill ran only sporadically, and it finally closed when health regulations were implemented that required the testing of corn before processing. (It is interesting that South Carolina is now the only state that exempts water-powered gristmills from health and regulatory inspections.)

The two-story mill has never been painted, although, as at other old wood-construction mills, some of the planks have been replaced, and the window shutters are new. The wood framing and supports for the structure were hewn by hand, and the exterior has always been covered with siding. The waterwheel is twenty feet in diameter, and the millstones weigh sixteen hundred pounds each.

Restoration work began in 1978, more than a decade after the mill was retired from commercial service, and continued into the late 1980s. The waterwheel was replaced, and the restored mill first became operable in 1996. Today, water is pumped to within eighty feet of the waterwheel before it empties into a headrace, and a wooden flume carries it the final distance to the wheel.

The family donated the mill and the property to Pickens County in 1971. Today the historic structure is part of the Hagood Mill Historic Site, which includes pioneer cabins and other educational exhibits. A collection of interesting millstones is displayed along the pathway to the mill.

A particularly interesting feature of the mill property is the Hagood Creek Petroglyph Site, where more than thirty ancient petroglyphs are displayed in situ on a rock outcropping. This area is set up as a museum with limited hours, so check the schedule on the Hagood Mill website before planning a visit. Along with the petroglyphs, a large soapstone boulder with five-thousand-year-old carvings can be seen on the property.

The visitor center provides a gift shop at which mill products and other local and related items can be purchased and activities can be planned.

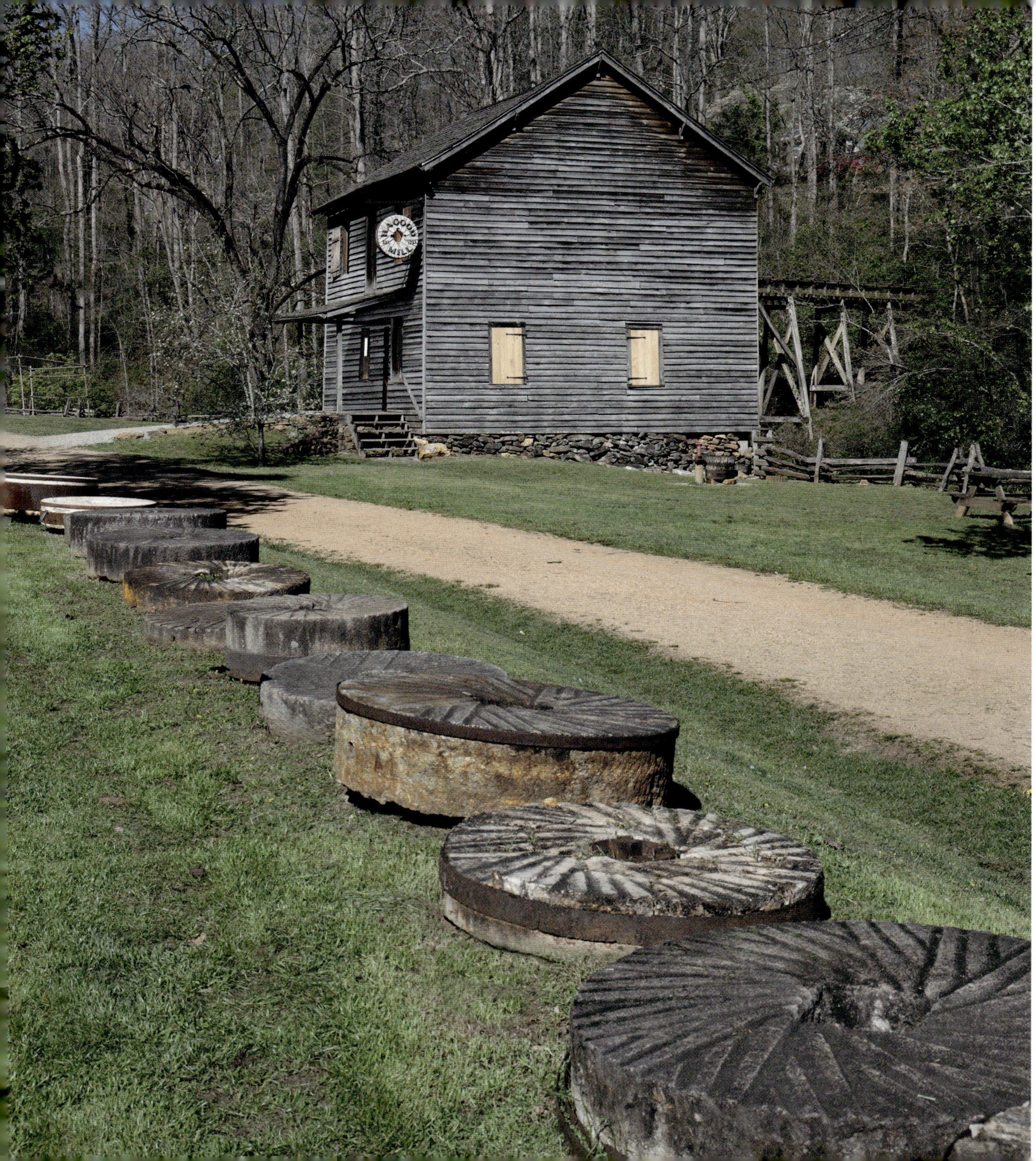

The walkway leading to Hagood Mill is lined with an interesting collection of historic millstones from across the area.

Suber's Corn Mill

Greer, South Carolina, 1908

THE SUBER FAMILY has been involved with milling in the Greer area since before the Civil War, when they first set up a sawmill and corn whiskey still at a site along the nearby Enoree River. The family has operated Suber's Corn Mill since 1908; most recently, James Suber, a fourth-generation miller, ran the business and welcomed visitors and customers to the mill. With his passing in 2025, his children, Lisa Suber Quick and Brian Suber, now carry on the dedication, passion, and tradition that have made the Suber name respected across South Carolina's Upcountry.

The 1908 mill replaced an 1860s mill that was located at the present site or at a site nearby. Initially, the mill was built to saw lumber, and the corn mill followed.

Today the historic mill stands as one of the last water-powered gristmills operating in the Palmetto State. The mill grinds cornmeal, grits, and animal feed with the same waterwheel and millstones used over a century ago. Visitors are welcome every Saturday to watch and learn about the ancient skill and art of milling corn.

The highlight for first-time visitors has to be the twelve-foot-by-sixteen-inch steel and wooden waterwheel. The waterwheel is located behind the mill buildings and is not visible from the road or the main mill entrance.

Waterpower for the wheel comes from a millpond on Princess Creek up the hill from the mill. A wooden flume delivers water over the final stretch of the trip from the pond, releasing it over the top of the wheel.

Most country watermills are housed in a single structure, but this mill consists of a series of small attached buildings that form a line up the steep hillside. Over the years, several small structures were added to the original mill to provide storage and add a feed mill. Most of these were built in the mid- to late 1950s, with the last project completed in 1959.

The buildings are mostly wooden (though cinderblock and metal siding were also used as various additions became part of the mill), and all are painted white. Grinding occurs in a building near the center of the lineup, opposite the waterwheel on the back wall of the mill. The mill uses two sets of stones for grinding, with packaging and weighing done by hand.

Suber's Corn Mill is in a relatively quiet area but is only minutes from the hustle and bustle of Greenville and busy Interstate 85. A visit to the mill is an experience on several levels. In addition to mill products and an education in the milling process, the mill also offers a glimpse of life in the rural South not so long ago.

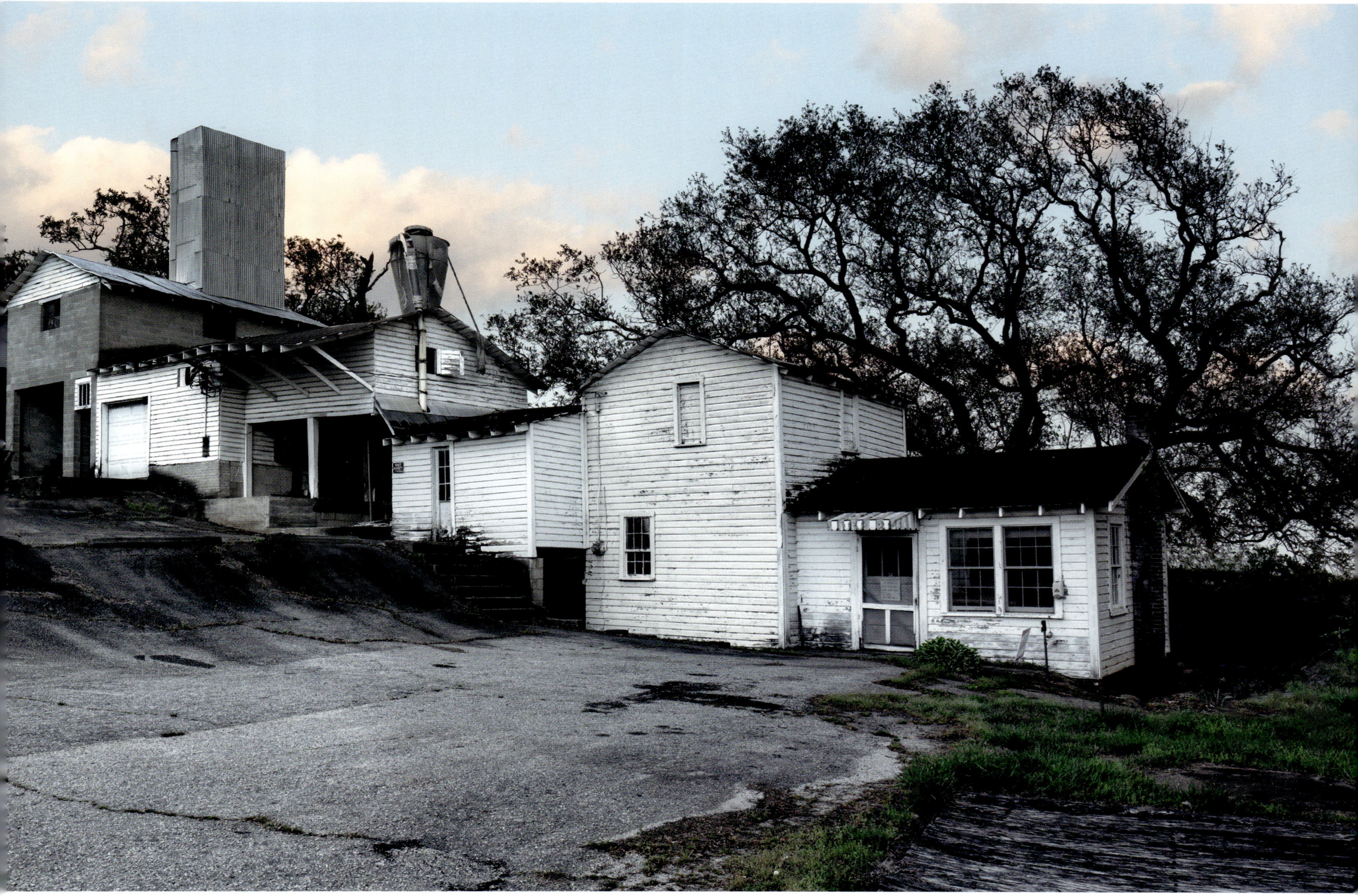

Suber's Corn Mill is unique in design and consists of a series of small white attached structures along a fairly steep hillside. The building at bottom right in the image is the mill office. The millstones and packaging area are located about halfway up the series of buildings.

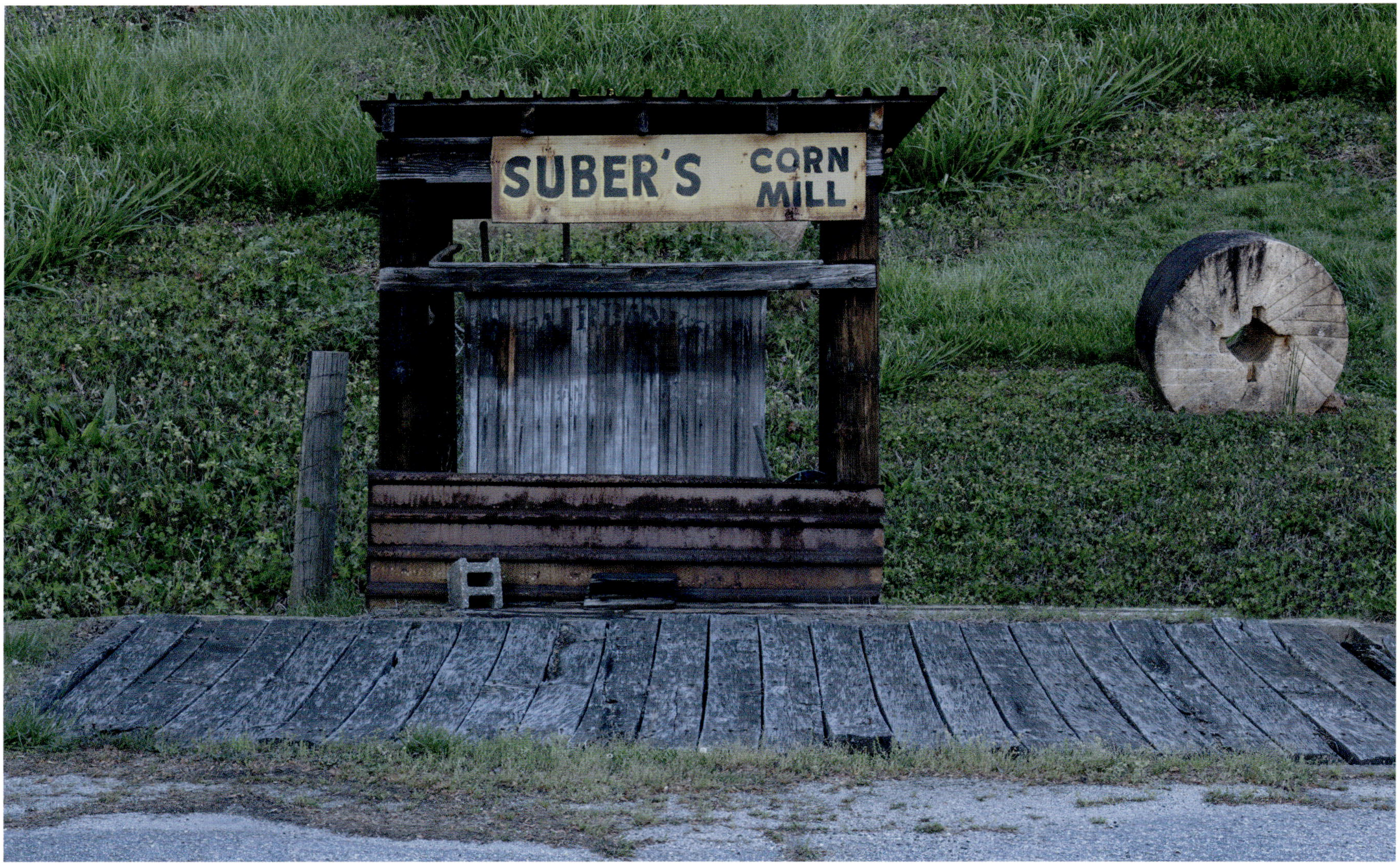

Although this interesting structure appears to be just a sign for the mill, with an old millstone displayed to the right, it is actually a scale for weighing trucks, trailers, and wagons.

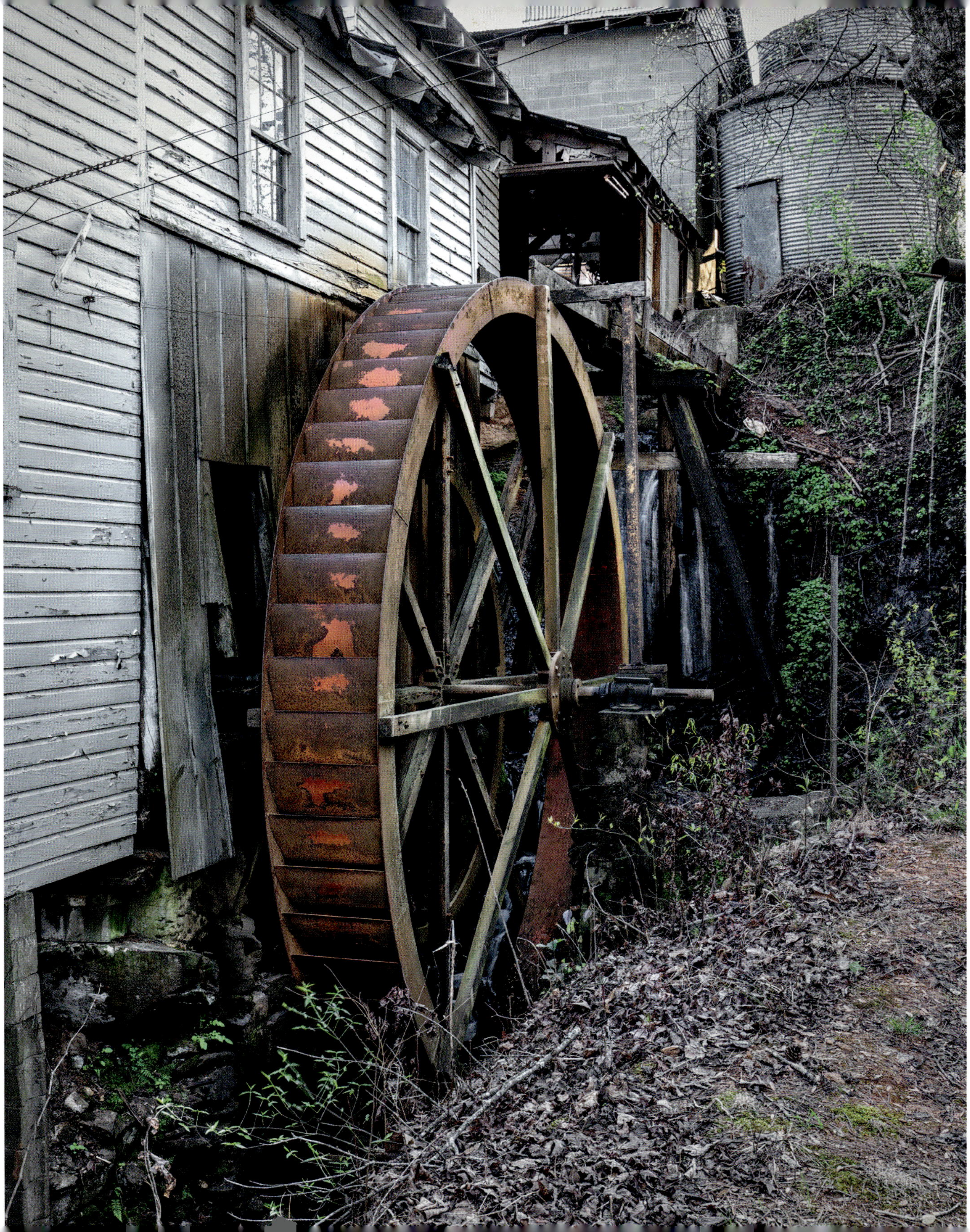

Most visitors, especially children and anyone with a mechanical fascination, find the large waterwheel and other historic features on the back side of the mill to be especially appealing.

Ballenger Mill

Landrum, South Carolina, 1820s / 1960s

BALLENGER MILL stands as the historic centerpiece of a forty-seven-acre environmental conservatory that has become a premier setting for destination weddings and other events. Situated in upstate South Carolina, only a short drive from Greenville and just a few hours from Asheville, North Carolina, this mill is in a picturesque and easily accessible location.

The mill has a storied past that dates back some two hundred years, long before its modern transformation. The gristmill and the site have been restored, rebuilt, or repurposed several times over the years. During that time, several other structures have been built on or relocated to the property.

George Wallace Huet Lewis Dickey constructed the mill in the 1820s using enslaved workers. Waterpower came from the south prong of the Middle Tyger River.

Dickey, who also operated a water-powered sawmill nearby, was a wealthy plantation owner with several thousand acres of land across North and South Carolina. In 1868, soon after the Civil War was over, he died. Over the next one-hundred-plus years, the mill passed through the hands of several owners.

The Greens acquired the site following Dickey's death and named the mill Green's Mill, which stuck until 1960. After almost a century of operation, the mill had fallen into disrepair and was purchased by the Ballenger family, hence its current name. The Ballengers restored the mill and held onto it until the early 1970s.

Morgan Graham was the mill's next owner, and he undertook a total twentieth-century transformation of the site. He was a visionary and set about designing a beautiful environmental area for destination weddings, which he called Willow Creek Conservatory. By this time the property included a covered bridge, an arched footbridge, a plantation-style miller's house, a gazebo, and a one-room schoolhouse converted into a wedding chapel.

As Graham aged, Ballenger again fell into disrepair. In 2022, Alan and Jennifer Zordan bought the mill and the conservatory and began the most extensive restoration to date. What they have accomplished is a charming, inspiring, almost magical site that retains its historic and rustic character. They call it the Venue at Ballenger Bridge.

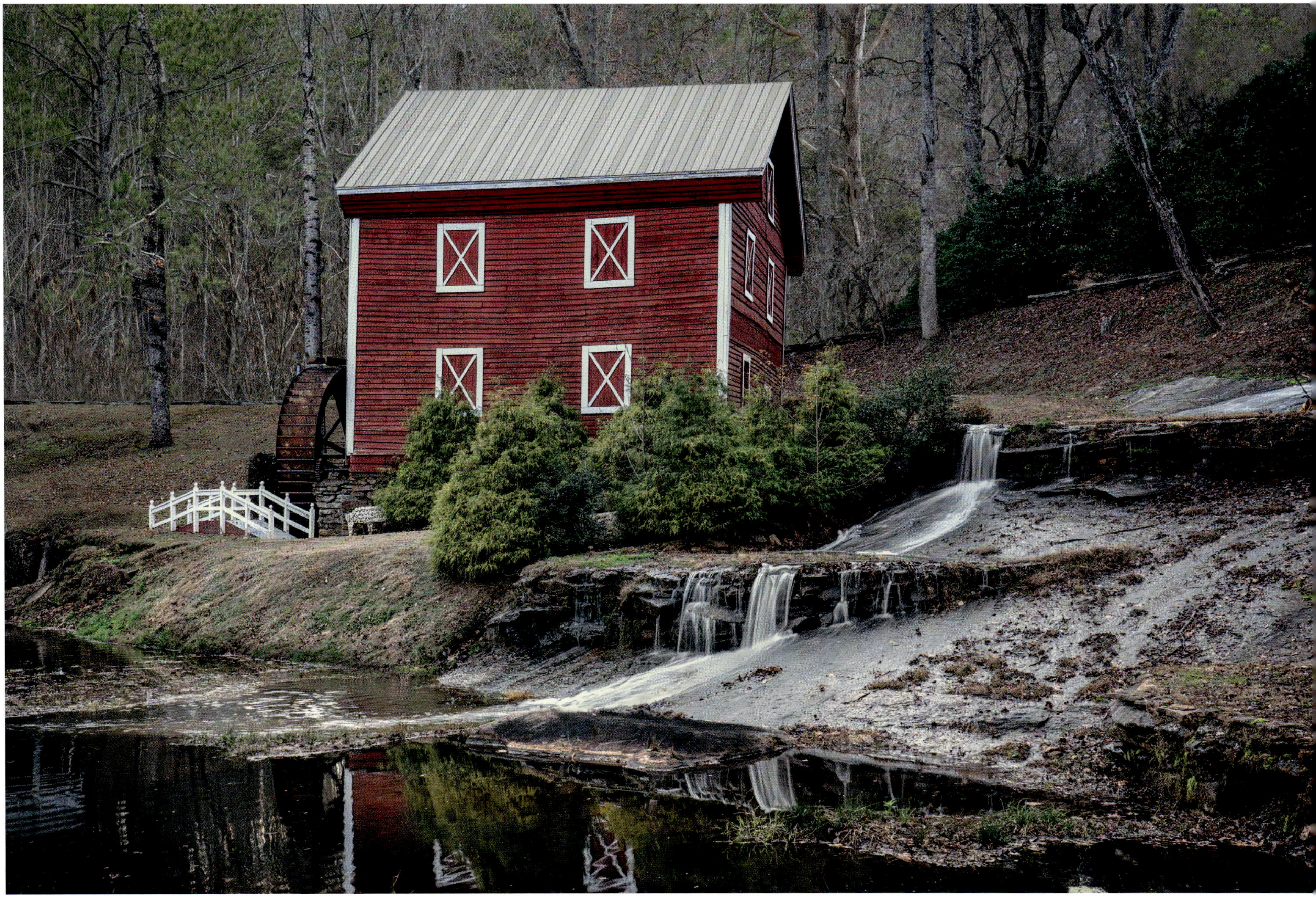

Featuring a red exterior with white trim and a large metal overshot waterwheel, Ballenger Mill is set within a beautiful and well-maintained environmental conservatory.

The Old Mill in Pigeon Forge

Pigeon Forge, Tennessee, 1830

THE OLD MILL in Pigeon Forge is one of the oldest continuously operated gristmills in the United States. It is also one of the most famous and photographed scenes in the country. Its location at the main western entrance to Great Smoky Mountains National Park has transformed this rural mill into a major tourist attraction over the past few decades.

The gristmill dates to 1830, but an iron forge adjacent to the mill began operation even earlier, in 1817, and produced iron bars until it was dismantled in 1885. For a time a post office operated out of the mill, and the miller was the postmaster.

The mill building is a three-story yellow poplar structure that is held together with hickory pegs and mounted on a substantial river rock foundation. The long one-story section of the building was added later as a sawmill. The mill has survived several torrential floods that washed away most other structures along the Pigeon River. In recent years, the foundation has been reinforced with concrete as a precaution, given the river's tendency to flood.

A waterwheel originally powered this mill. In 1889, the operation was updated with a water turbine and roller mills. A generator was also added to provide electricity for nearby houses and to provide another source of income for the mill.

The mill is still turbine powered today, but it also features a prominent twenty-four-foot red-orange breastshot waterwheel that powers the mill elevators. Several smaller wheels that are not as readily visible can provide supplemental power. The mill is equipped with a pair of four-thousand-pound French buhr stones but formerly used roller mills for commercial grinding, as the roller mills were faster than millstones and could keep up with the high demand for mill products at what was once a busy location.

Mill products sold in the gift shop on the main level are believed to be milled off-site (at least in part) due to more stringent government health regulations. A large and visually complementary restaurant has been added next door. Almost shockingly, commercial development around the mill has evolved, or more accurately exploded, over the last few decades.

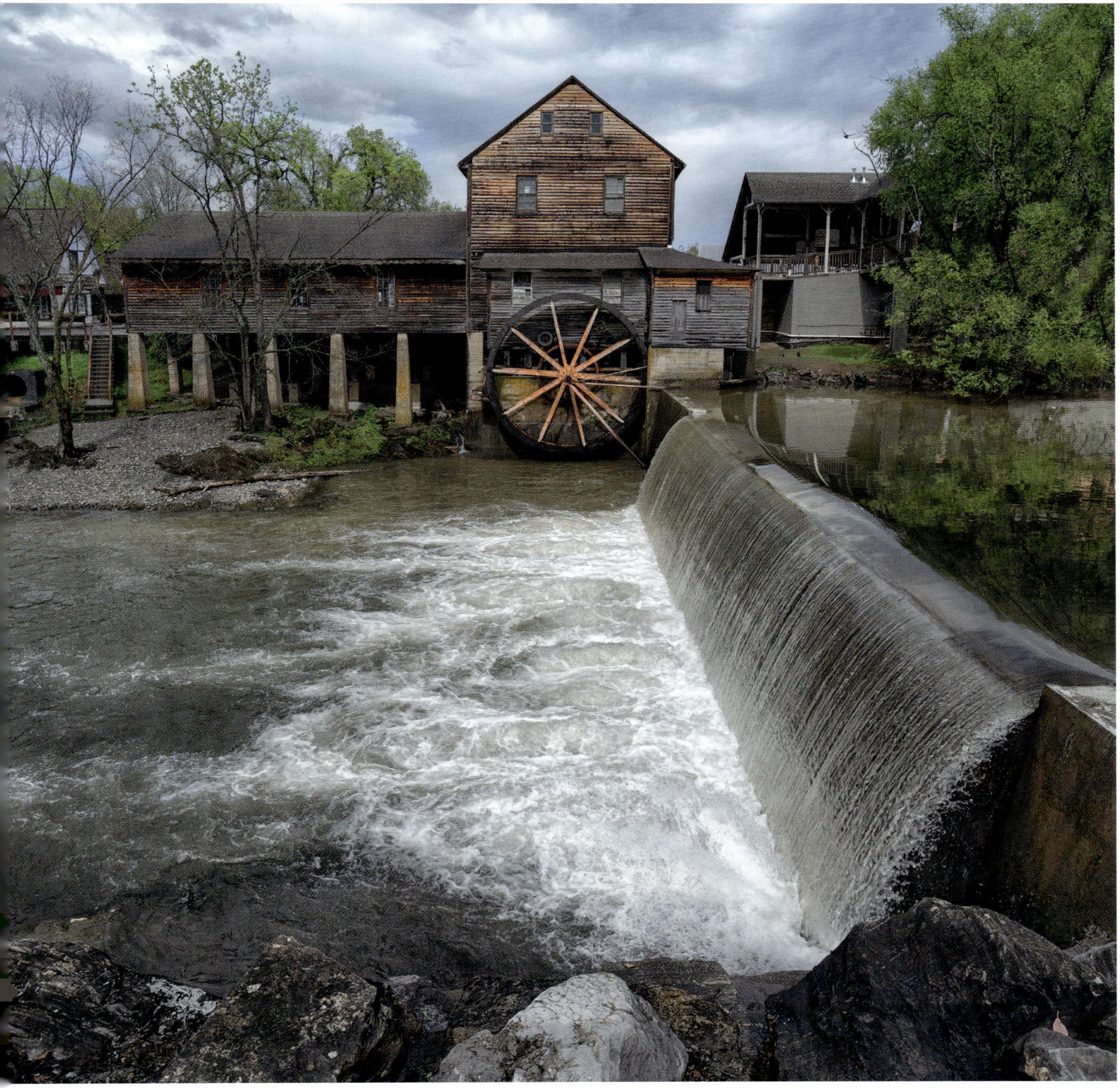

More than a million visitors to the Great Smoky Mountains see this scene every year. To the right of the Old Mill is the more recently built but matching Old Mill Restaurant. Behind the mill are artists' working studios, used especially for pottery.

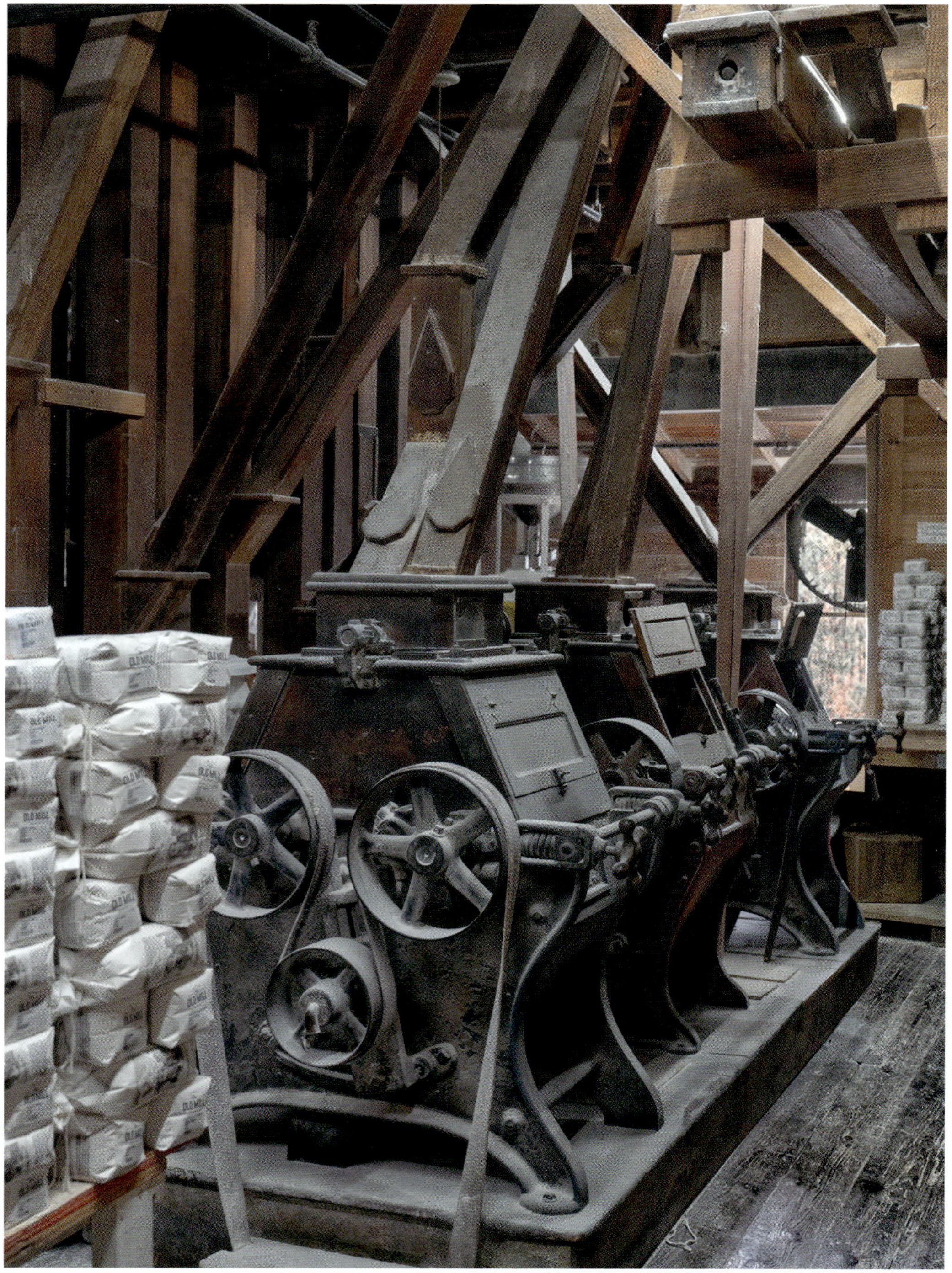

Although the working side of the Old Mill is not normally open to the mill's many visitors and customers, limited tours are offered most days. The three machines shown here are called "roller mills." As described more than once in this book, the slanted, square wooden structures above the mills automatically feed grain to the rollers.

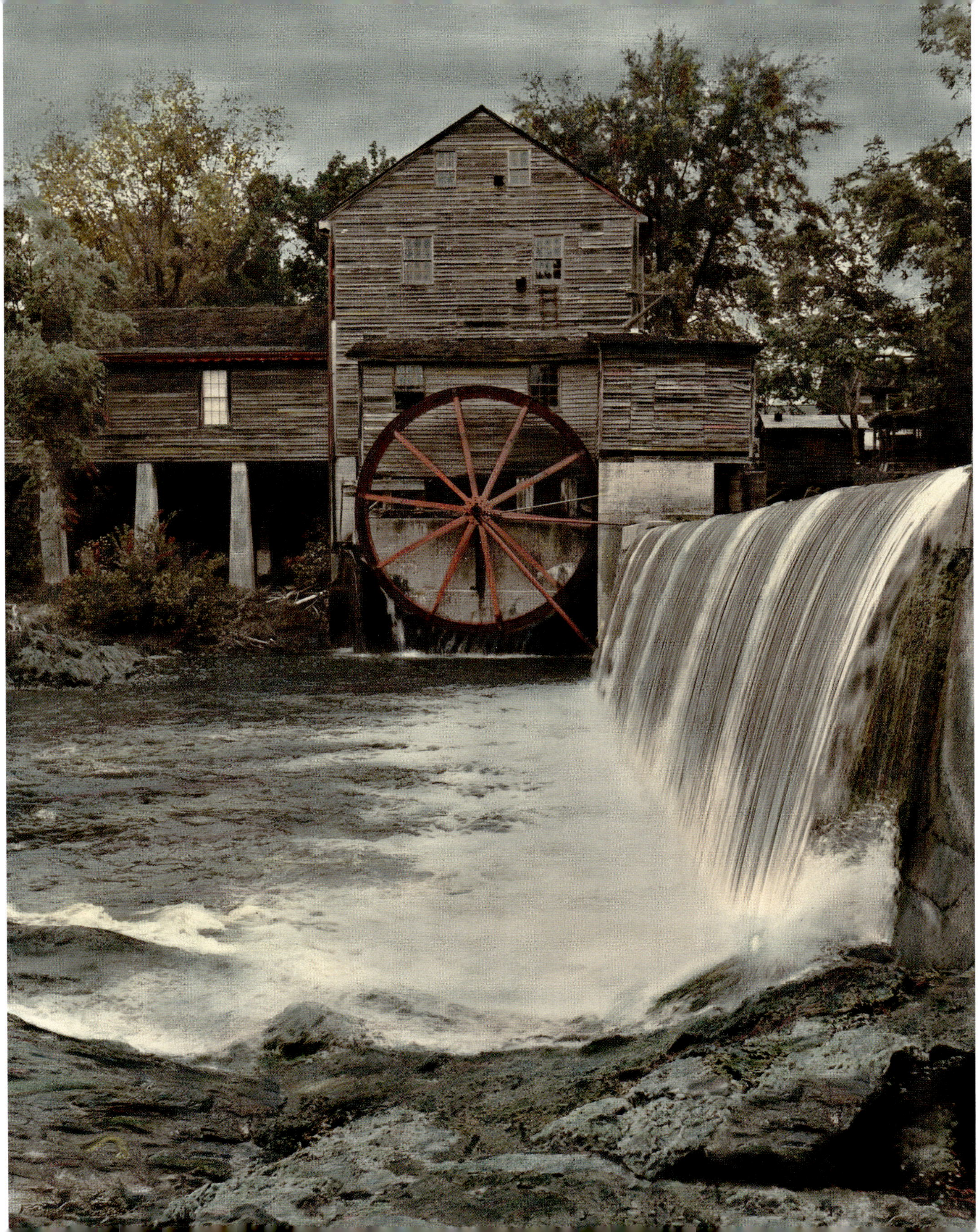

This vintage, hand-tinted image of the Old Mill in Pigeon Forge seems appropriate for such a historic landmark. The dam has been reworked several times since this image was made years ago.

Hacker Martin Grist Mill

Norris, Tennessee, 1790s / 2002

IN 2002, this watermill was relocated from the community of Boones Creek in northeastern Tennessee to Norris in East Tennessee. The move involved forty large truckloads of carefully identified and labeled parts— and, of course, the waterwheel and millstones.

The mill was known as Dove Mill when it began operating in the late 1700s. It was later sold and renamed Isenberg Mill and used in an operation to make, of all things, wooden waterwheels for other mills! SPOOM describes the mill as having been a sawmill during this era.

Today, the reassembled structure is a featured exhibit at the Museum of Appalachia, a premier pioneer and early American history museum in Norris. The mill is clearly visible from the main entry and parking area at the museum. In the area around the mill, a collection of millstones and other related items are on display as part of the educational exhibit.

The Museum of Appalachia is affiliated with the Smithsonian Institution and includes more than thirty historic buildings and thousands of authentic relics and folk items.

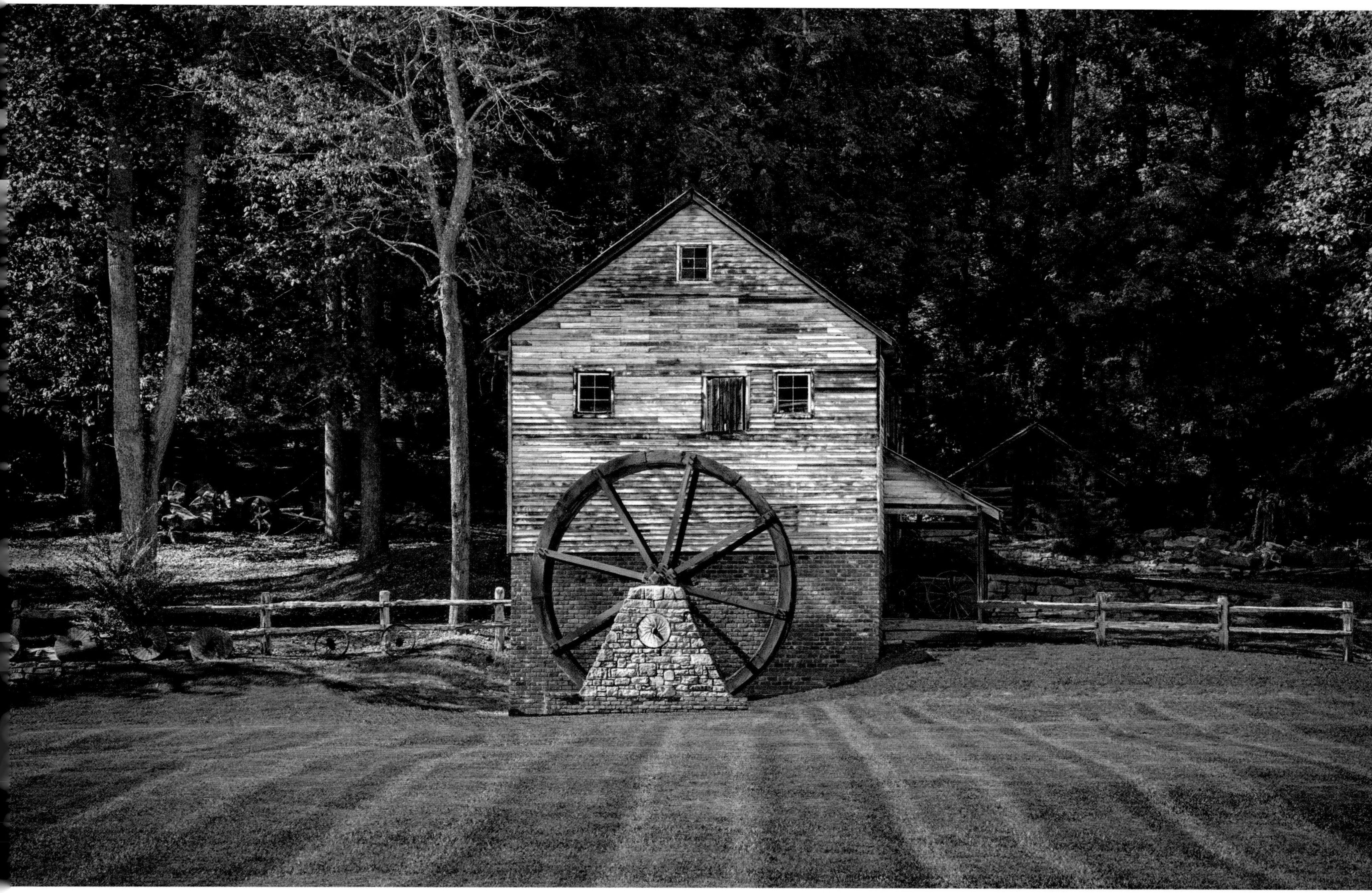

Hacker Martin Grist Mill is a relocated watermill at the Museum of Appalachia.

Falls Mill

Belvidere, Tennessee, 1873 / Ongoing Since Mid-1980s

FALLS MILL is a delightful, stately, and elegant structure adjacent to a small waterfall in a picturesque valley along Factory Creek in southern Middle Tennessee. A larger waterfall and an 1871 dam just upstream of the mill provide power to a waterwheel that is thirty-two feet in diameter and four feet wide, making it one of the largest operational waterwheels in the United States. The bricks used to construct the mill were produced and fired onsite.

John and Jane Lovett have continually made upgrades and improvements to the mill and the surrounding site since they became the owners in 1984. Their attention to detail is apparent everywhere, from the landscaping to the displays inside the mill. The site includes the owners' home as well as a bed-and-breakfast facility, outdoor historical exhibits, and several picnicking areas. The ruins of an 1876 gristmill, the Mann and David Flour Mill, are visible nearby on private property.

Today Falls Mill operates as the nonprofit Museum of Power and Industry, with educational exhibits spread across the three floors of the building. The mill was originally constructed as a cotton and woolen factory and cotton ginning operation, with several women and children tending the equipment. The mill was later repurposed as a woodworking shop before being converted into a fully functional water-powered gristmill, which operated from 1970 to 2015. It has now been returned to its roots and is the only water-powered nineteenth-century textile factory with period machinery in the South.

Falls Mill is always a pleasant, friendly, and welcoming experience for visitors. Step inside the doorway, and you are met by a long old wooden work counter, where someone can explain a little about the mill and help with typical mill purchases of meal, grits, and flour, as well as honey, jellies, jams, and items that are unique to Falls Mill.

Step to the left and see a functioning line shaft system powered by the waterwheel at the opposite end of the building. Some of the machinery is operational, and other interesting pieces, including rare vintage textile machinery, are displayed with signage that explains the function of each item. Beyond the first floor is a porch from which the big waterwheel can be observed from the top and the valley and creek below the mill can be surveyed.

Walk back near the front door, past the front counter, and take the stairs to see the textile equipment and other exhibits on the second and third floors. This is where the Power and Industry Museum is at its finest!

Note that all belts, gears, and machinery are operated by the waterwheel; this is not an electrically powered mill, unlike some old watermills today.

As with other historic and operational watermills, be sure to check the schedule on the Falls Mill website before planning a visit, as the mill closes for some of the winter season.

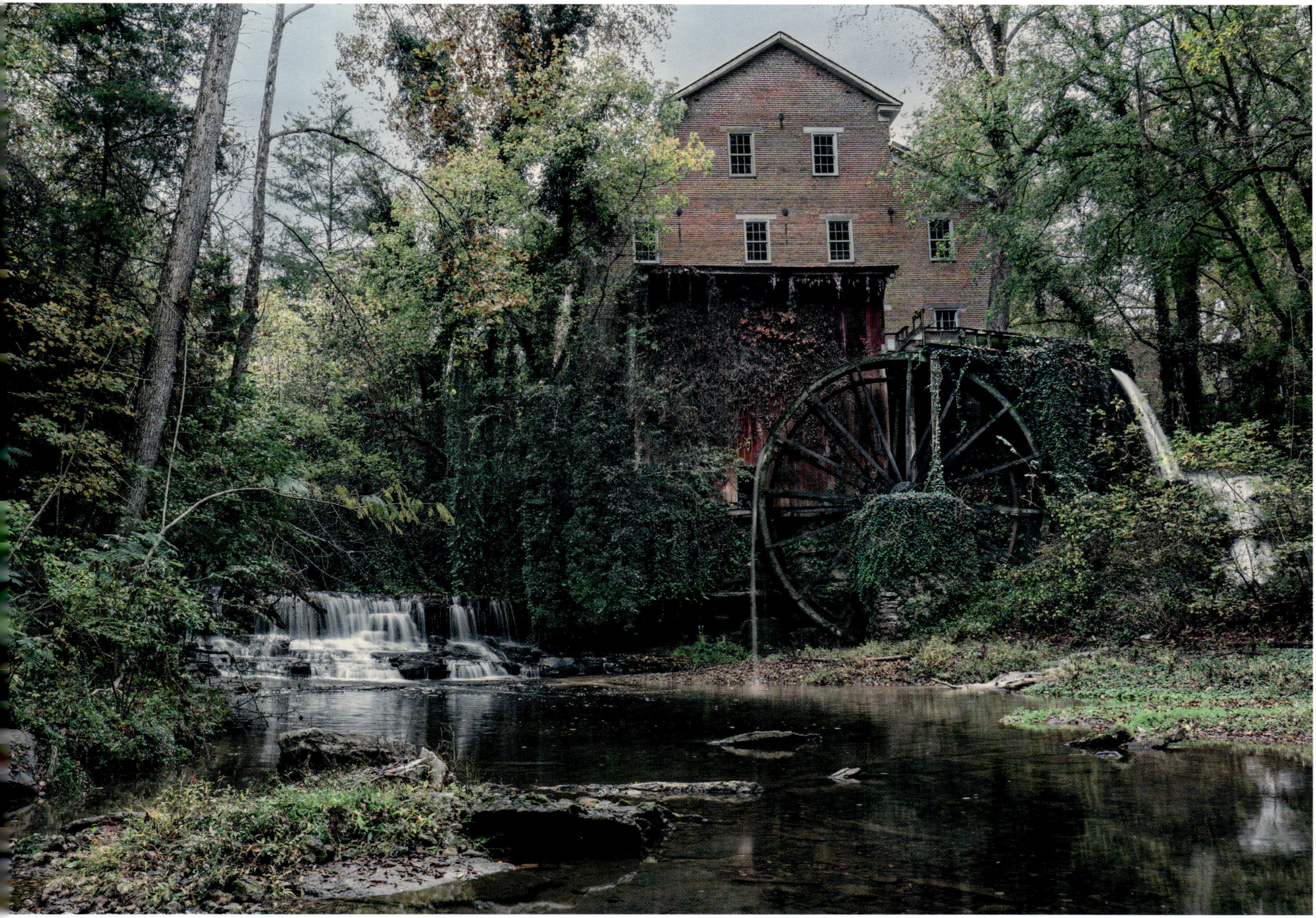

Falls Mill with its massive thirty-two-foot overshot waterwheel is an impressive and elegant sight, especially with its image reflected in Factory Creek.

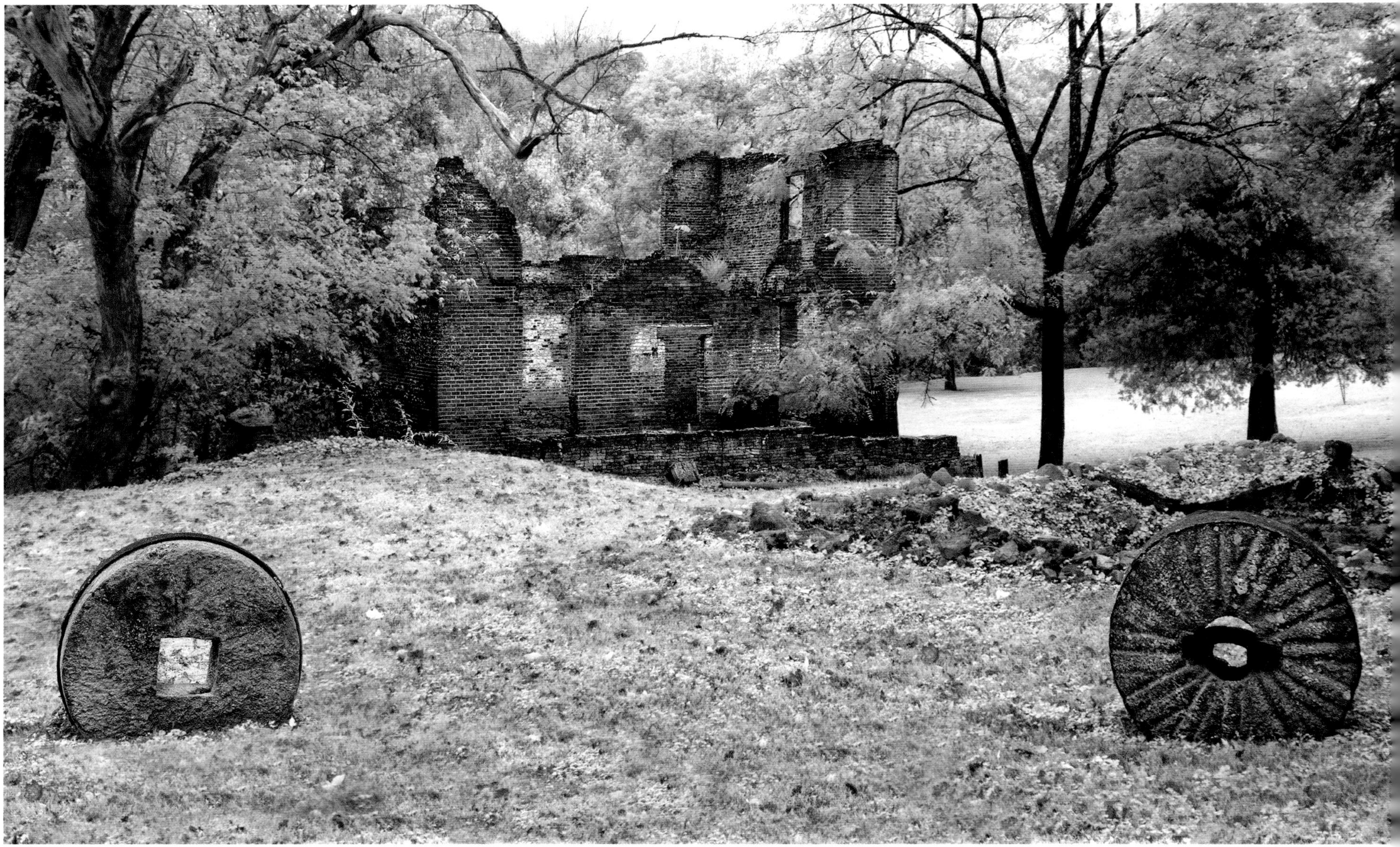

The ruins of the Mann and David Flour Mill (built in 1876) are located on private property near Falls Mill. Both mills are of brick construction.

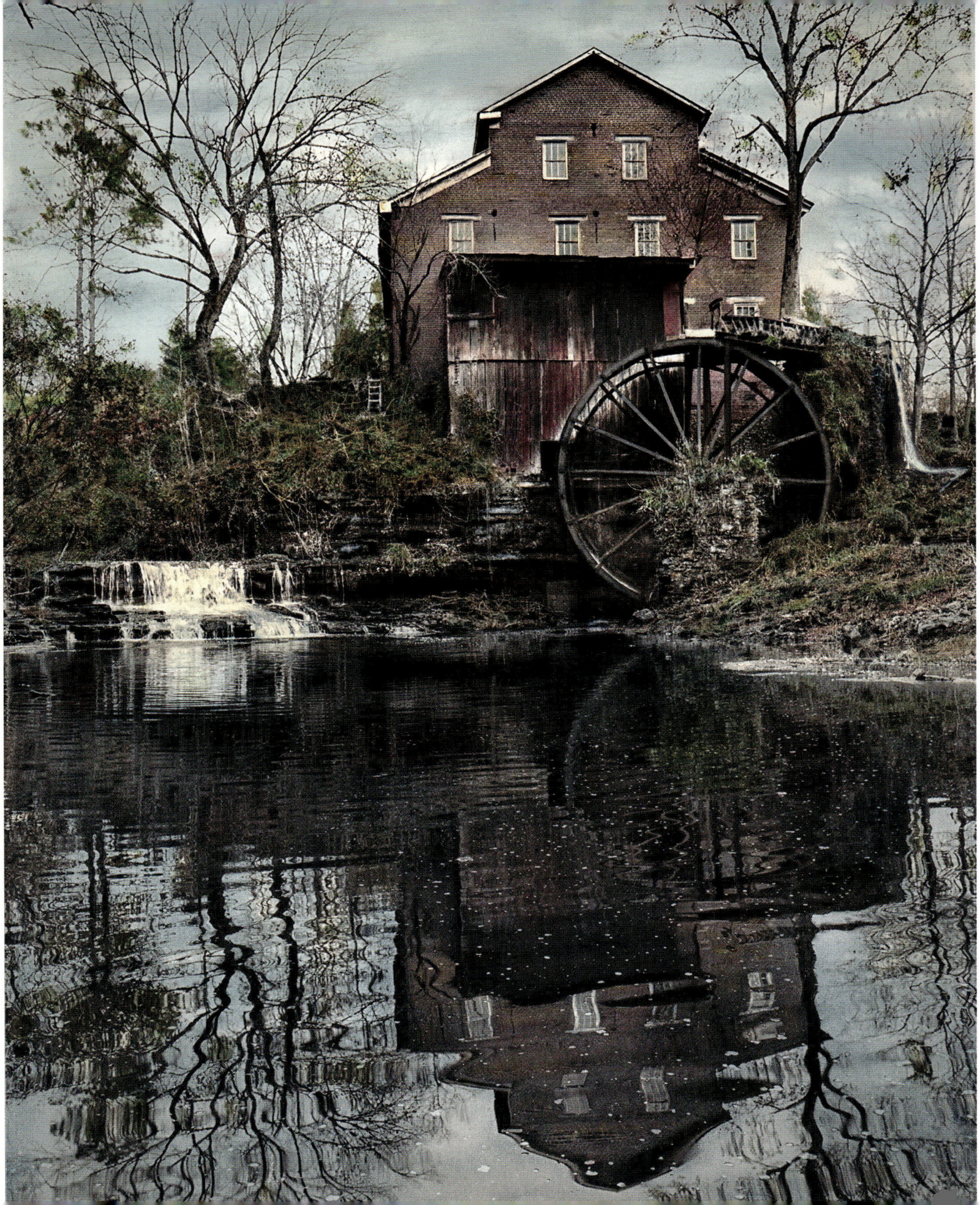

This hand-tinted photograph of Falls Mill dates from the early 1980s.

K. C. French Mill

Dandridge, Tennessee, Late 1790s / 1877 / 1990

IN AN EARLIER INCARNATION around the turn of the eighteenth century, this mill was part of an ironworks called the Dumplin Bloomery Forge. The site, which is about twenty miles east of Knoxville, was purchased by Perkins Iron Works in 1796, and the mill was built between 1798 and 1800. The ironworks failed in the mid-1820s, and about five years later, the local sheriff sold off the property.

The two-and-a-half-story mill was subsequently converted to a gristmill, with two sets of forty-two-inch granite millstones. Oliver Evans's automated milling system was introduced. Much of the current structure dates to 1877.

Over the course of the nineteenth century, a number of people bought and sold the mill. Each owner, of course, brought renovations and modifications, some good and some not so good.

In 1910, the mill was "upgraded" from millstones to more "modern" roller mills, and the aged wooden gearing was replaced with metal pulleys and leather belts. In 1949, equipment from another watermill was added. In 1950, an eighteen-foot steel overshot waterwheel replaced an old and rotting wooden wheel. Similarly, in 1953 the wooden millpond dam was replaced with a concrete impoundment.

Many of these upgrades were made by James French, who had purchased the mill from the Kite brothers in 1920. He and later his son ran the business until 1979, when a major flood washed out structural supports under the roller mills and misaligned the drivetrain. The mill then shut down, and it remained closed until 1990, when Freeman Lambright purchased the site and began restorations.

The mill and property were recently sold at auction. Today the mill is operational with waterpower from Dumplin Creek.

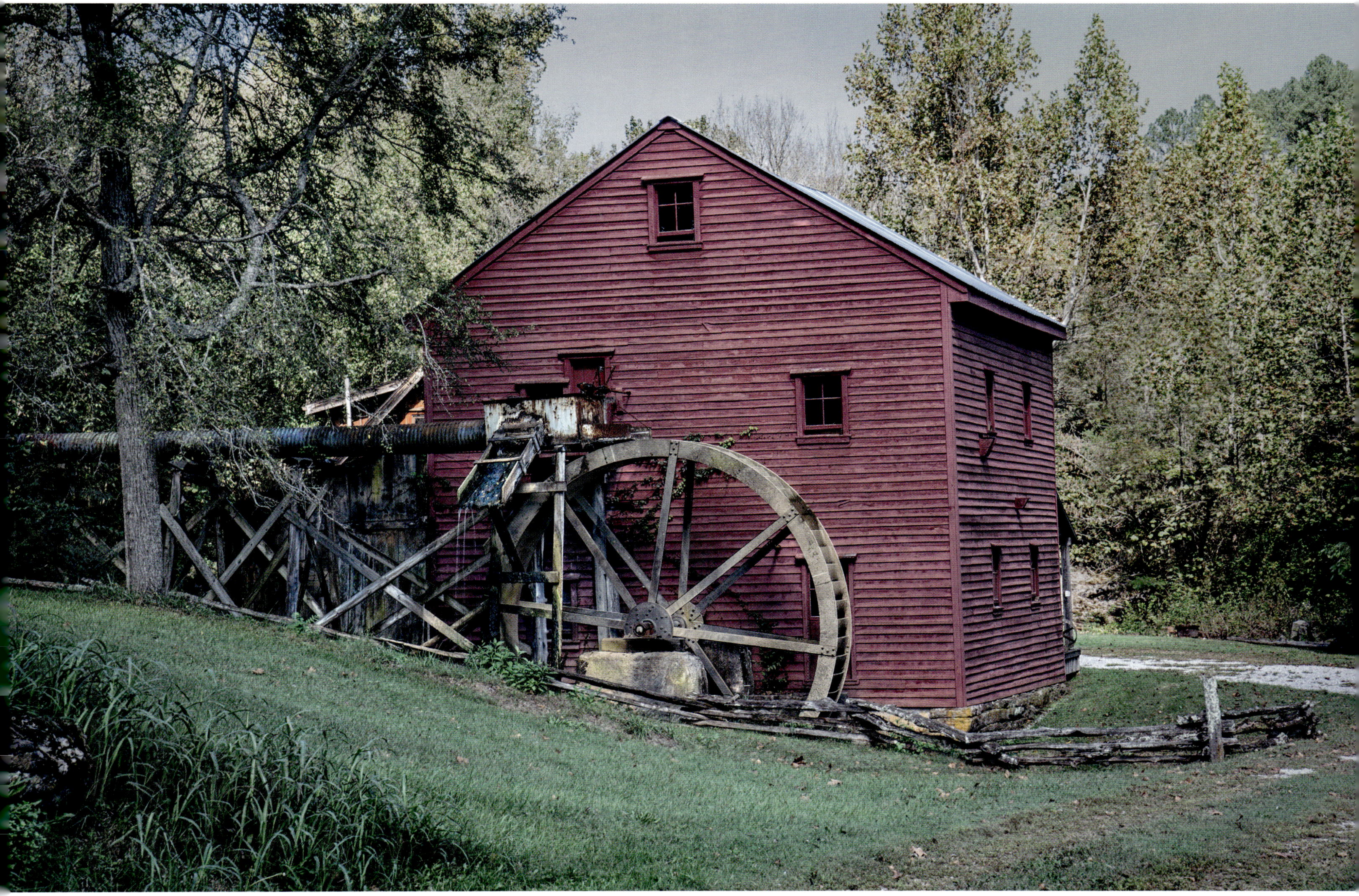

K. C. French Mill is a picturesque gristmill that was originally part of an iron forge. Only the mill building from an early industrial site remains today. The overshot waterwheel is eighteen feet in diameter.

Ketner's Mill

Whitwell, Tennessee, 1882 / 1977

JUST A SHORT AND SCENIC drive west from Chattanooga, Ketner's Mill is located at a site on the Little Sequatchie River, not far from the Tennessee River. The mill operated full-time until 1992, when Clyde Ketner, the grandson of the original owners, passed away. Ketner was a highly respected miller and millwright across the industry. He and his mill were often considered a definitive standard by other millers. He welcomed and hosted educational groups, including groups led by the author of this book, who came to experience his mill.

The mill hosts the region's largest country fair each fall. The concept for the fair dates to 1977, when it began as a celebration of a major restoration of the mill by the Ketner family. The mill has recently been sold, but plans are to continue the annual fair.

In 1824, the Ketner family began operating another mill and blacksmith shop at a site several miles from the current mill. During that same period, other settlers in the area completed a dam and mill at the site of the current Ketner's Mill. In 1868, the Ketners bought that site, including the dam, a gristmill, and a sawmill, and relocated their business.

Construction of the current brick mill was completed in 1882. The new mill retained a water-powered sawmill operation, and a few years later, the family moved a woolen carding machine to the site from another location. The sawmill operated until 1955. The McDonald family actually owned the mill for many years, but they allowed Clyde Ketner to continue his milling operation there.

Today the site includes the main brick mill building and two wooden structures. One of the other structures is a large open-sided pavilion-type building that houses a display of historic sawmilling machinery. All of the buildings stand on tall wooden beams for protection from high water, although the mill has still experienced serious flooding at times. Most days, when the water is at a normal level, visitors can have a look below the mill and observe some of the shafts and linkage that deliver power from a turbine in the water to the millstones inside the mill.

Each millstone weighs between 1,200 and 2,000 pounds and measures forty-two inches in diameter; these are typical horizontal gristmill stones. Each stone was cut from conglomerate rock and includes both coarse grains and smooth white flint pebbles to ensure excellent grinding performance.

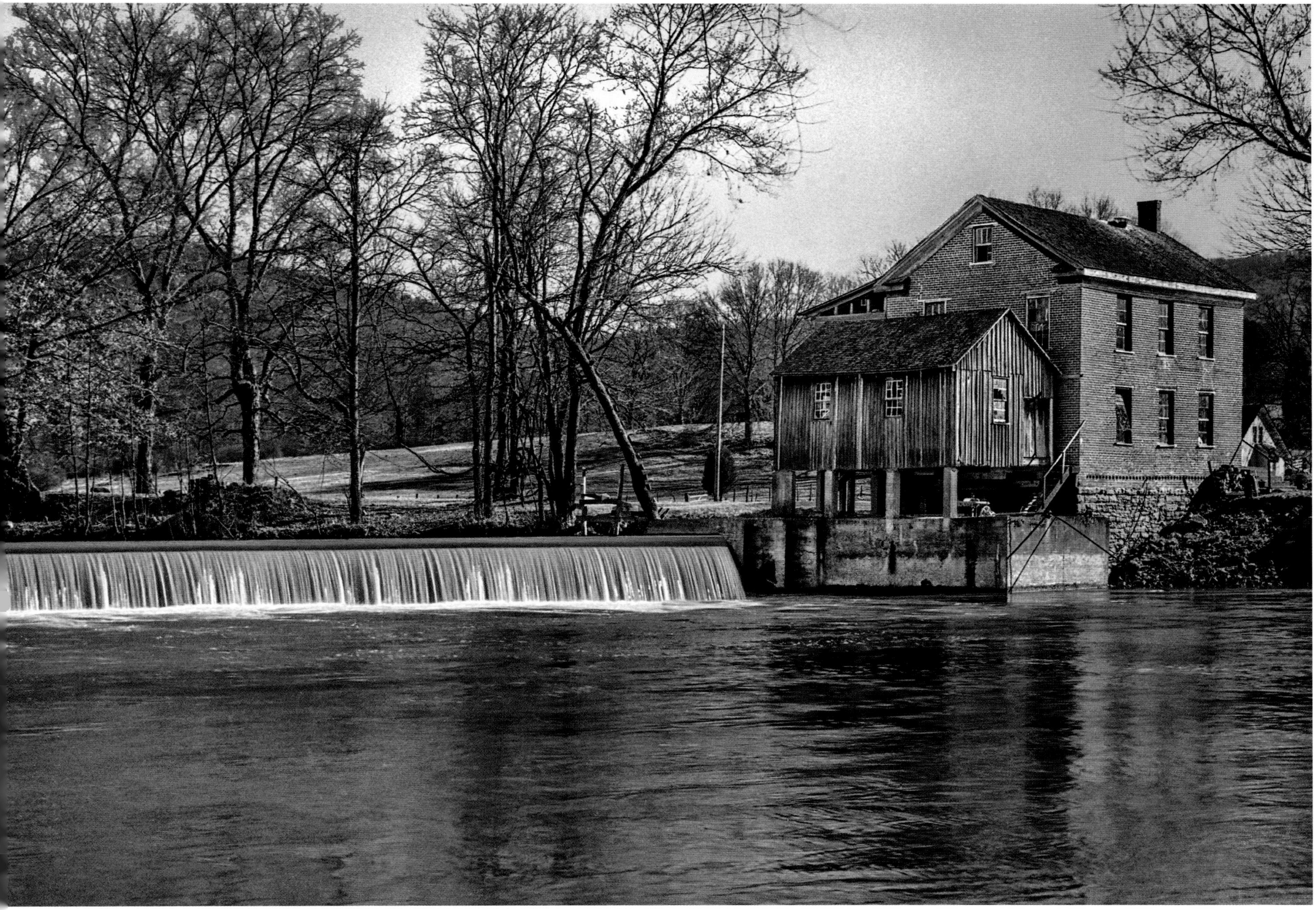

This photograph shows Ketner's Mill as it looked some forty years ago from across the Little Sequatchie River.

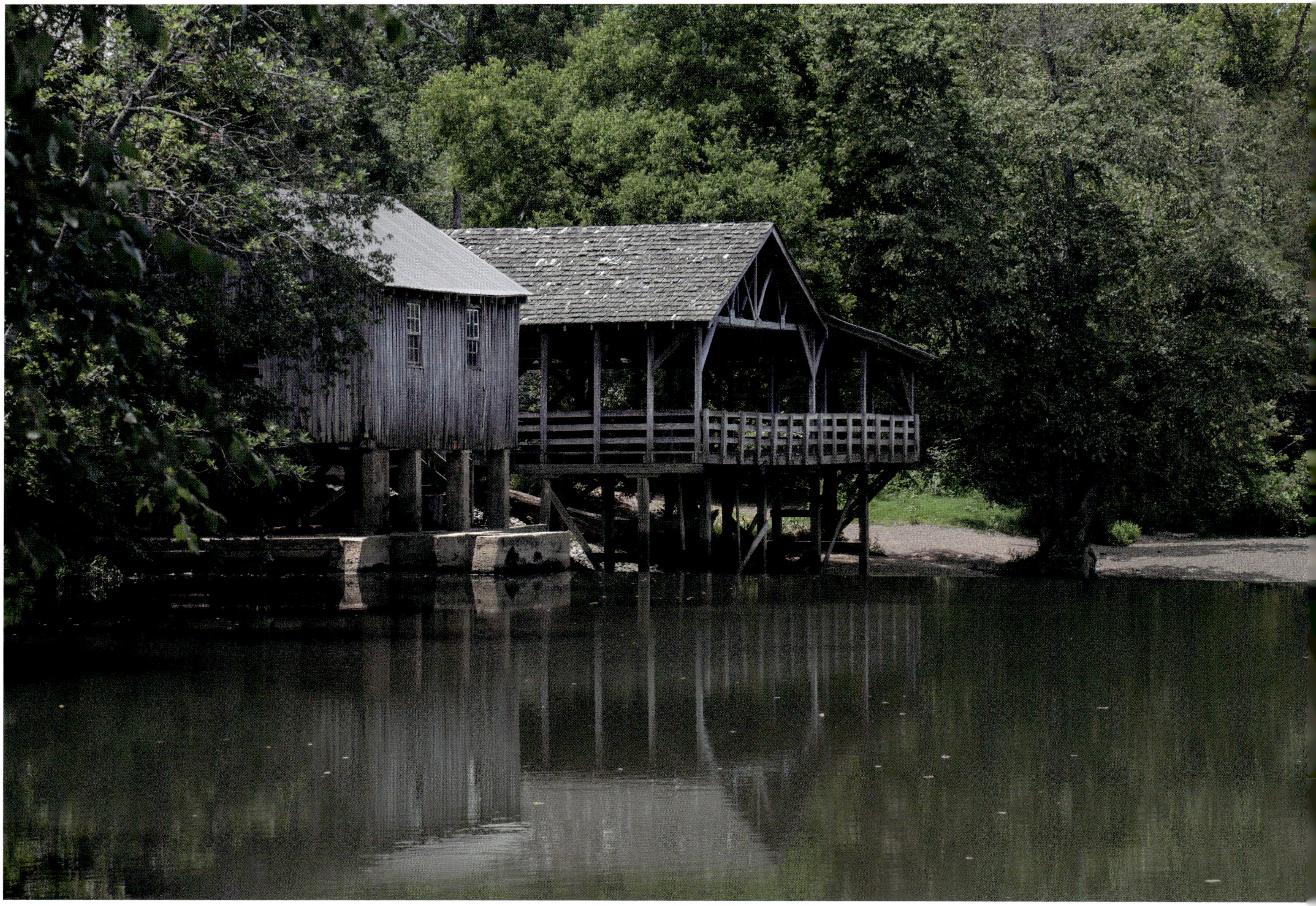

From upstream above the millpond, the concrete water intake to the turbine can be seen along the waterline at the left. The open pavilion structure with the wooden guardrail houses a museum of historic sawmilling machinery.

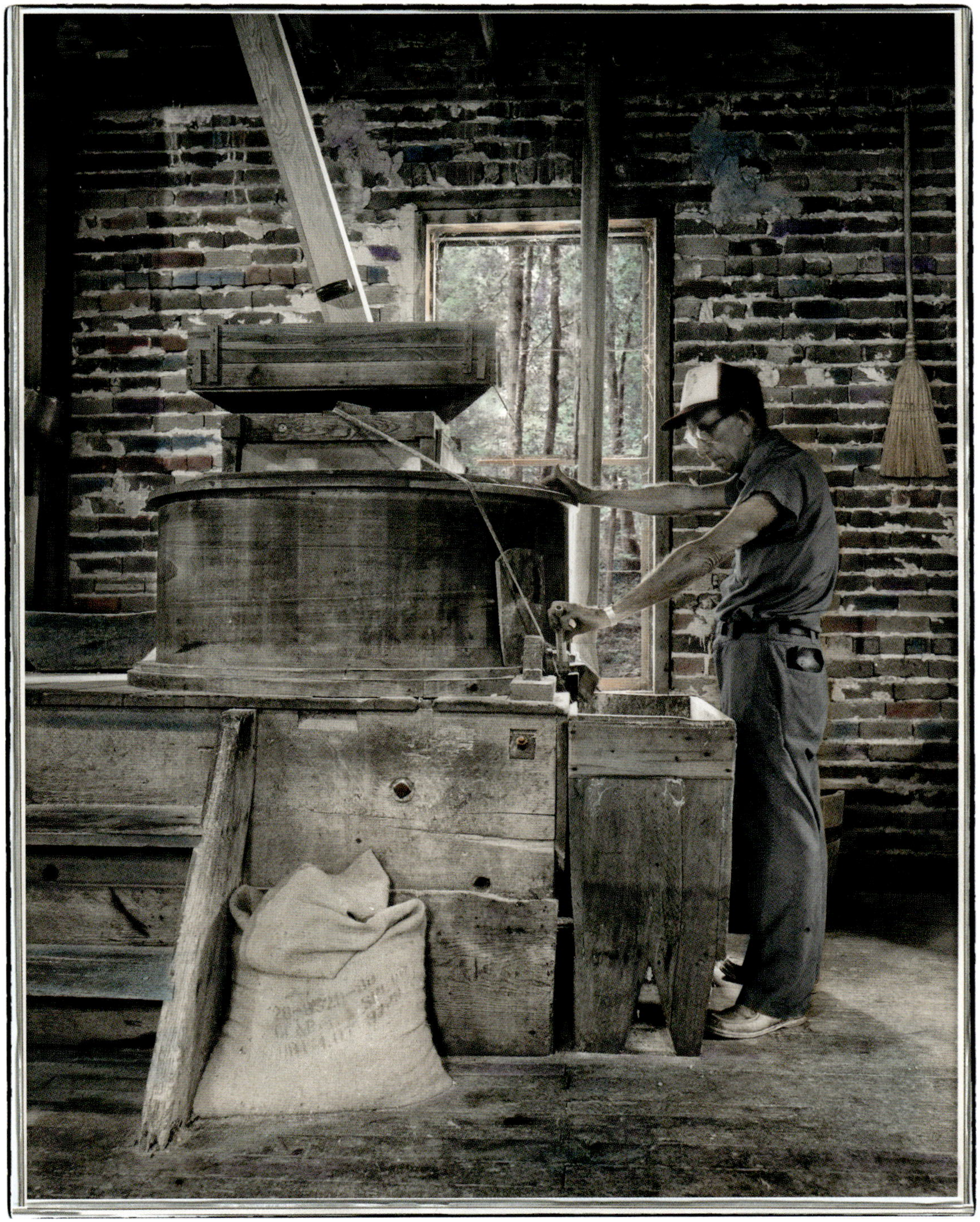

This photo taken by the author in the 1980s shows Clyde Ketner grinding corn in his mill. Ketner passed away in 1992.

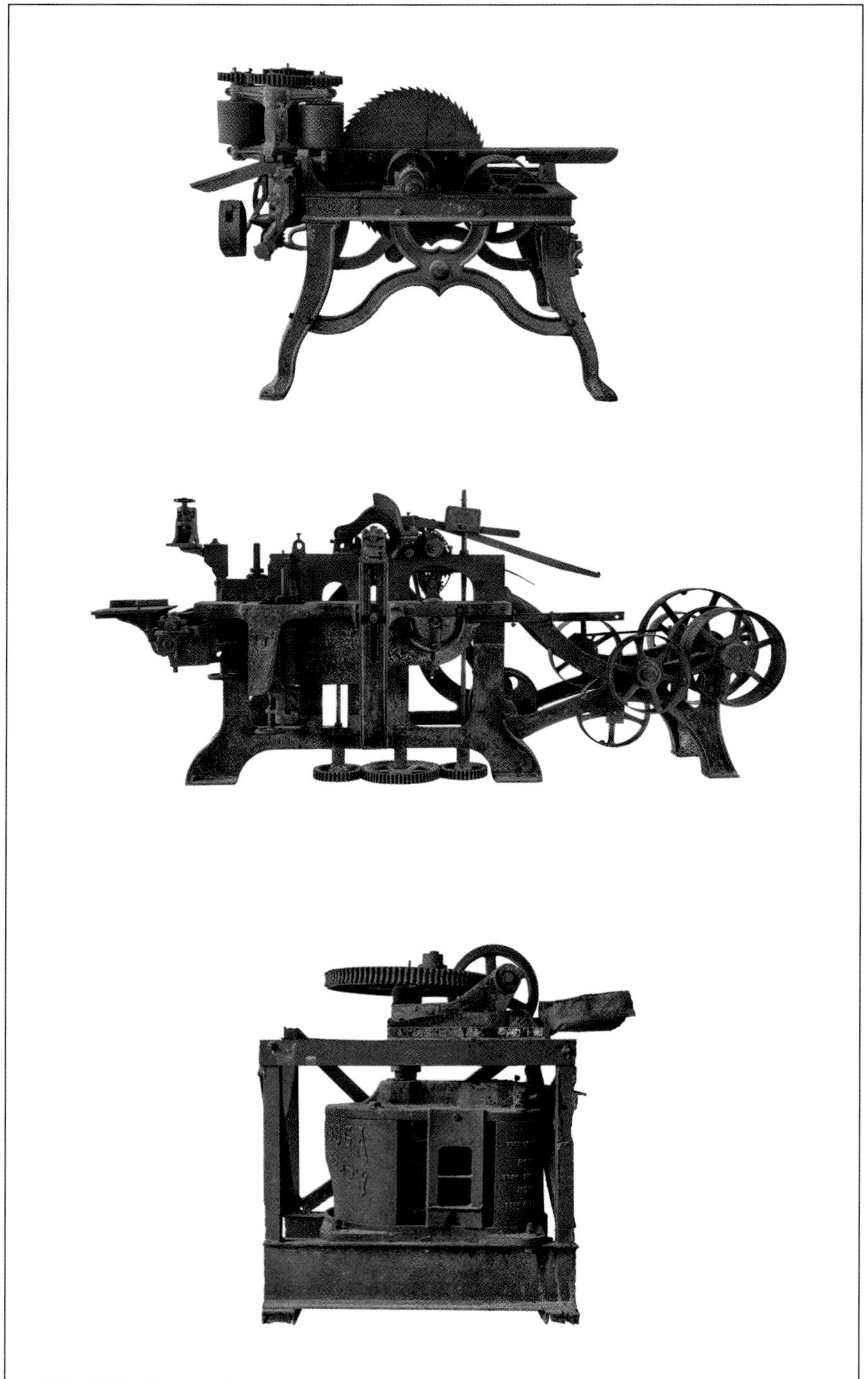

A collection of rare and historic milling machinery, including an early water-powered circular saw, is preserved and displayed at Ketner's Mill.

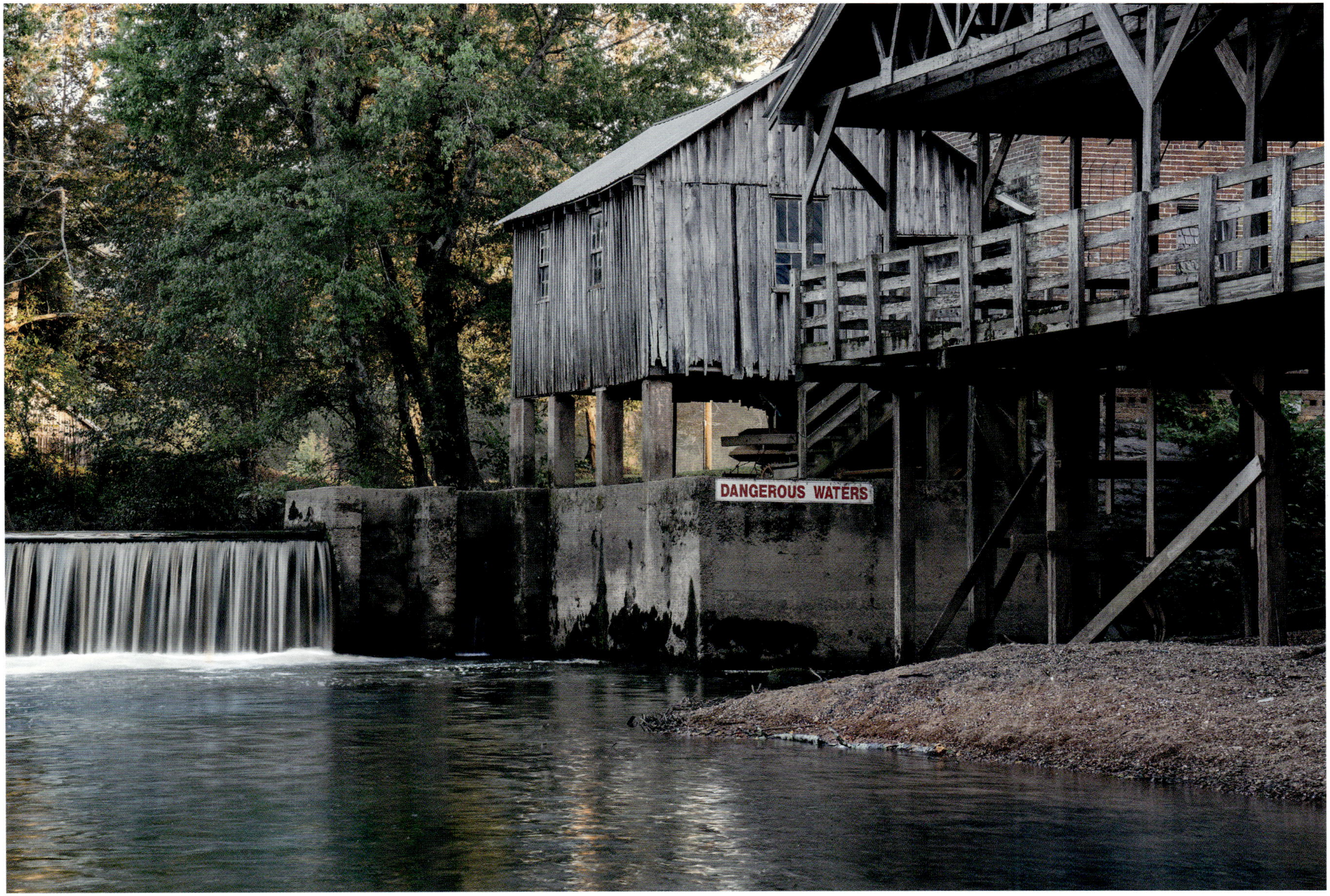

The Little Sequatchie River is dangerous for swimmers, especially around the dam, and several people have drowned here over the years (as has happened at many watermill sites). The hydraulics below the dam create powerful currents and whirlpools. Warning signs are posted around the mill and dam.

G. T. Wilburn & Son Grist Mill / Fall River Mill

Lawrence County, Tennessee, 1865 / 1885

IN 1823, a gristmill and general store opened in the Fall River community of Lawrence County, and these businesses were soon followed by a blacksmith shop, a tannery, a sawmill, a sorghum mill, a country doctor, two churches, and a cemetery. During the Civil War, the mill was burned by the invading Union army, but it was rebuilt beginning in 1865.

In 1885, the Wilburn family purchased the mill, built a new dam, installed a water turbine, and began a multigenerational business that would span seventy years. By the 1920s, the Wilburns' grinding skills were so renowned that farmers traveled as far as fifty miles to have their wheat and corn ground on the Wilburn millstones! By 1950, the mill was no longer grinding wheat flour, but the demand for cornmeal and animal feed remained high. Operations continued until the mill finally closed in 1956. The general store continued in business until 2000, and then it too went out of business. By this time, the last generation of Wilburn millers had passed away, and the mill soon went up for sale.

Today the mill, store, and homeplace are considered to be among the most historic properties in Tennessee. The mill is situated along a one-lane road in a scenic but remote and somewhat difficult-to-find location in Lawrence County. The mill is about seven miles off U.S. Route 64, between Pulaski and Lawrenceburg.

In 2020, Chris Belland and Piper Smith purchased the property. Smith described seeing the mill as "love at first sight!"

Under the leadership of Belland and Smith and with support from the Fall River community, the old mill is coming back to life. The equipment was still operational when the mill closed. The general store is being rebuilt, and a music festival was held on the grounds in October 2023.

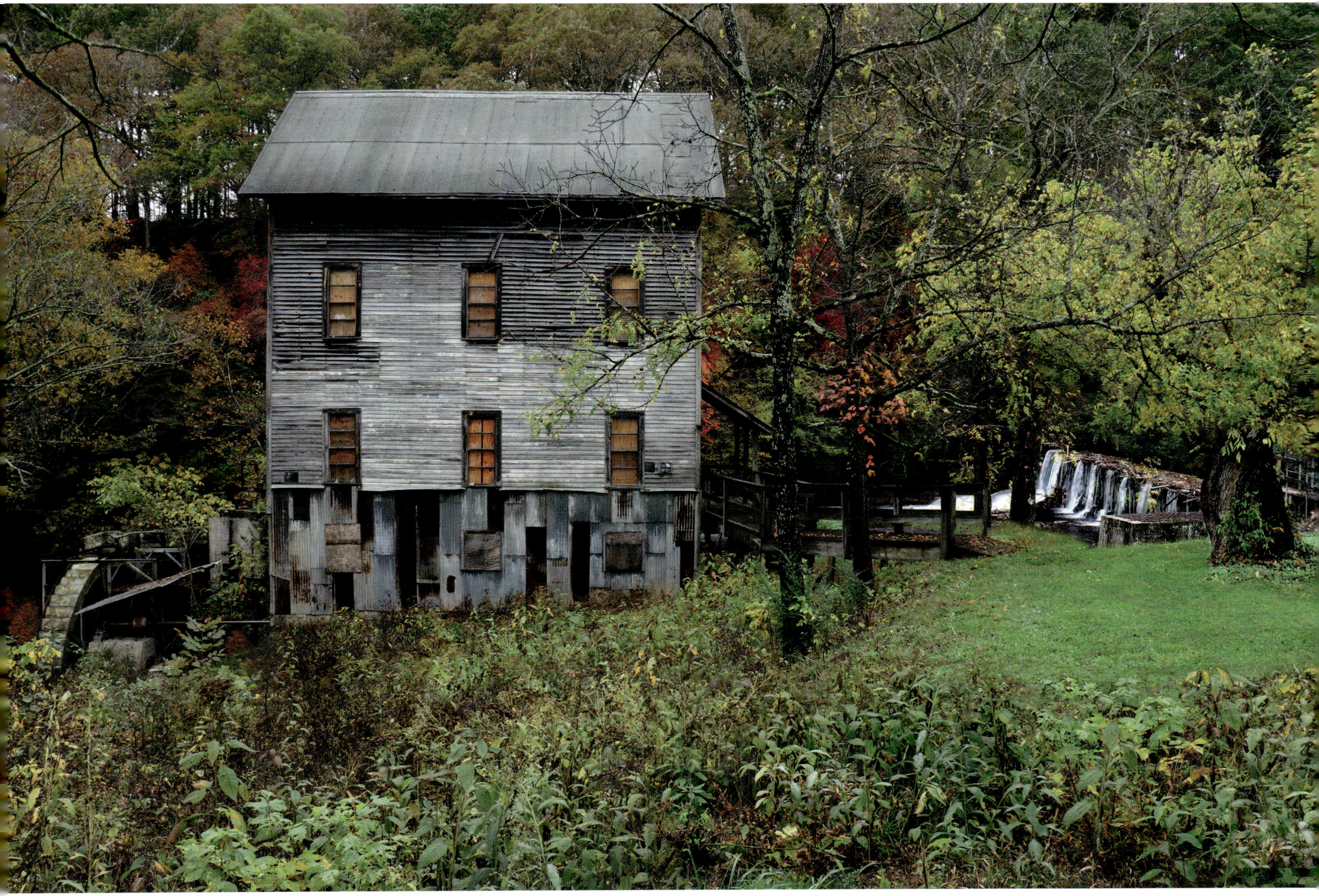

G. T. Wilburn & Son Grist Mill is shown here alongside Clear Creek, with the waterwheel at left and the mill dam at right. Today this beautiful mill site is coming back to life.

Tennessee's Historic Rural Tub Mills

Great Smoky Mountains National Park, 1800s

WHILE MANY PEOPLE TODAY envision the classic watermill as a magnificent edifice in an enchanting and romantic setting, with a tall waterwheel and a picturesque millpond, in reality most early Southern watermills were small, primitive, and rudimentary structures. These mills were built by struggling pioneer, frontier, and farming families using only locally available materials.

A tub mill usually produced just enough grain for a family or small extended community. Most tub mills and other small farm mills did not survive into the modern era. The examples shown here have been preserved or restored by the National Park Service in Great Smoky Mountains National Park.

With origins dating back some two millennia, the tub mill had been known in some European cultures as the "Norse mill." Unaware of this heritage, early Southern farmers constructed tub mills by the hundreds along small mountain streams throughout Appalachia and in other locations where small, somewhat manageable, rapidly flowing mountain streams could be applied as sources of energy.

Significantly, these crude, humble mills were vitally important for deprived and isolated folks in need of daily food. Tub mills were often constructed near the headwaters of streams, where the water flow was swift but low in volume. This type of water source was fine when rainfall was abundant, but the streams might dry up at other times of the year.

Everything used to build a tub mill came from the immediate area around the stream and farm. That means the construction was wooden except for the millstones, which usually came from the streambed. The waterwheel was often about the size of a wagon wheel and installed horizontally, with a shaft, usually a tree trunk, leading directly to the millstones mounted on the mill floor in a modest structure just above the stream.

The use of the term "tub" is interesting. The most popular explanation is that the waterwheel was usually mounted inside a partially enclosed wooden housing for better utilization of the available water. The enclosure was something like a tub, hence the name. Another story is that the mills were so small and inefficient that the quantity of grain involved was limited to a bushel or less—that is, the amount that would fill a small tub.

As mentioned earlier, because of the construction materials and techniques and exposure to the elements, tub mills were often short-lived, and most did not survive into the last half of the twentieth century. The Noah Ogle and Alfred Reagan mills are two reconstructions in an area that was once home to at least fourteen tub mills.

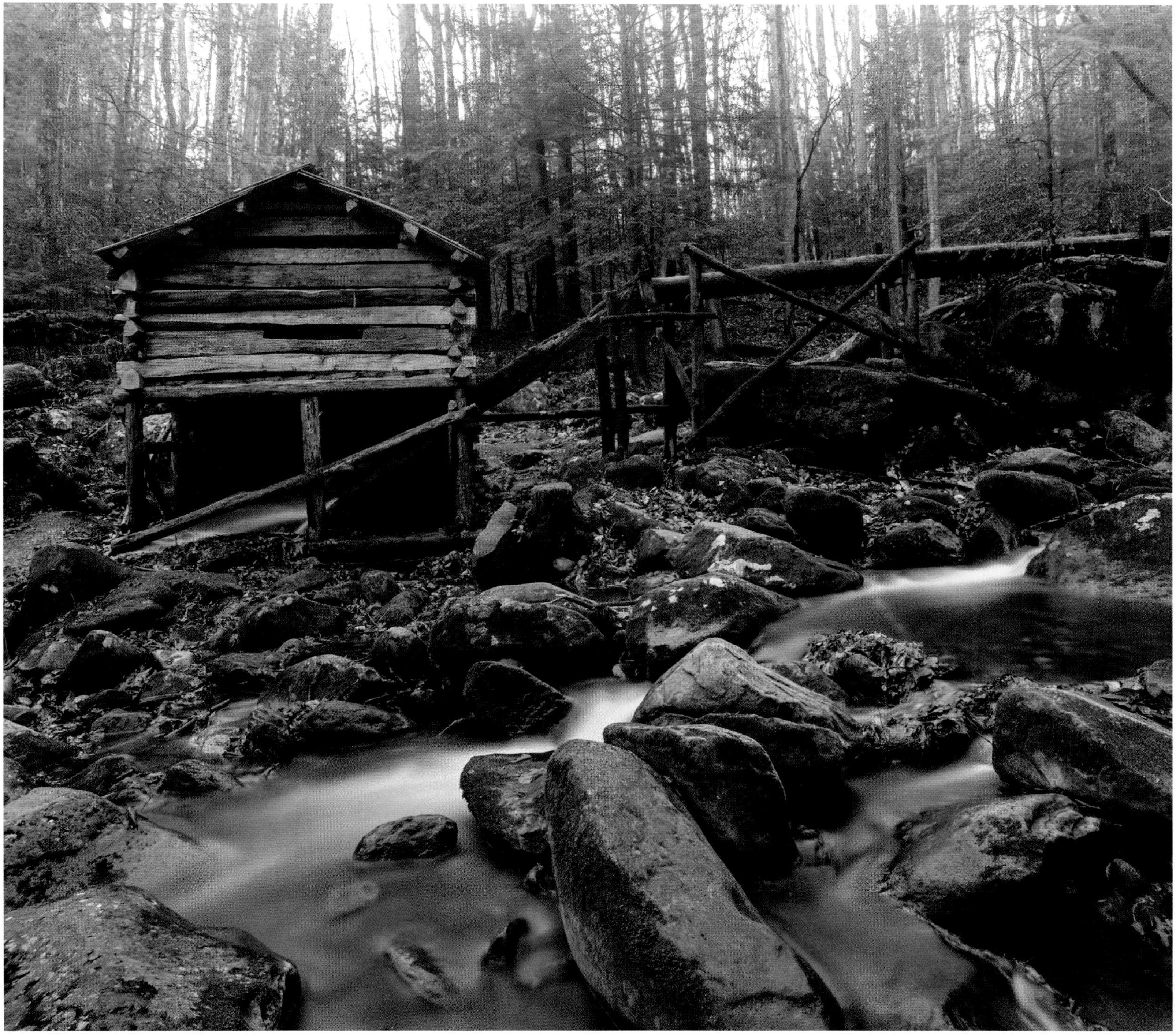

The Noah Ogle Tub Mill dates to 1880 and is a classic example of a simple pioneer watermill.

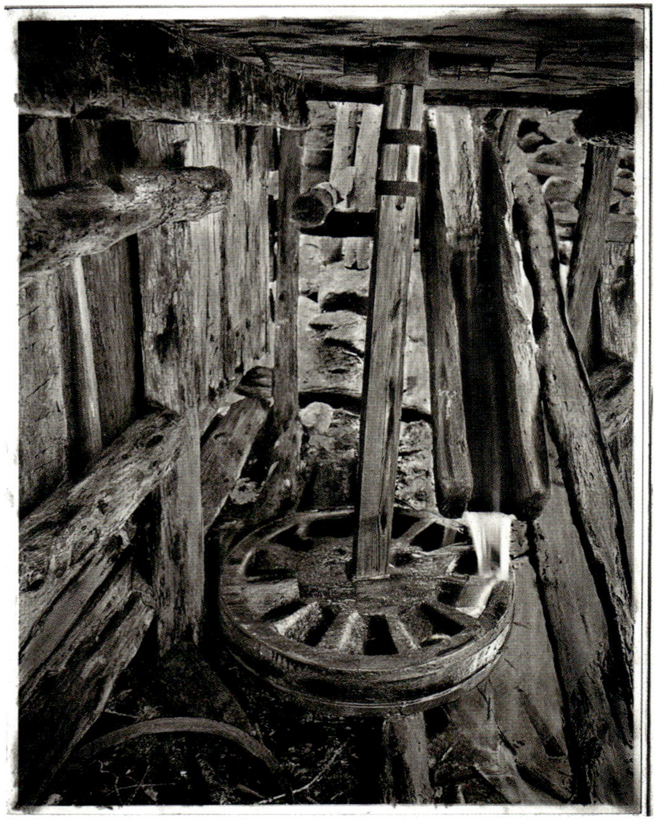

This reconstructed horizontal waterwheel is fitted beneath the Noah Ogle Tub Mill. The square vertical shaft leads directly up to the millstones on the main floor.

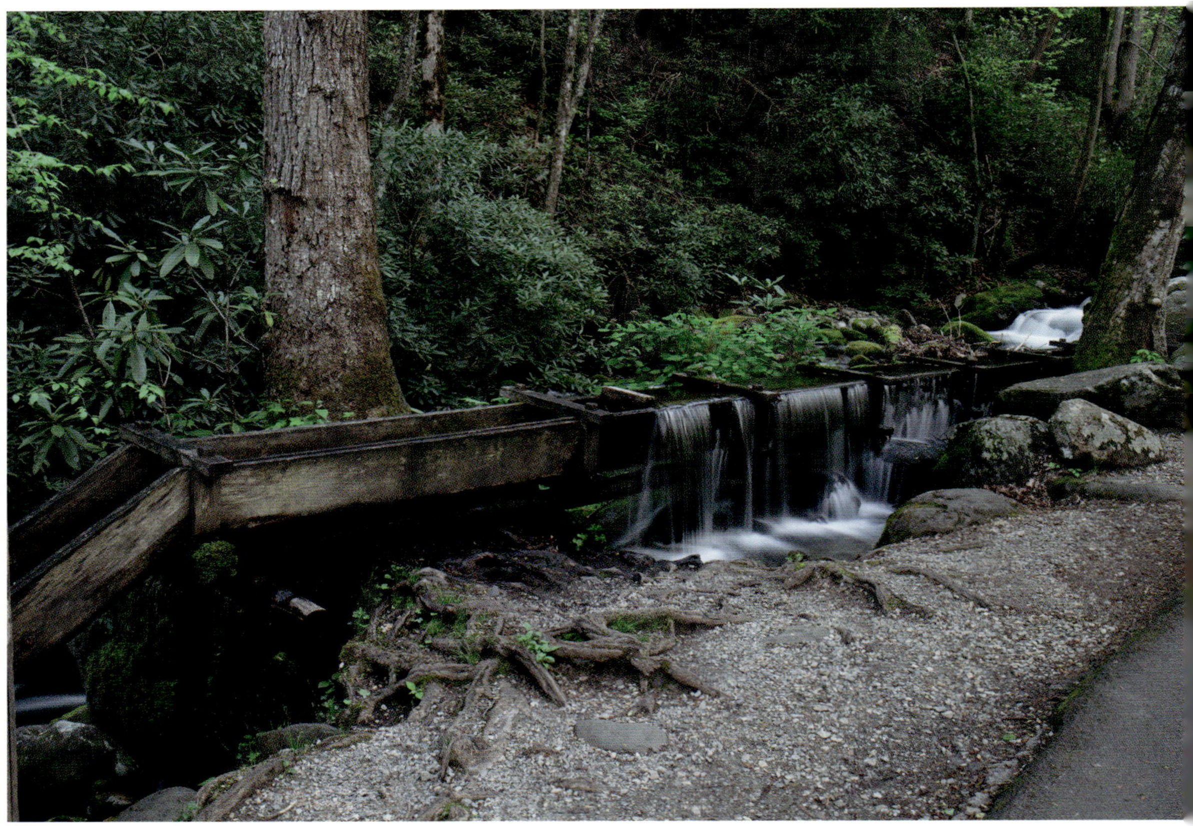

A simple hand-hewn trough, or "flume," was used to deliver the water that powered the millstones in the Alfred Reagan Tub Mill.

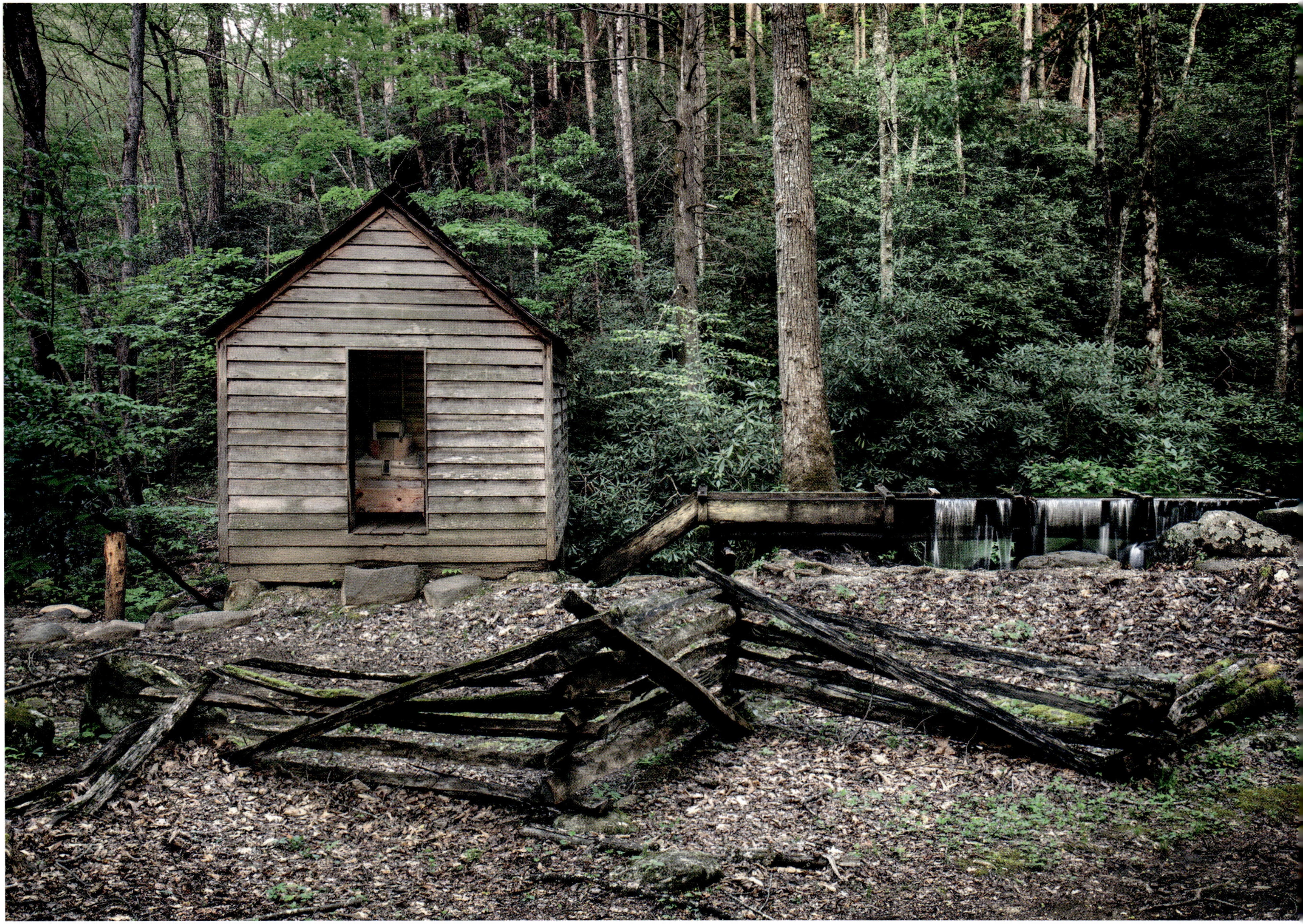

The simple millstone housing can be seen through the open doorway of the Alfred Reagan Tub Mill (ca. 1895). The structure has been rebuilt, and very little of the mill is original, but the image provides a glimpse into the realities of nineteenth-century life in the Tennessee mountains.

Cable Mill

Great Smoky Mountains, Tennessee, 1870 / 1930s

IN THE LATE 1800S, several watermills operated in the isolated Cades Cove community of East Tennessee. When John Cable began operating his mill here in the early 1870s, it was the largest one in the area. He also built a sawmill nearby, but it is not preserved.

As was common among millers of this era, Cable was also a farmer. When someone arrived to have grain ground or lumber sawed, they rang a bell to summon him from the fields. The toll for grinding was one-sixth of any wheat and one-eighth of any corn.

Waterpower for the mill was and still is supplied from Mill Creek, with supplemental water drawn from Forge Creek during dry periods. (Simple, descriptive names were often given to the streams at Southern watermills.) The water is managed through several cleverly designed water gates to better regulate the flow to the waterwheel. A large, elevated wooden flume carries the water the final distance of about 230 feet to the wheel.

The wooden overshot waterwheel is eleven feet tall and five feet wide. It is fitted with forty buckets to catch the water and rotate the wheel. The current waterwheel was installed in 2018. Wheels of this type typically last for eight to twelve years before needing replacement or significant maintenance.

The interior of this two-level mill is relatively small—about eighteen by twenty feet. The left side of the single room on the main floor is elevated on a platform to provide vertical space below for the gears, shafts, and linkages between the waterwheel and the millstones. The mill originally ran with two sets of stones, one for wheat and one for corn. Corn stones were more important for a Southern farming community. The stones at Cable Mill are still operational today.

The gears on the lower level of the mill change the direction of shaft rotation from horizontal at the waterwheel to vertical at the millstones. In addition, the gears increase the speed of shaft rotation. The millstones turn thirty-five times for each rotation of the waterwheel. This mill uses six gears that originally were all wooden but today are mostly metal.

The mill can process about three bushels of corn in an hour, and the park generally operates the mill between April and October as a demonstration. It ceases operation during the winter months. Preservation and maintenance of the mill site are ongoing.

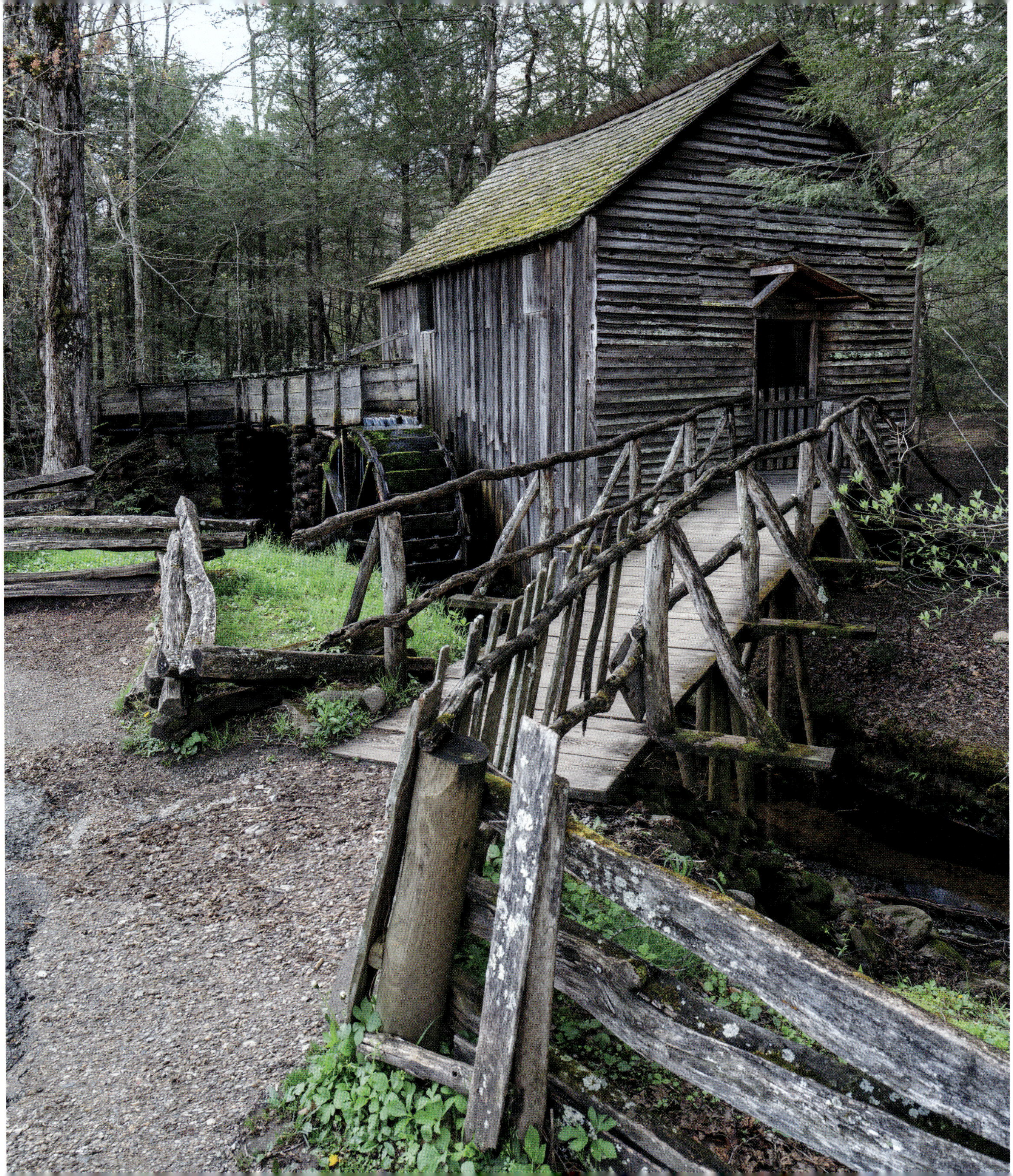

Cable Mill is well maintained and preserved as an example of an early wilderness mill at Cades Cove in Great Smoky Mountains National Park. The mill was rebuilt in the 1930s, and restoration of the site continues.

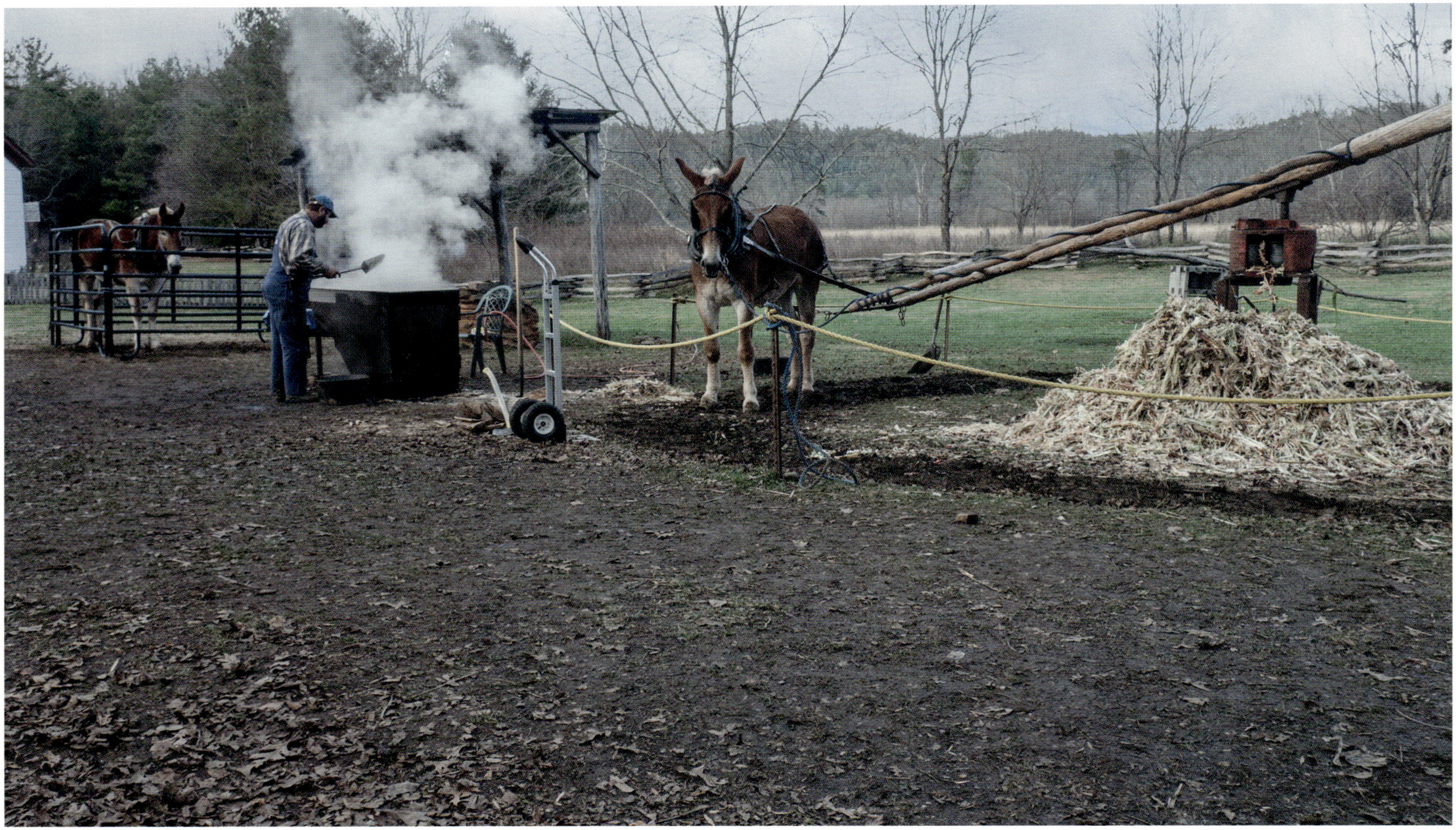

In the fall season, a mule-powered sorghum mill sometimes operates as a demonstration near Cable Mill.

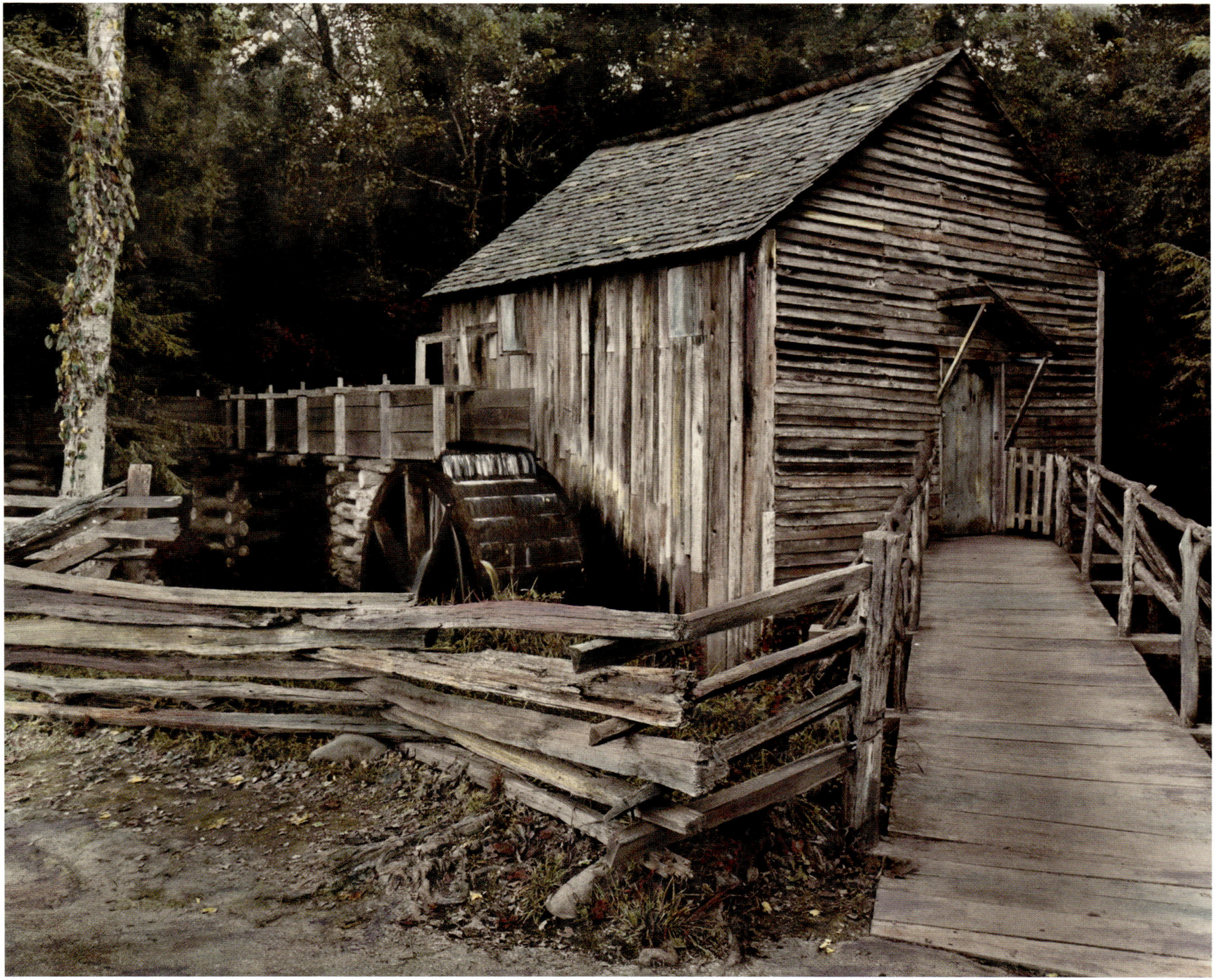

This hand-tinted image shows Cable Mill with water flowing over the waterwheel. The waterwheel has been replaced at least twice since this photograph was taken in the early 1980s.

Bush Mill / Bond Roller Mill

Nickelsville, Virginia, 1896 / 1986

A WATERMILL was first built on this site in 1831. That mill burned in 1895 and was replaced a year later with the current mill, a three-story log and timber frame construction. Tragically, the son of one of the builders was killed on the waterwheel in 1902.

That wooden waterwheel was replaced with a twenty-four-foot steel overshot wheel in the 1920s. The mill continued to operate with the steel wheel until well into the 1950s, and it served as a social center and community gathering place during most of that period. The mill, especially the interior, changed little for much of the twentieth century. The roller mills, bolters, sifters, hoppers, spouts, and belts are all still in place today, and the wooden pulley and connecting shafts remain intact in the basement.

In 1986, the old yellow poplar siding was replaced with new western cedar, which has since aged to look like the original siding. In recent years, the mill has been restored and is again operational, with a thirty-foot-diameter and four-foot-wide metal overshot waterwheel that generates approximately thirty horsepower at maximum operating capacity.

The Nickelsville Ruritan Club occupies a second structure on the mill property. This building dates from 1920 and was moved to its present location in 1974. In 2008, Scott County transferred ownership of the mill to the Southwest Virginia Community Foundation. Today the Ruritans operate it for special events, including an annual Bush Mill Day festival.

The structure is listed on the Virginia Landmarks Register and the National Register of Historic Places.

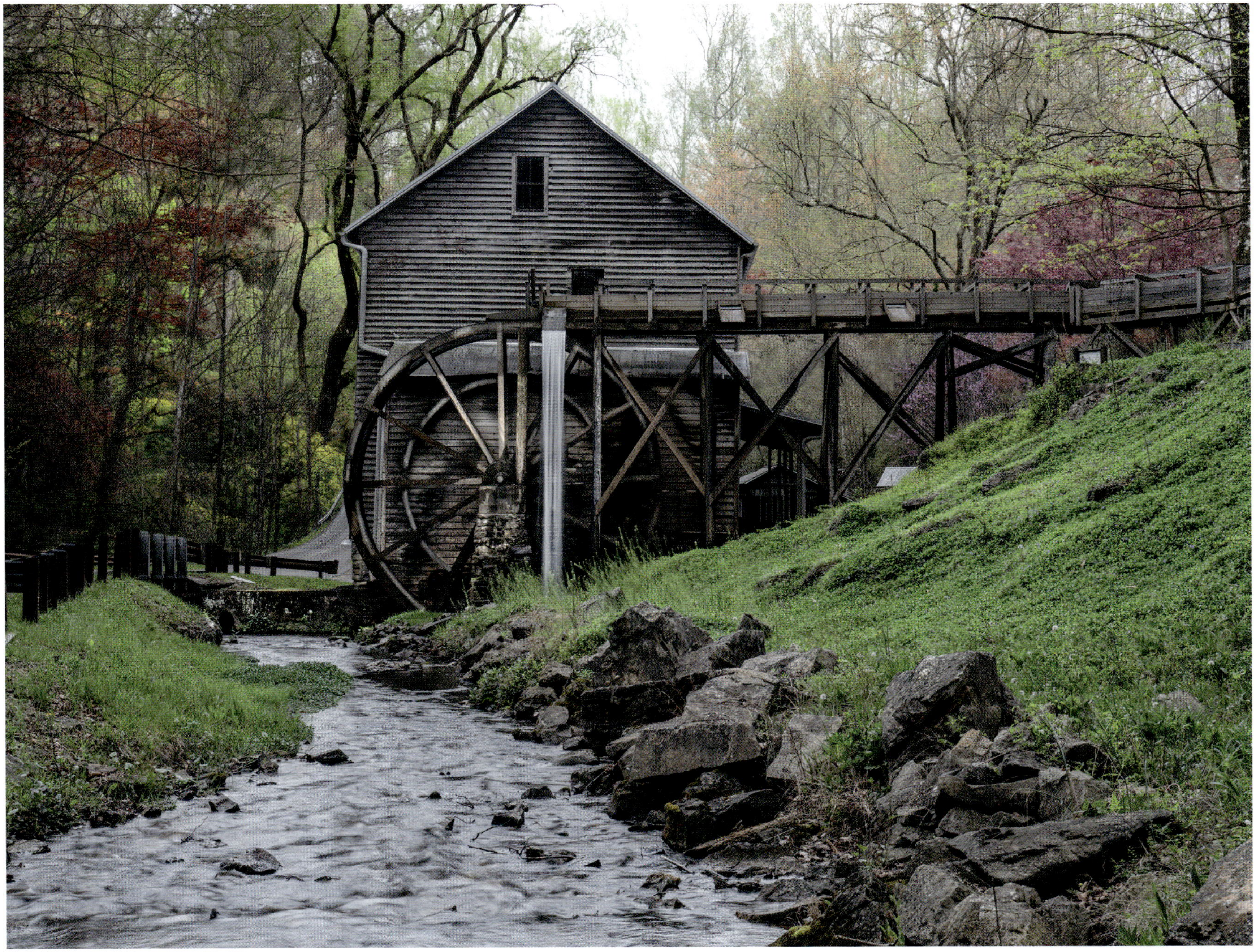

Bush Mill is located in a small, quiet park along Amos Branch near Nickelsville. The flume structure is designed for the release of water or excess water just before the water goes over the wheel.

Meadow Run Mill

Albemarle County, Virginia, 1797 / 1974

MEADOW RUN MILL AND GENERAL STORE is a completely restored gristmill and merchandise establishment in the Michie Tavern complex just south of Charlottesville. The mill and store building is about half a mile from Thomas Jefferson's Monticello estate and was moved to this location in 1974. Several other structures have also been relocated to this site along the route in and out of Monticello.

Historically, a gristmill that dated to about 1784 was part of the Michie plantation, which stretched for several miles across northern Albemarle County. As part of the redevelopment plans for the Michie Tavern area, the relocation and restoration of that original mill was considered, but it was just too deteriorated structurally to be saved. As work on the Michie Tavern area proceeded, a gristmill was needed. The current mill was located and purchased and moved from Laurel Hill, Virginia, where it had operated from 1797 until 1958. Any missing parts and other needed components were added from the old Drumheller's Mill in nearby Augusta County.

As workers disassembled the mill in Laurel Hill, they meticulously numbered and cataloged every part. Then the entire mill was loaded onto trucks for the fifty-mile trip to the current location. An extensive two-and-a-half-year rebuilding and reconstruction process ensued to assure a completely operational, historic, and authentic mill for the tavern property. The reconstructed mill is spectacular, as evidenced by the images shared here.

The mill is said to feature one of the best-preserved examples of the "I-X-L steel overshoot water wheel," which was completed before the I-X-L company changed its name to the more familiar Fitz Water Wheel Company in 1902. In the mill's current location, a wooden flume delivers water over the wheel, but the water has to be pumped and recycled for operation.

Inside, the milling machinery is neatly arranged, laid out, and labeled for both operation and display. The adjoining store offers items related to the mill as well as assorted gift items and some snack foods.

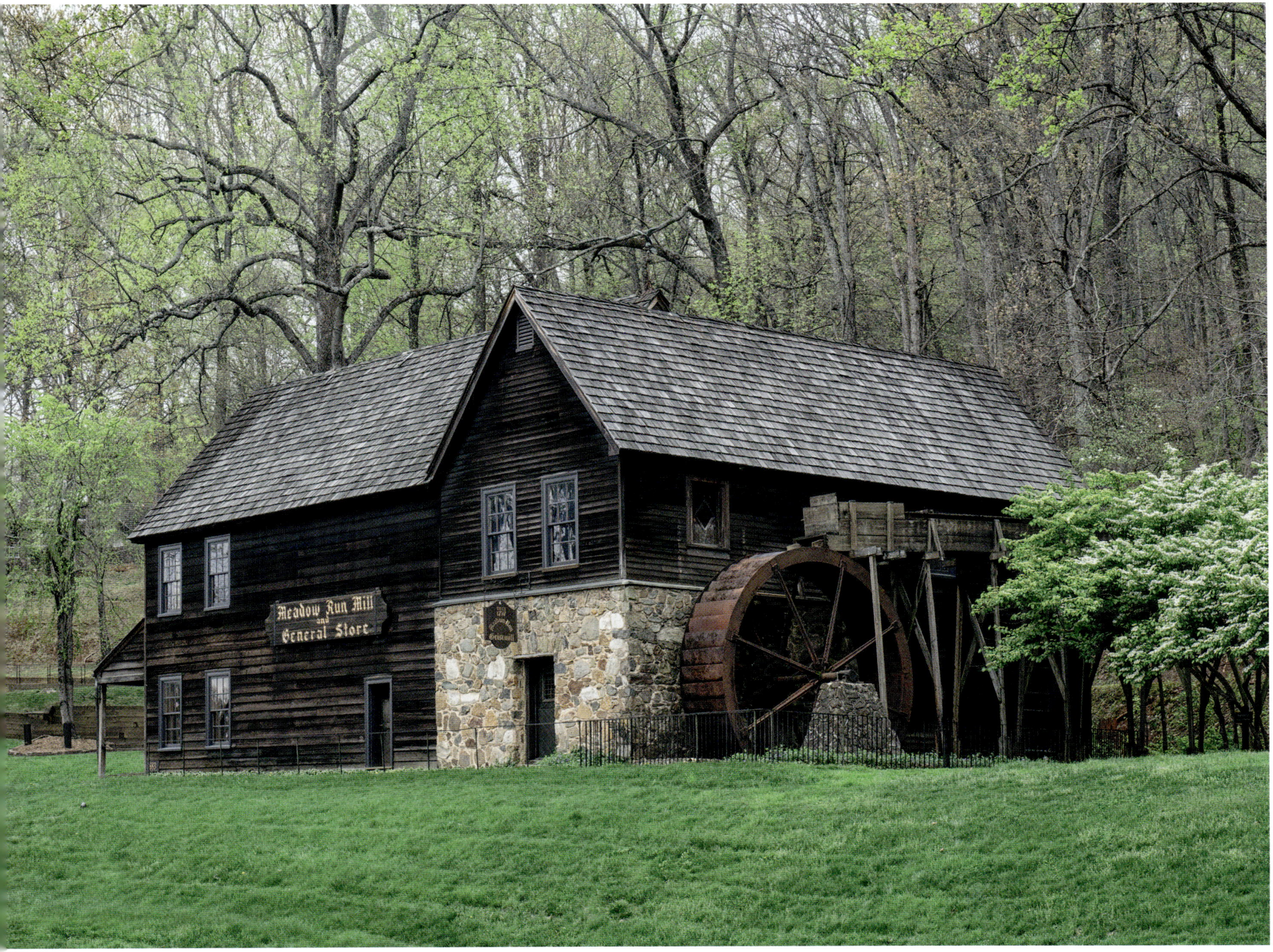

The beautifully rebuilt and restored Meadow Run Mill is sited on the Michie Tavern property.

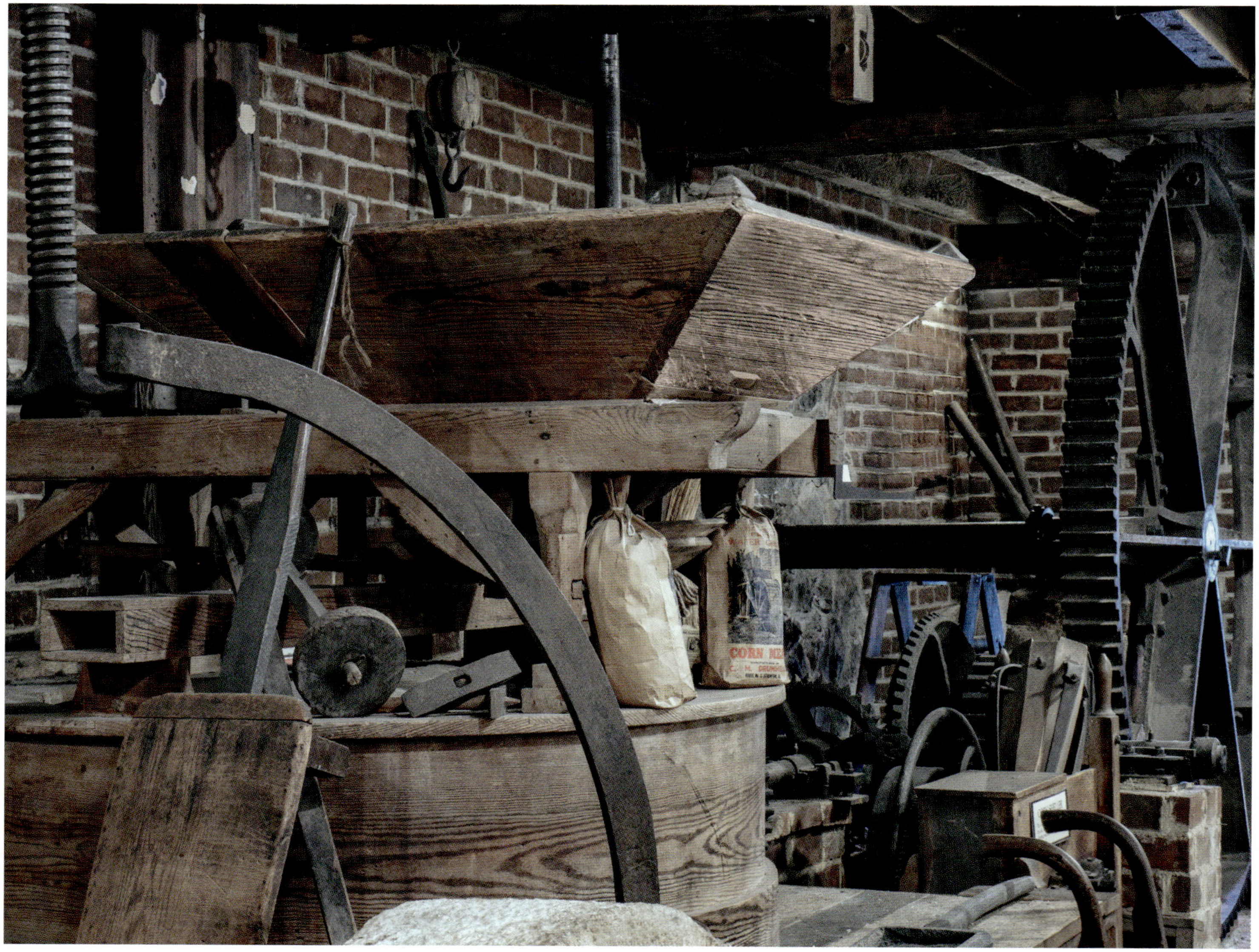

This mill operates as a general store, gift shop, and gristmill. The circular wooden millstone housing with the wooden hopper and mechanism for feeding grain into the millstones is beautifully displayed. The gears that connect the millstones with the waterwheel can be seen in the background.

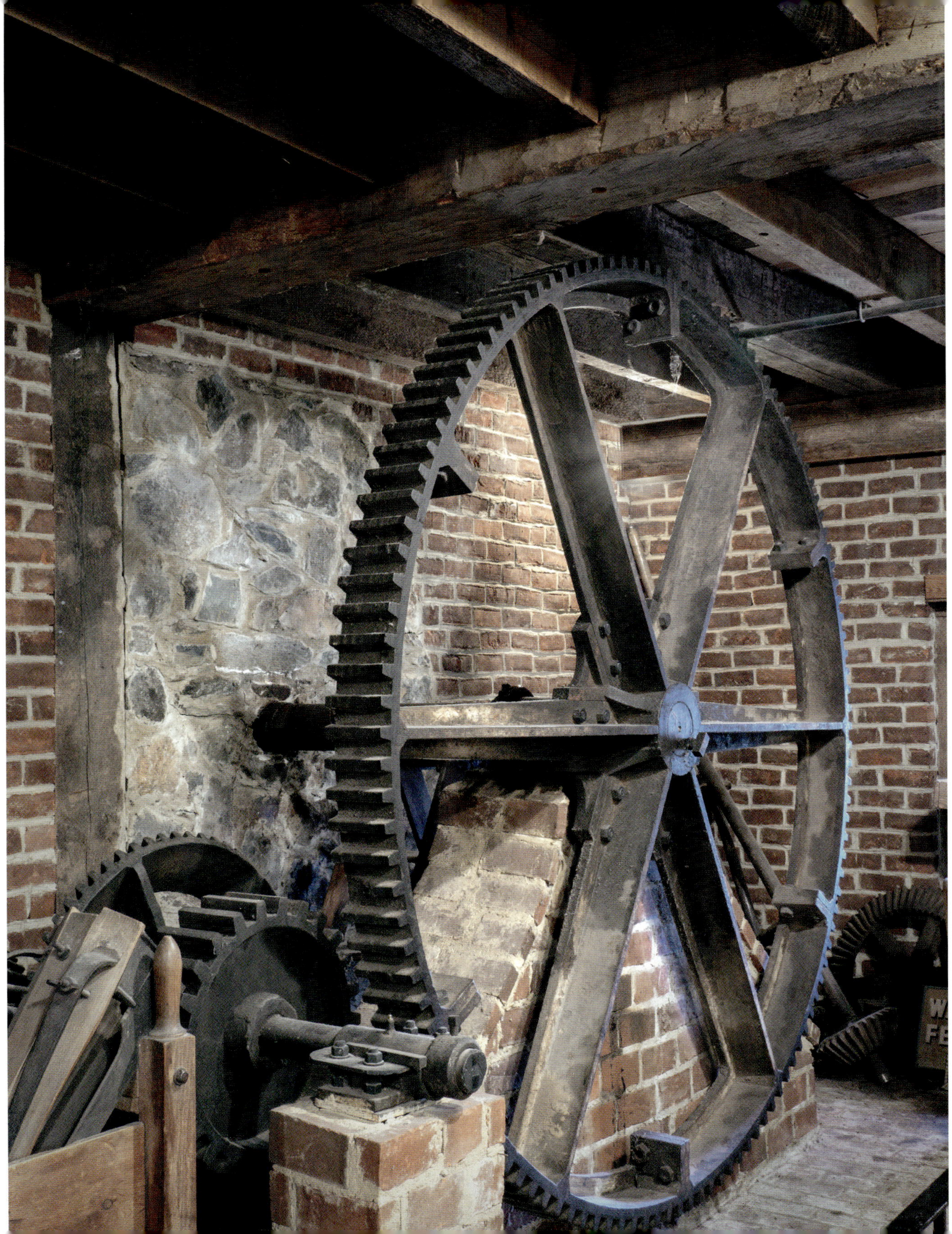

This large gear, called a "face" gear, is located inside the mill, just opposite the location of the waterwheel on the outside of the mill. A short shaft directly connects the gear to the waterwheel.

Mabry Mill

Meadows of Dan, Virginia, 1903 / 1942

MABRY MILL is one of the most famous watermills in the world. It is located along the Blue Ridge Parkway and operated by the National Park Service. This mill is both a gristmill and a sawmill. Ed and Lizzie Mabry began work to build the mill in about 1903. It is believed that the gristmill was in operation by 1905, though some accounts say 1908. Around 1910, a sawmill was added behind and on the side of the gristmill.

The gristmill uses quartz conglomerate stones from Brush Mountain Quarry in Blacksburg, Virginia. The sawmill was designed around a circular saw blade with a mechanism to slide logs into the blade. Overall, the mill was designed to run on belts and shafts rather than gears and cogs. Everything is driven by a sixteen-foot wooden overshot waterwheel.

Throughout its years of operation, the mill often suffered from an insufficient water supply and as a result was slower to grind or saw than some other mills in the area. In later years, two kerosene engines were added to speed its operation.

Lizzie was the principal miller. She completely took over operation of the mill in later years when Ed's health declined. After he died in 1936, the mill deteriorated, and she soon closed the mill. In 1938, the National Park Service acquired the property and completed restoration in 1942.

The site also includes a restaurant and gift shop as well as other historic structures and exhibits for folks traveling along the parkway.

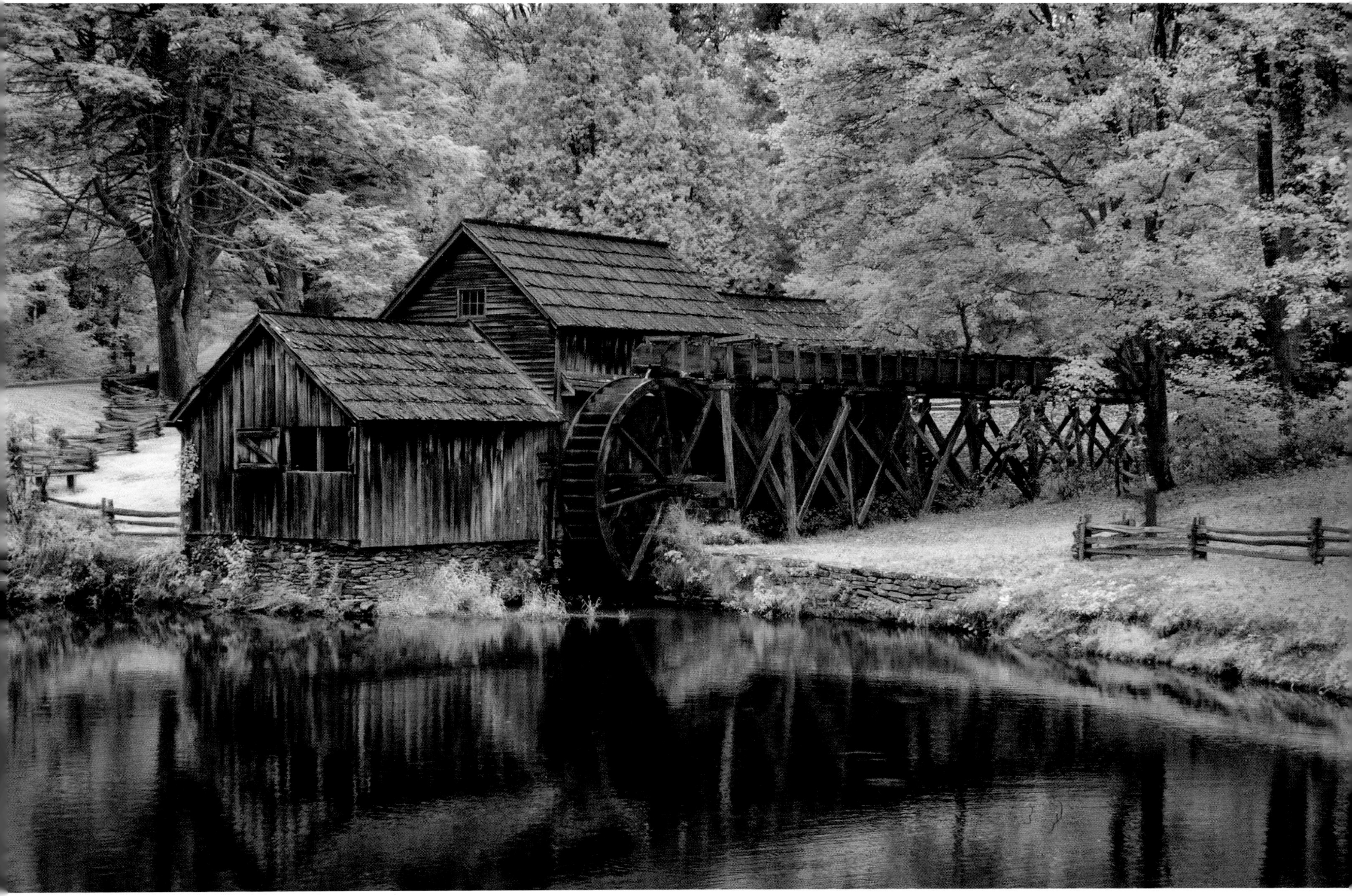

Mabry Mill is a classically beautiful structure. Over the years the split-rail fencing, reflective pond, and graceful wheel with truss-mounted flume have been admired by millions of travelers making their way along the Blue Ridge Parkway.

McCormick's Grist Mill

Raphine, Virginia, Circa 1800 / 1930s / 1993

THIS MILL IS ESPECIALLY FAMOUS because it is located on the farm of the nineteenth-century inventor Cyrus McCormick, who is known for designing the mechanical reaper. His mill, built by his grandfather, is a classic example of so many mills of the time that were constructed of logs and stood on simple stone foundations. The drivetrain is a wooden cog, peg, and gear design. The waterwheel is 17 feet in diameter, and water is delivered about 525 feet from an impounded millpond.

Because mill machinery, especially wooden drive machinery, vibrates and shakes intensely, the log structure of the building is isolated from the milling floor; the milling machinery stands on its own separate foundation. This dual-foundation concept was often implemented in gristmills across the South to protect the buildings from damage caused by excessive vibration.

A matching blacksmith shop sits next to the mill; this is the building in which metal parts for McCormick's reaper were forged. McCormick was aided in his design projects by the previous work of his father and by Jo Anderson, an enslaved friend and a highly creative collaborator on the reaper. McCormick and Anderson grew up together and have been described as being more like brothers than enslaver and enslaved. Anderson was recently credited for his contribution to work on the mechanical reaper and other McCormick inventions, credit that was not allowed in his own time.

The gristmill includes two runs of stones, one set for small grains like wheat and the other adjusted for grinding larger corn kernels. The millstones are housed on the upper level of the mill, with the gearing and drive mechanisms on the lower level. This is the standard watermill layout at many sites.

Until the late 1800s, the mill served the McCormick farm and the Steeles Tavern and Raphine communities along Marl Creek. After that, the mill suffered from neglect and was basically abandoned. In the 1930s, the mill was restored because of its historical significance, and the waterwheel was replaced.

Significantly, in 1954 the farm was deeded to Virginia Polytechnic Institute, and the mill now receives ongoing and much-needed attention. Some of the old wooden parts have been replaced or reinforced with metal. By 1993, a new shingle roof was needed to maintain the historic appearance of the structure, and the waterwheel was rebuilt again. The mill dam, raceway, and flume have been restored, work that was financed by a grant from SPOOM and by investment from a descendant of the McCormicks. The mill is fully operational today, and admission to the park and buildings is free of charge.

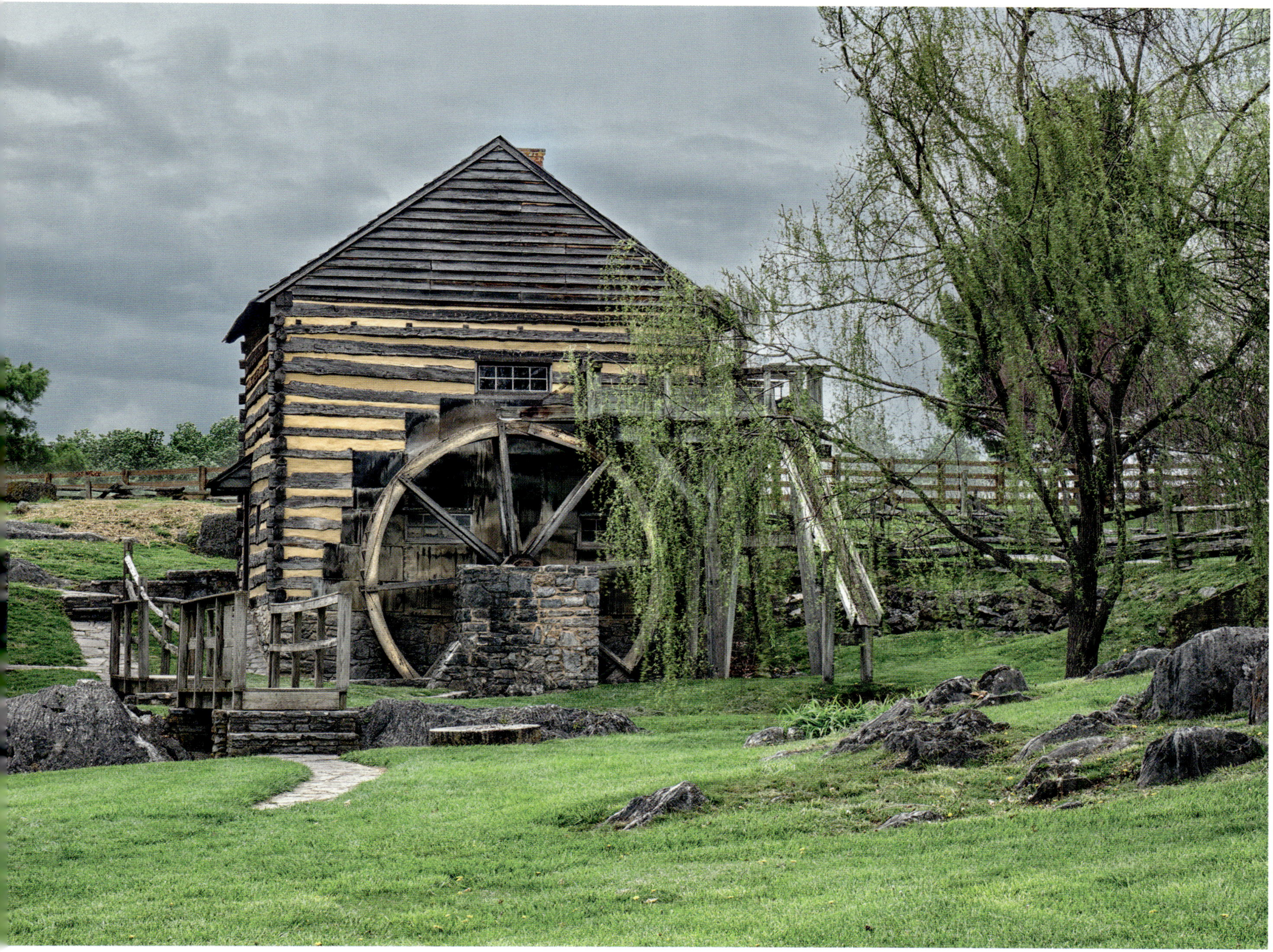

McCormick's Grist Mill is a two-and-a-half-story log structure with a stone foundation. The mill is powered by a seventeen-foot wooden overshot waterwheel that is fed by water from nearby Marl Creek.

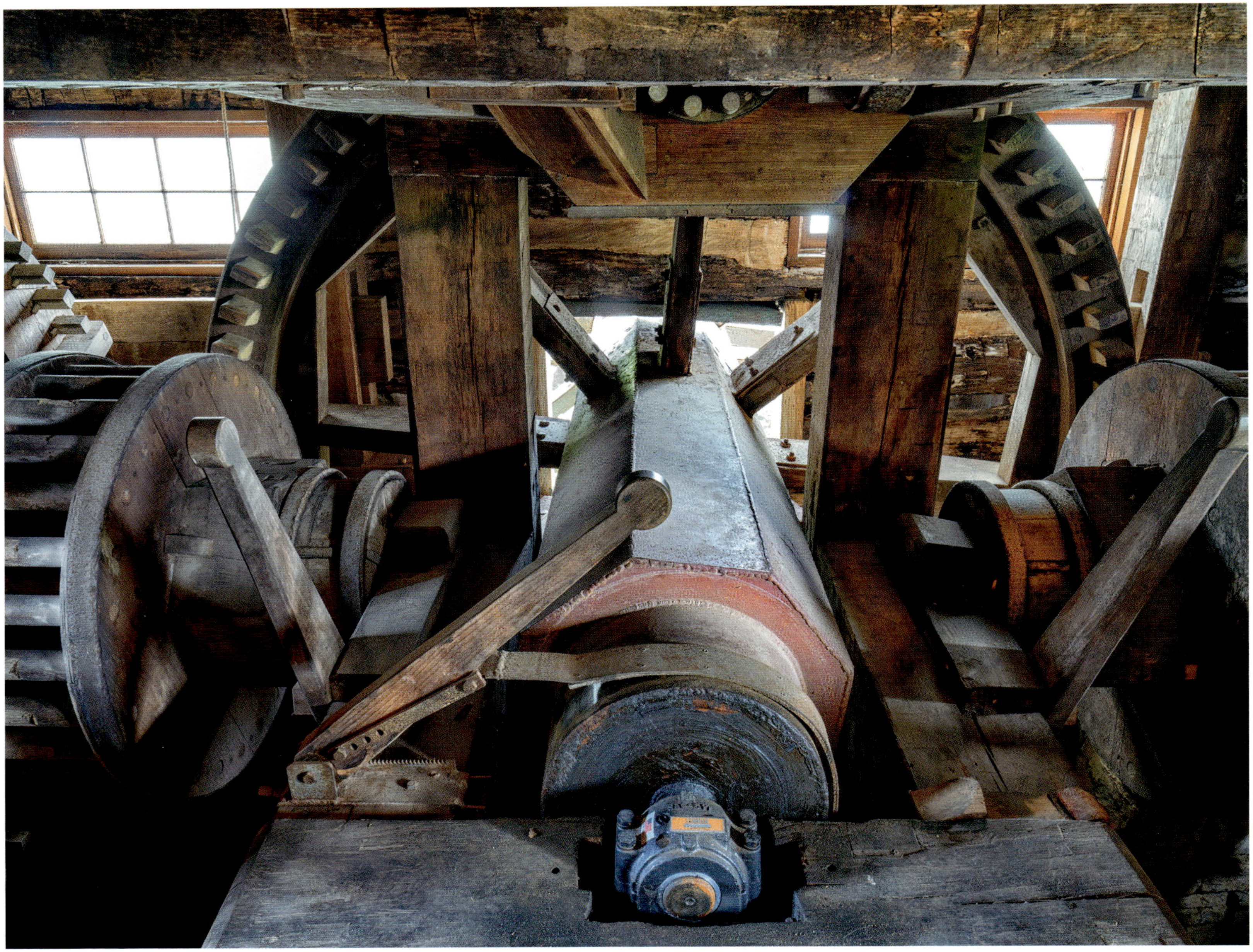

In this image from the lower level of McCormick's Grist Mill, the massive driveshaft can be seen at the center of a symmetrical design. Masterfully conceived and set, wooden gearing at the right and left of the driveshaft powers the two sets of millstones on the upper level of the mill. The waterwheel is outside and directly behind and beyond the central driveshaft.

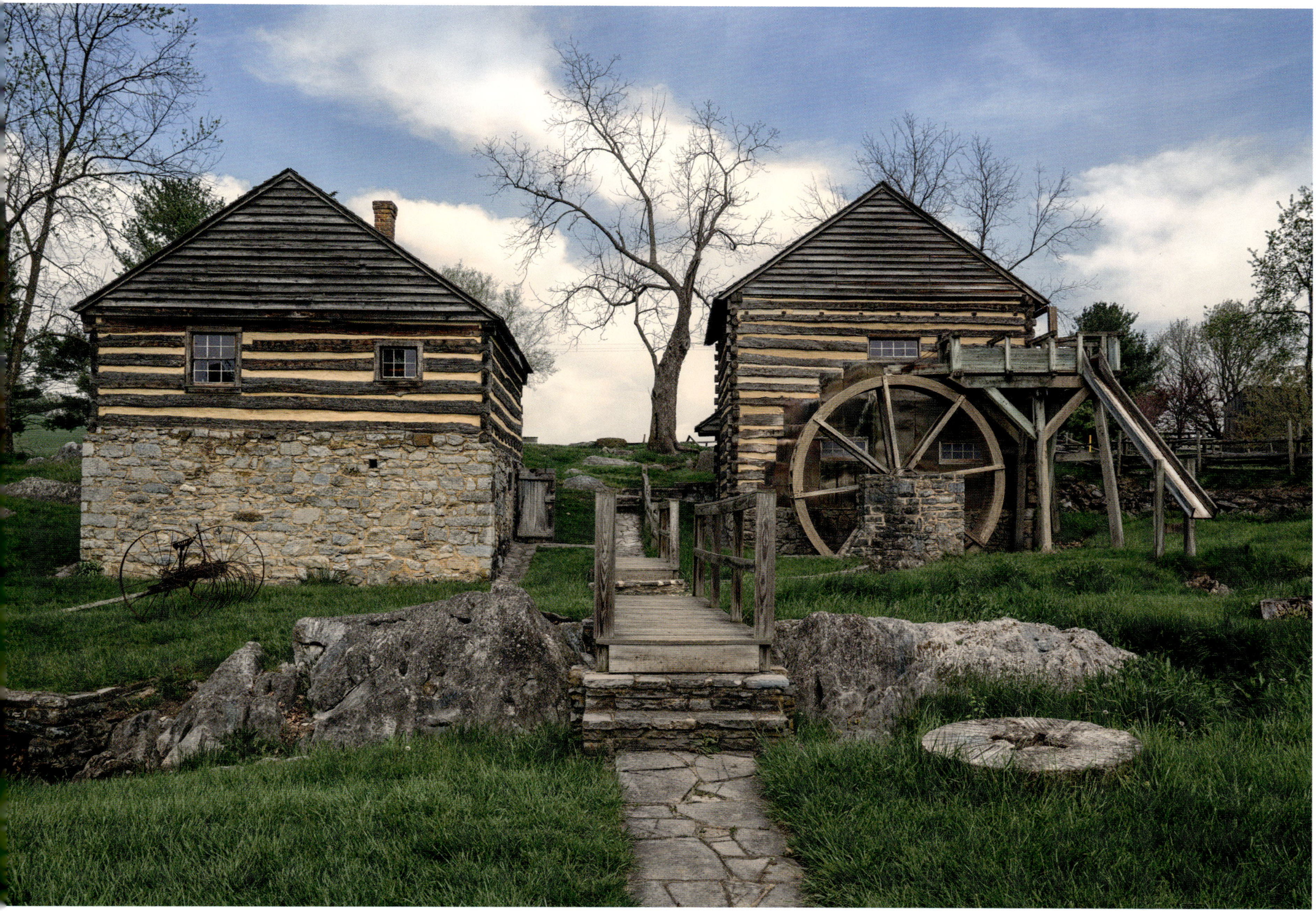

At the McCormick farm, a matching historic blacksmith shop sits adjacent to the watermill.

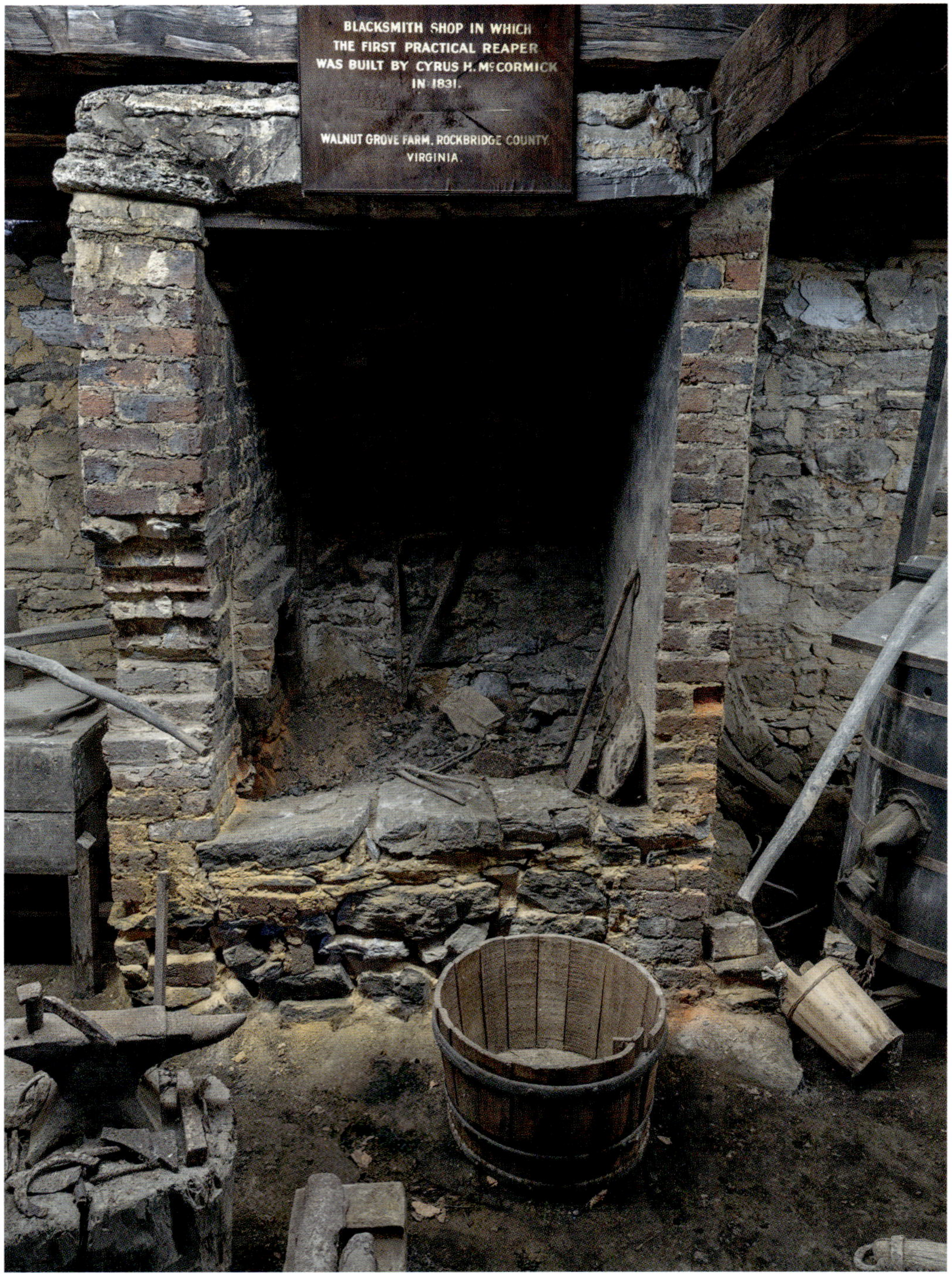

Cyrus McCormick and his enslaved partner and best friend, Jo Anderson, used the forge and blacksmith shop to fabricate their famous mechanical reaper.

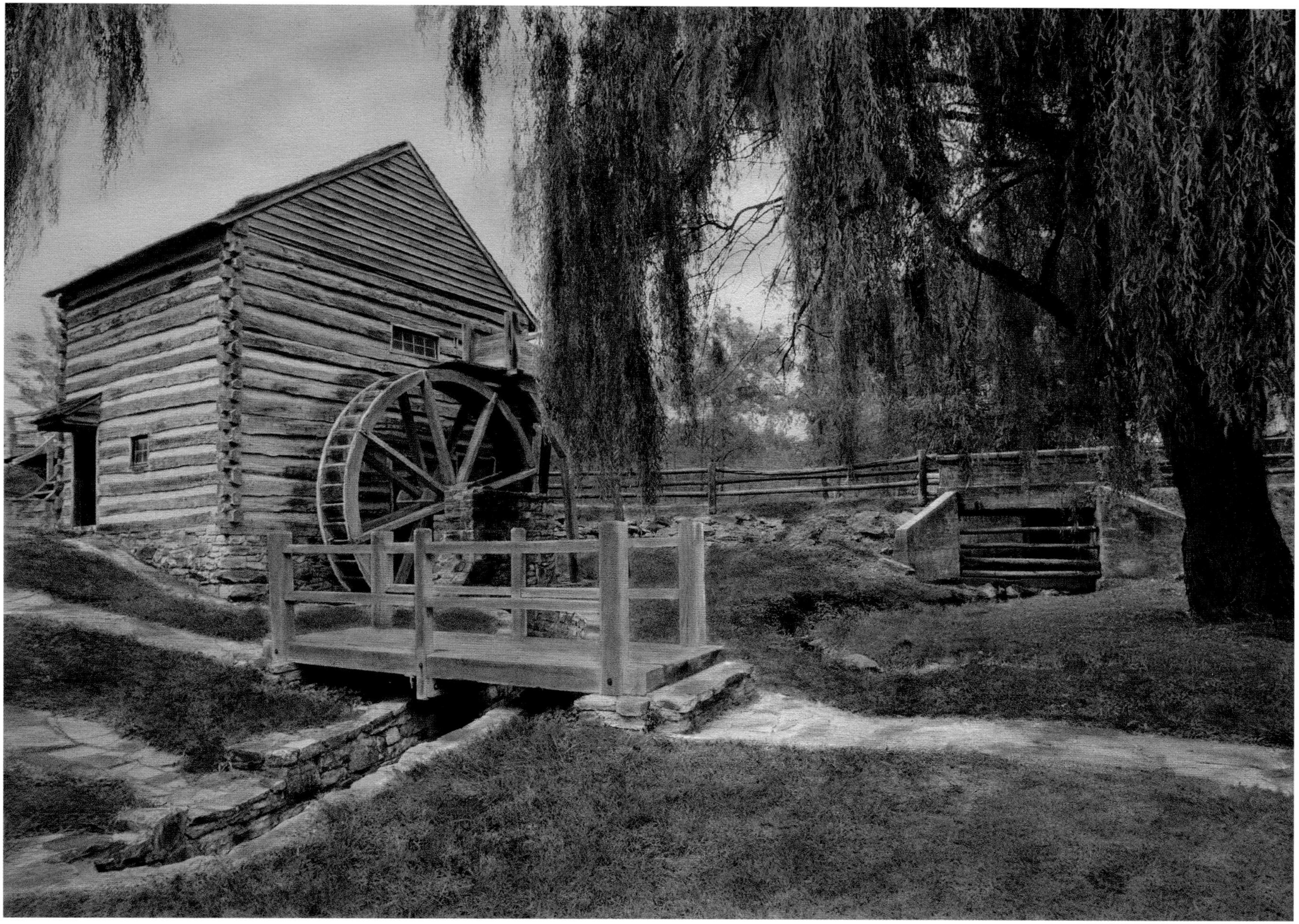

This vintage hand-tinted image by the author shows McCormick's Grist Mill more than forty years ago, when it was still outfitted with an earlier waterwheel, and before other restoration work was undertaken.

Osceola Mill

Vesuvius, Virginia, 1849

WITH ITS STRIKING twenty-eight-foot red-orange overshot waterwheel, Osceola Mill in the world-famous Shenandoah Valley is impressive to anyone with a fascination for watermills. It is one of two remaining watermills on Marl Creek. This mill dates to about 1849 and, like McCormick's Grist Mill, was once part of the McCormick estate, where five additional watermills operated. Osceola Mill is located on a lower stretch of the creek and was always the largest of the estate's mills.

A classic white Victorian-style miller's house overlooks the mill grounds from across the main road through the valley. The house, built in 1875, has a colorful history of its own, originally serving as a McCormick homeplace, later as a mill inn, then as a Virginia orphanage, and now as a private residence apart from the mill.

The mill sits on a substantial limestone foundation and is constructed of chestnut beams and timbers that have survived in good condition for almost two centuries. Over its operational years between 1849 and 1969, the mill was set up first with millstones, then with roller mills, and finally with a hammer mill. The mill closed after Hurricane Camille tore through the area in 1969.

For the last fifty years or so, Osceola Mill has served as an inn and a restaurant under several different owners and with varying levels of commercial success. Recent owners added gray vertical siding to the exterior of the mill and constructed a "honeymoon cabin" nearby. The gray siding has received generally negative reviews from area residents and historians. The mill is equipped with a modern commercial kitchen that as of 2024 was not in use.

As with many old mills, colorful stories about strange sounds, ghosts, and unusual phenomena at Osceola are fairly common. At least ten people have died or been murdered at the mill or in the immediate vicinity, adding to the folklore. There are additional accounts of missing persons and unexplainable events at the site. Ghost hunters and paranormal investigators have visited the mill, and their reports are said to confirm elevated energy levels and perhaps a presence in the mill itself.

Today, the appearance of the old miller's house adds to the overall eerie experience around the mill. It looks like a scene from a Hitchcock thriller. The place is overgrown, the white paint is faded and badly flecking, and shutters have fallen off some of the windows. When the house served as a state orphanage, disease, sickness, and unexplained circumstances occasionally claimed the lives of children, some of whom are likely buried on the mill property. This adds still another level of intrigue to the mill experience and enhances its popularity with visitors and guests.

As of 2024, the mill is set up with self-check-in but has sketchy Wi-Fi and no cell service. All alone through the long hours of the night, guests at the creaky and ghostly old mill can feel a little isolated in this lonely area of Virginia. Some thrive on the experience, while others are ready to pack up and get out!

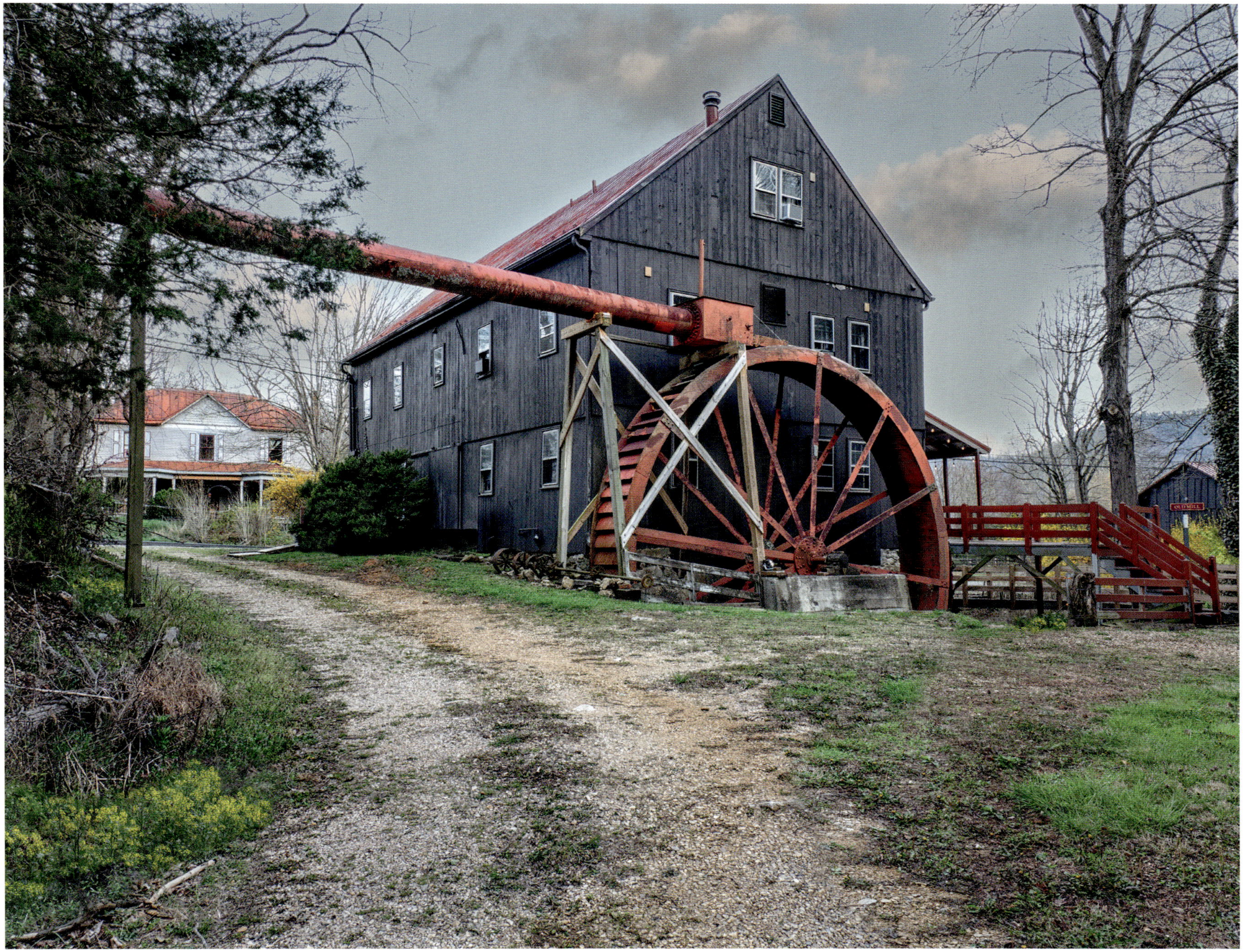

Osceola Mill is located in a picturesque valley only a few miles from McCormick's Grist Mill. Before the Civil War, five additional watermills were located on Marl Creek between the Osceola and McCormick mills. The Union army burned those other mills during the war, but locals disguised Osceola Mill as a barn. This action saved it from the torch!

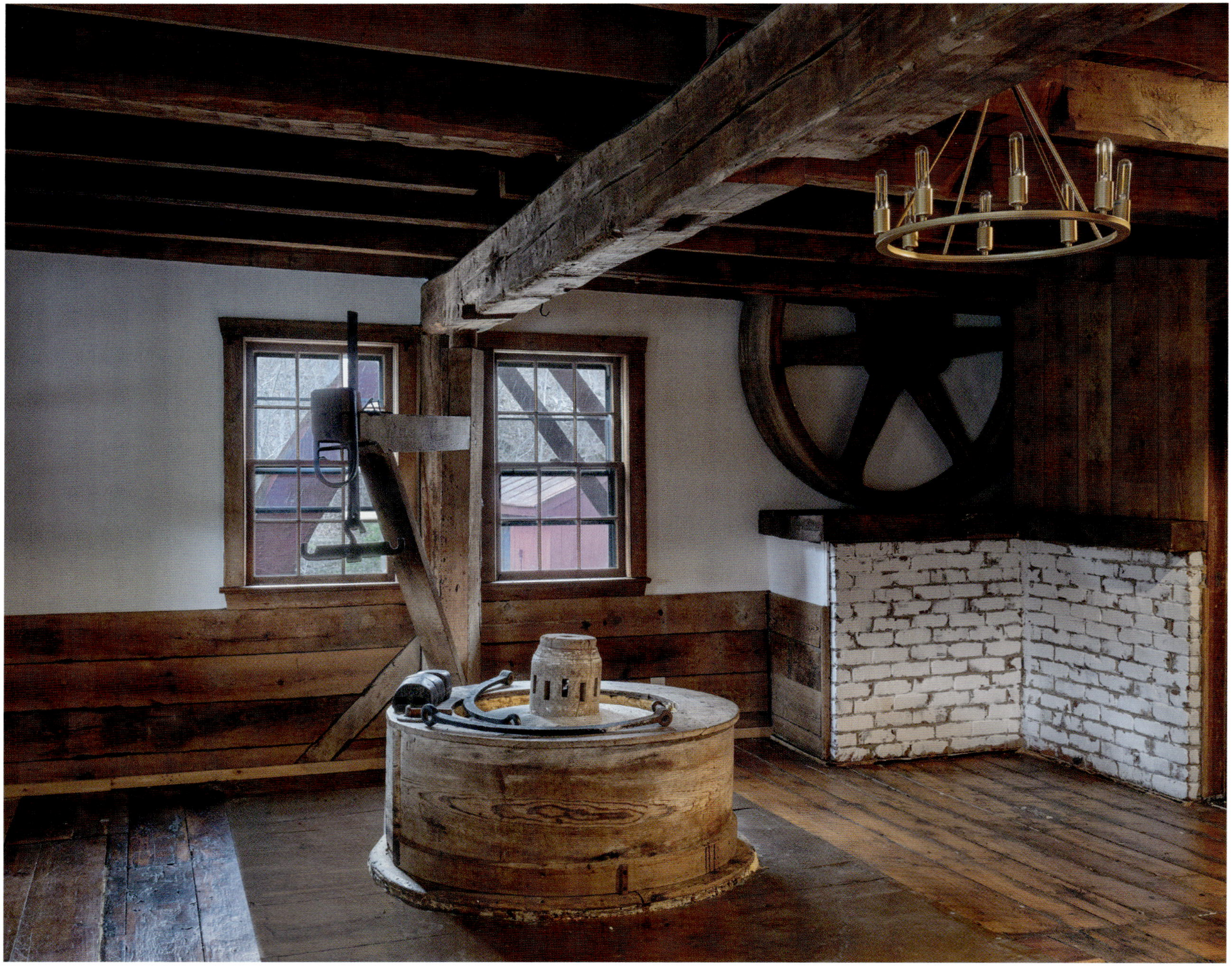

The interior of Osceola Mill has been converted into a country inn and restaurant, but some milling equipment has been retained as a beautiful educational display for guests.

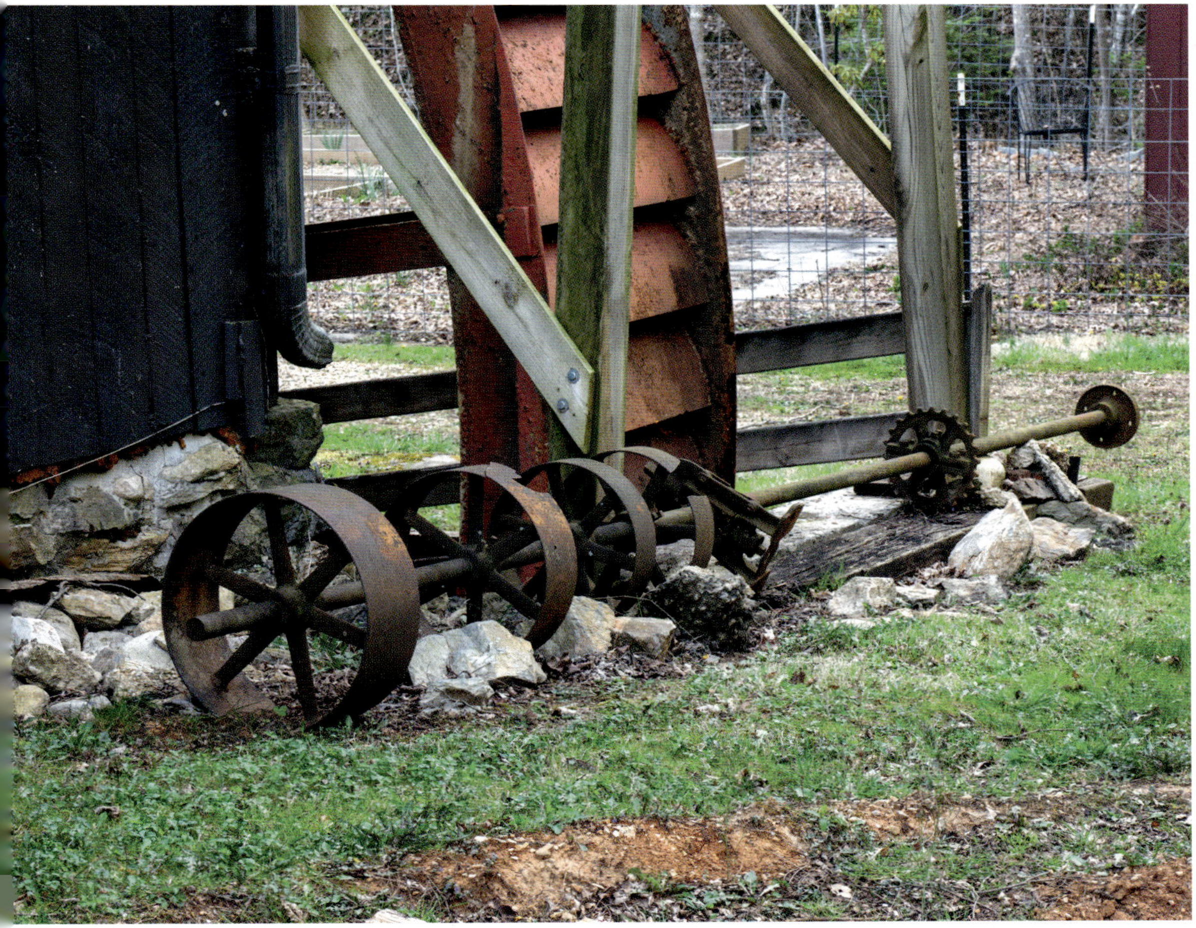

This long shaft with a gear at one end and four pulleys further along its length was used to power various pieces of equipment when the mill was operational. The shaft is now displayed outside, near the waterwheel.

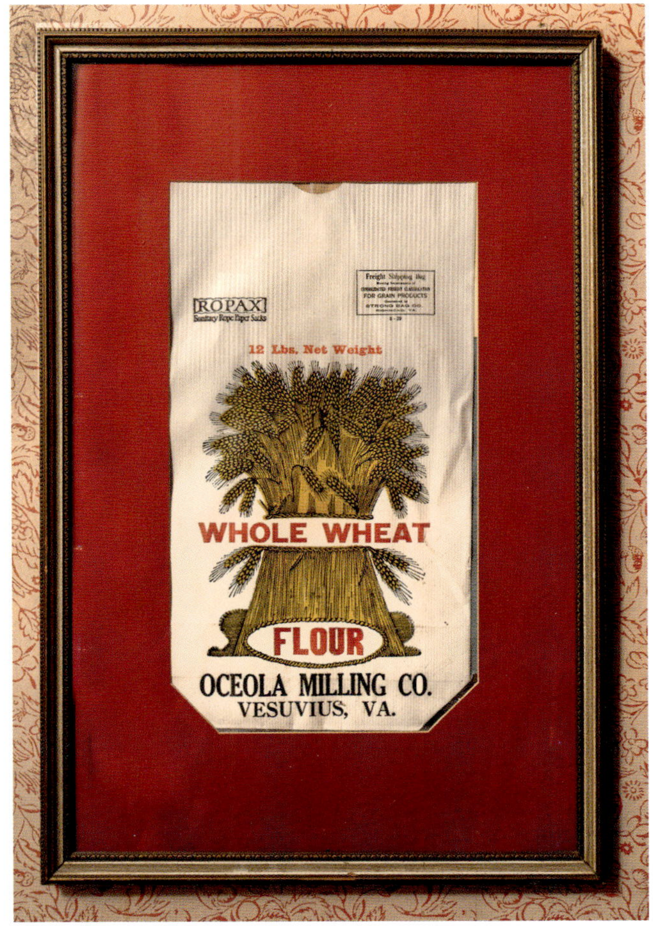

Flour and meal bags can be quite beautiful as art. Notice that the mill's name is spelled without an "s" ("Oceola") on this vintage bag that is framed and displayed on the upper level inside the mill.

Pine Creek Mill / Nolen Mill / Shelor Mill

Floyd County, Virginia, 1913 / 1996

PINE CREEK MILL has all the charm of a classic Southern watermill. The ancient wooden siding has never been painted. The mill sits along a narrow gravel country road and is powered by a twenty-foot waterwheel clearly visible beside the mill. The building is a two-and-a-half-story wood and timber frame construction with various sizes of windows installed.

By the time the mill was built in 1913, turbines had proven to be generally more powerful and have lower maintenance requirements than waterwheels, and thus a turbine was initially installed at Pine Creek. However, from the beginning, the turbine experienced problems with "silting up" and filling with debris, and it was soon removed and replaced with a more traditional waterwheel. A three-thousand-foot in-ground waterway provides water for the wheel.

A breastshot waterwheel was chosen as the replacement for the turbine. These waterwheels are generally about 50 percent efficient, but this installation proved to provide adequate power for the single set of millstones at Pine Creek. Significantly, this was the last watermill in Floyd County to commercially grind cornmeal for local farmers. Over the years, the mill also produced corn whiskey.

Pine Creek has benefited from some restoration and appears to be structurally sound. The wooden flume was replaced in 1996.

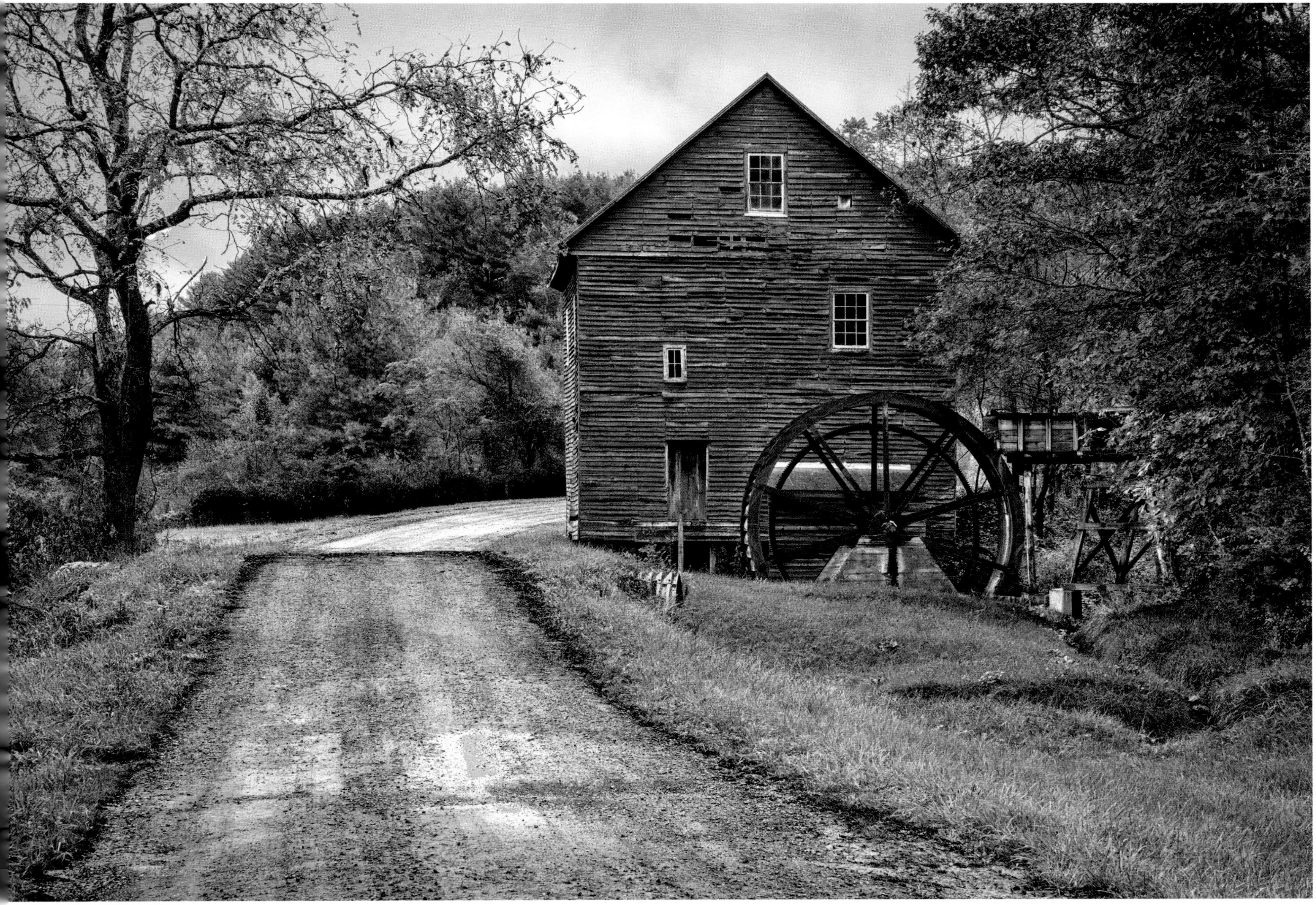

Pine Creek Mill makes for a classically charming scene in rural Floyd County, Virginia. Floyd County is home to at least thirteen historic watermills.

Lee's Mill at Stratford Hall

Montross, Virginia, 1740s / 1939

SITUATED ABOVE THE 165-foot cliffs and beaches along the Potomac River, Lee's Mill at Stratford Hall, also known as Stratford Mill, was a horse ride away from the main house on the sixteen-thousand-acre plantation estate. When the mill was built on this site in the 1740s, the area along the river was primarily a shipping wharf for plantation tobacco and other crops.

Thomas Lee constructed the one-story mill building to provide grain for the residents living and working on the plantation. They were primarily an enslaved population who did most of the work on the estate, including construction of the original mill. The millers were also likely skilled enslaved workers.

The current mill is a reconstructed timber and plank structure on the original 1740s brick foundation. Relatively authentic, the eighteen-foot wooden waterwheel is covered and protected by a lean-to shed and connected with authentic wooden gearing to grind cornmeal, grits, and pancake flour. Some of the milling machinery for the restoration came from a mill in Maryland.

Students of American history will be interested to know that two signers of the Declaration of Independence, Richard Henry Lee and Francis Lightfoot Lee, lived at Stratford Hall. The Robert E. Lee Memorial Association has owned the estate since 1929. The association has closed the sandy and dangerous cliffs and the area beyond the mill for safety reasons and to protect the rare Miocene Epoch fossils that remain along the river. The ten-million-to-seventeen-million-year-old Miocene fossils are found in only a few locations around the world.

During the spring season, the millpond is lined with beautiful wildflowers and flowering vines. In the fall, the hillside opposite the access road is illuminated with the colors of the season.

The plantation is often linked with General Robert E. Lee, but he never owned Stratford Hall. He was born there in 1807 but lived there for only a few years as a young child.

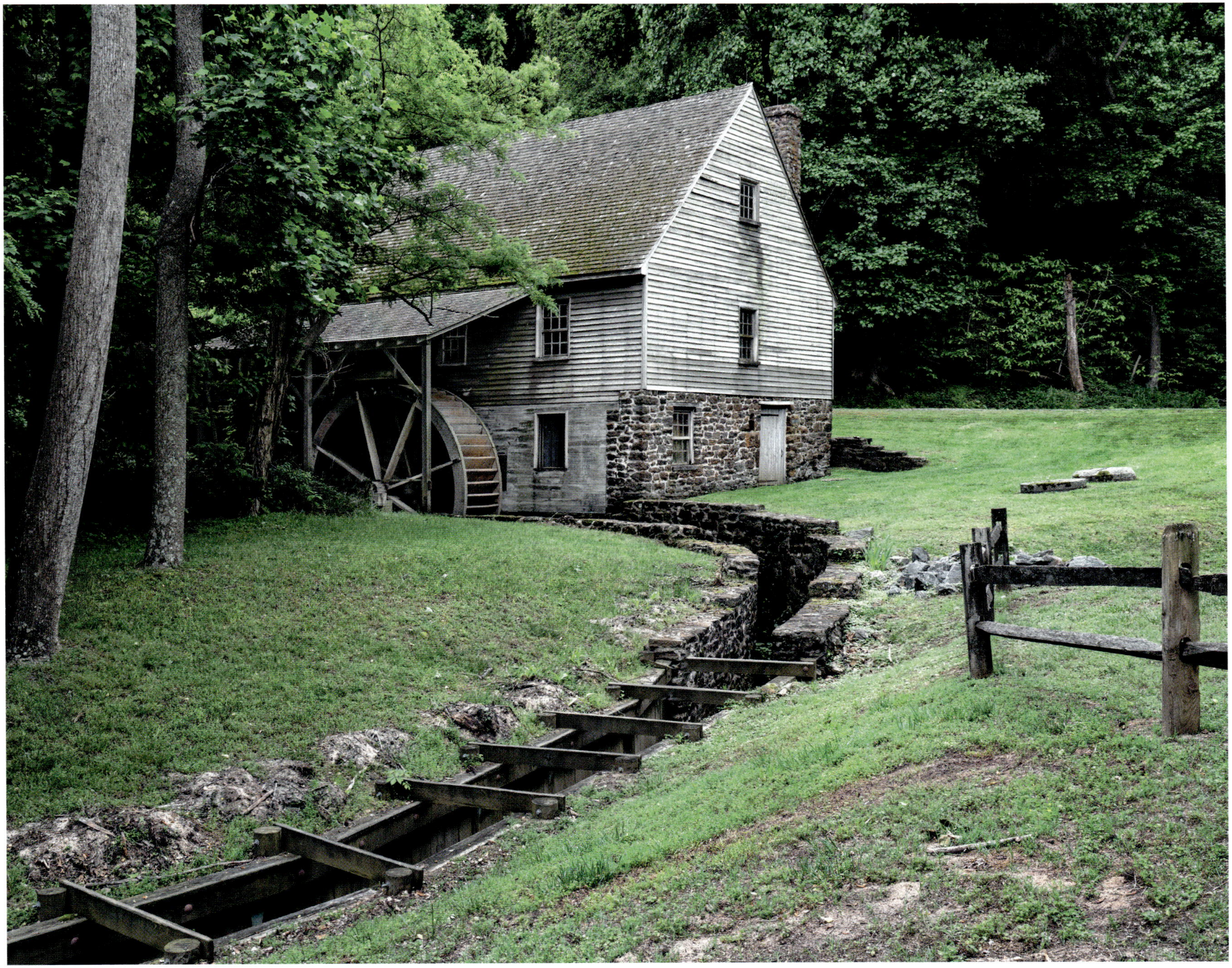

The eighteen-foot-diameter wooden waterwheel at Lee's Mill is covered by a lean-to shed that protects the wheel and main driveshaft from the weather—especially important during the winter months along the Potomac River, when snow and ice could damage the works.

Kennedy-Wade Mill

Raphine, Virginia, 1750 / 1990s

LOCATED IN THE picturesque and celebrated Shenandoah Valley, the Kennedy-Wade Mill is the oldest continuously operating watermill in Virginia. The mill has an admirable and prolific history; it dates to about 1750, when Captain Joseph Kennedy from the north of Ireland had the structure built on a 113-acre site and powered it with water from what he named Captain Kennedy's Mill Creek.

For more than a hundred years, the Kennedy family ran the mill. The mill survived the Civil War and was transferred to new owners in 1867, but it suffered fire damage soon after, in 1873. Fire was a constant risk in the dusty environment of watermills. Before electricity, window light was supplemented by oil lamps and lanterns, which in many mills were essential to seeing and working; however, this arrangement heightened the risk of fire or explosion.

The mill was subsequently rebuilt and resumed operations. It was later bought by the Wade family, hence the name "Kennedy-Wade Mill." For four generations—well into the twentieth century, when many watermills were going out of business—the Wades ran the mill.

In 1991, more than a century after it had been rebuilt, the mill needed operational, structural, and cosmetic restoration. The Young family purchased the mill and the surrounding properties and began updates, including fairly major and much-needed exterior work. When John and Karen Siegfried acquired the site in 2016, becoming the mill's seventh owners, they vowed to uphold the rich tradition and heritage dating to the time of Captain Kennedy.

Part of the deal for the Siegfrieds' purchase included lessons and training by the Youngs for proper operation, care, and maintenance of the historic landmark. John Siegfried did not grow up in the watermill business. In fact, the Siegfrieds relocated from England to begin an amazing new life in western Virginia. Today, a visit to this mill is an engaging, educational, and enlightening experience. The Siegfrieds are warm and welcoming and are always excited to share their experiences with guests.

The Kennedy-Wade Mill, sometimes known simply as the Wade Mill, is a three-story, thirty-five-by-forty-five-foot stone and wooden structure. The mill is powered by a twenty-one-foot metal overshot waterwheel. The main millstones were installed more than a hundred years ago, although electrically operated stones have since been added and are used today for most commercially produced mill products.

The mill regularly grinds yellow and white corn, hard and soft wheat, rye, and buckwheat into a menu of products. Popular items include grits, polenta, a variety of flours, and mixes for pancakes, bread, cornbread, and hush puppies. The grains are all locally sourced. White grits are the most popular item, with "anything to do with pancakes" coming in second!

In the old days, watermill sites were often host to parties, socials, and other events throughout the year. Even today, the Kennedy-Wade Mill hosts an annual festival each October at which visitors can buy a jar of old-fashioned apple butter that was slow-cooked in a 130-year-old copper pot!

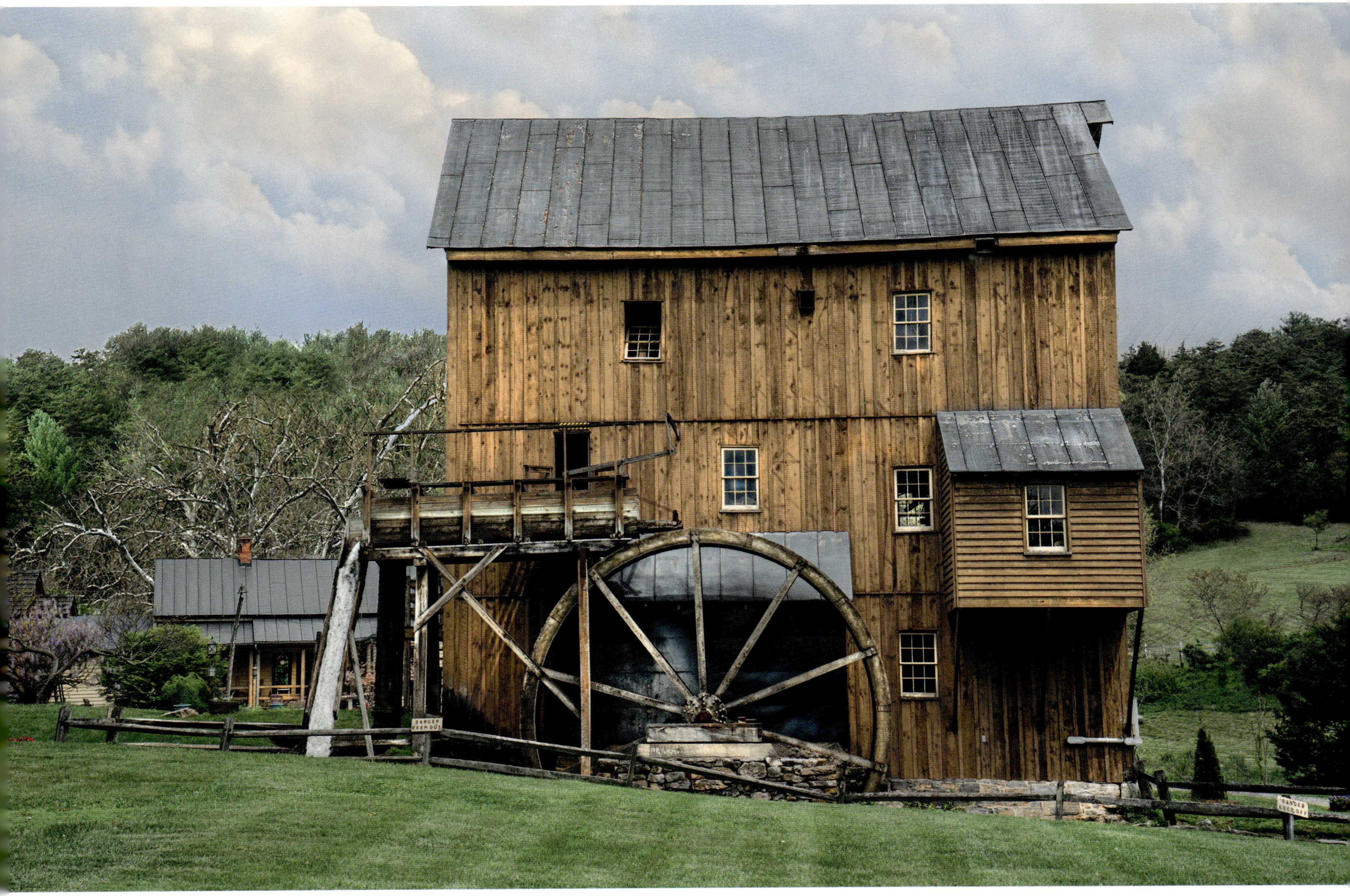

This recent image shows the Kennedy-Wade Mill following restoration work. The twenty-one-foot overshot waterwheel is prominent in the foreground, with water being released through a discharge just before the wheel. The miller's house can be seen behind and to the left of the mill.

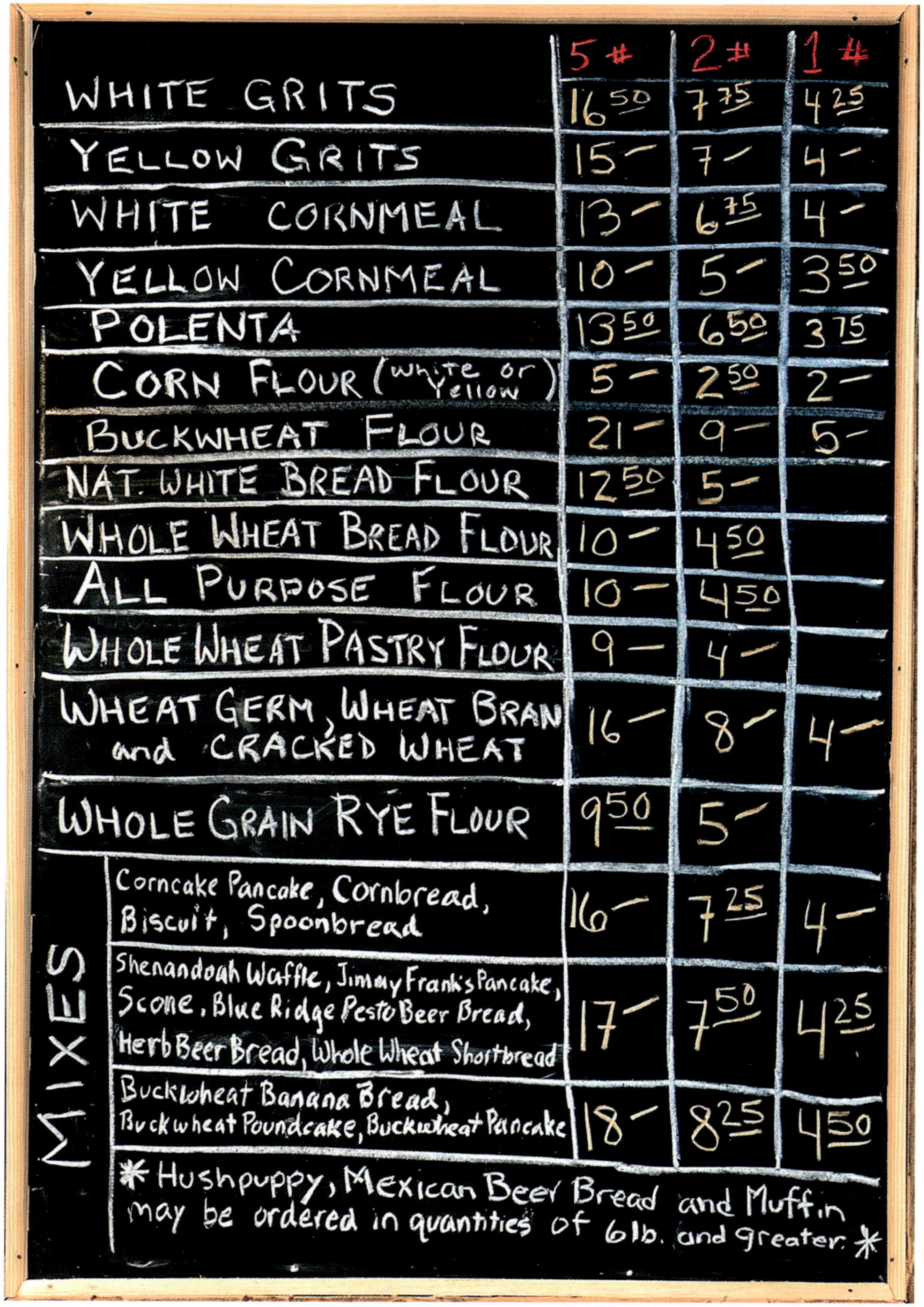

Customers and visitors to the Kennedy-Wade Mill are drawn to the chalkboard describing the extensive list of products offered for sale.

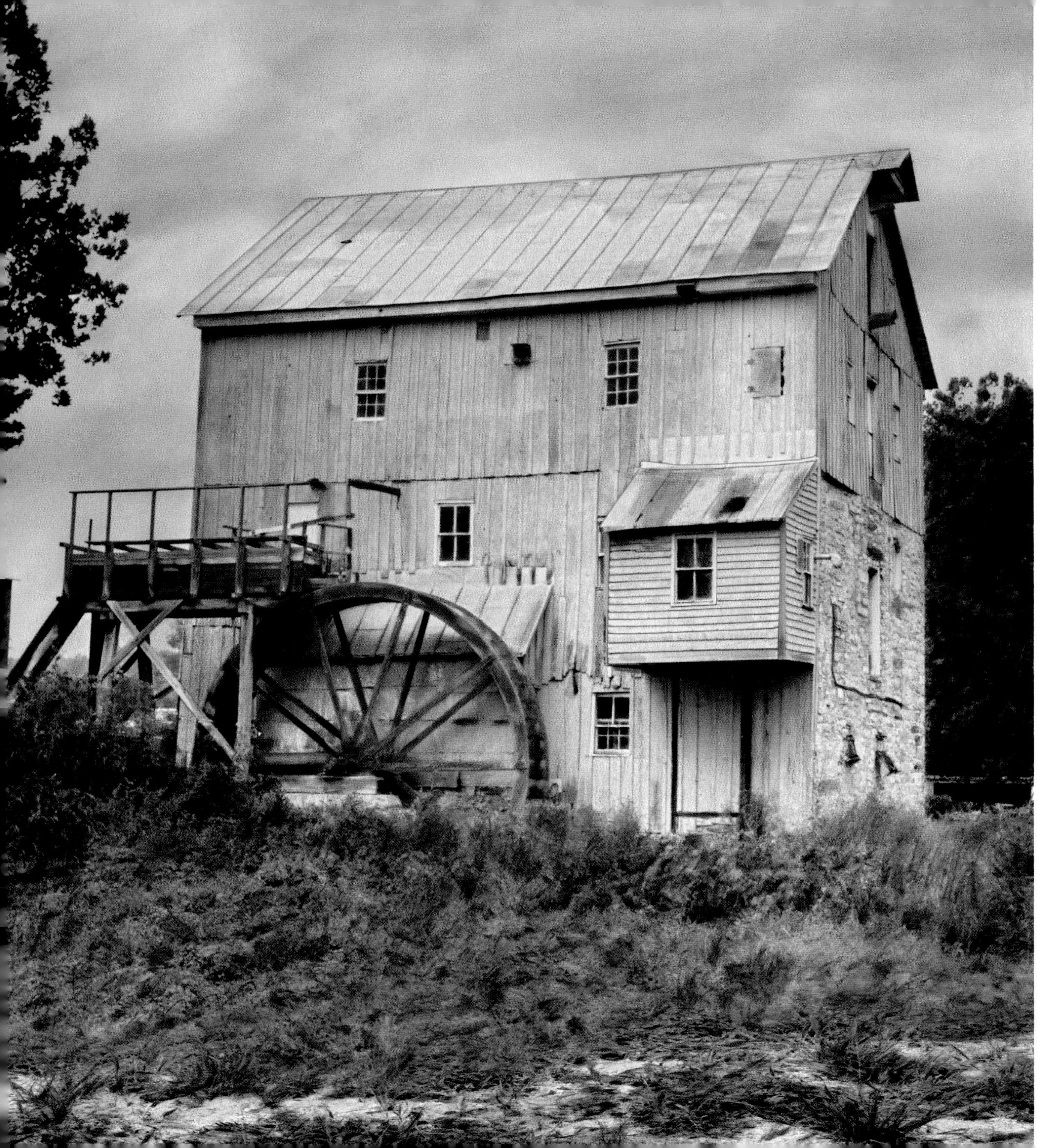

This historic photo of the mill was taken by the author before restoration began. The mill was still operational at this time but was in need of mechanical, structural, and cosmetic work.

Colvin Run Mill

Great Falls, Virginia, 1802–1811 / Restorations in Recent Decades

THIS BEAUTIFULLY REESTABLISHED, preserved, and operational watermill is located on property that was once owned by George Washington. Washington, a miller himself, recognized the site as a good location for a mill, noting the sloping terrain and available water supply, the surrounding ancient forests for providing building materials, and the site's proximity to the Potomac River for shipping grains. The mill was a commercial-scale operation, although the merchant millers also provided grinding services for local farmers.

Today Colvin Run Mill, with its massive twenty-four-foot wooden waterwheel, is one of the country's best-kept examples of an early nineteenth-century automated gristmill. However, this mill was not without its period of peril; after miller Addison Millard died and the family closed the mill in 1934, it sat abandoned and fell into disrepair. The Fairfax County Park Authority took over the site in 1965, and restoration efforts took place in 1968 and 1972 at a cost of about $1 million. One of the most significant projects was to reestablish the water supply, which had been disrupted by highway construction after the mill closed in the 1930s.

A new waterwheel and wooden flume were installed in 2003–4, and the immense six-thousand-pound white oak driveshaft has since been replaced. The mill features three sets of millstones.

The property includes a miller's house, a general store and gift shop, an educational center, picnic grounds, gardens, and nature trails.

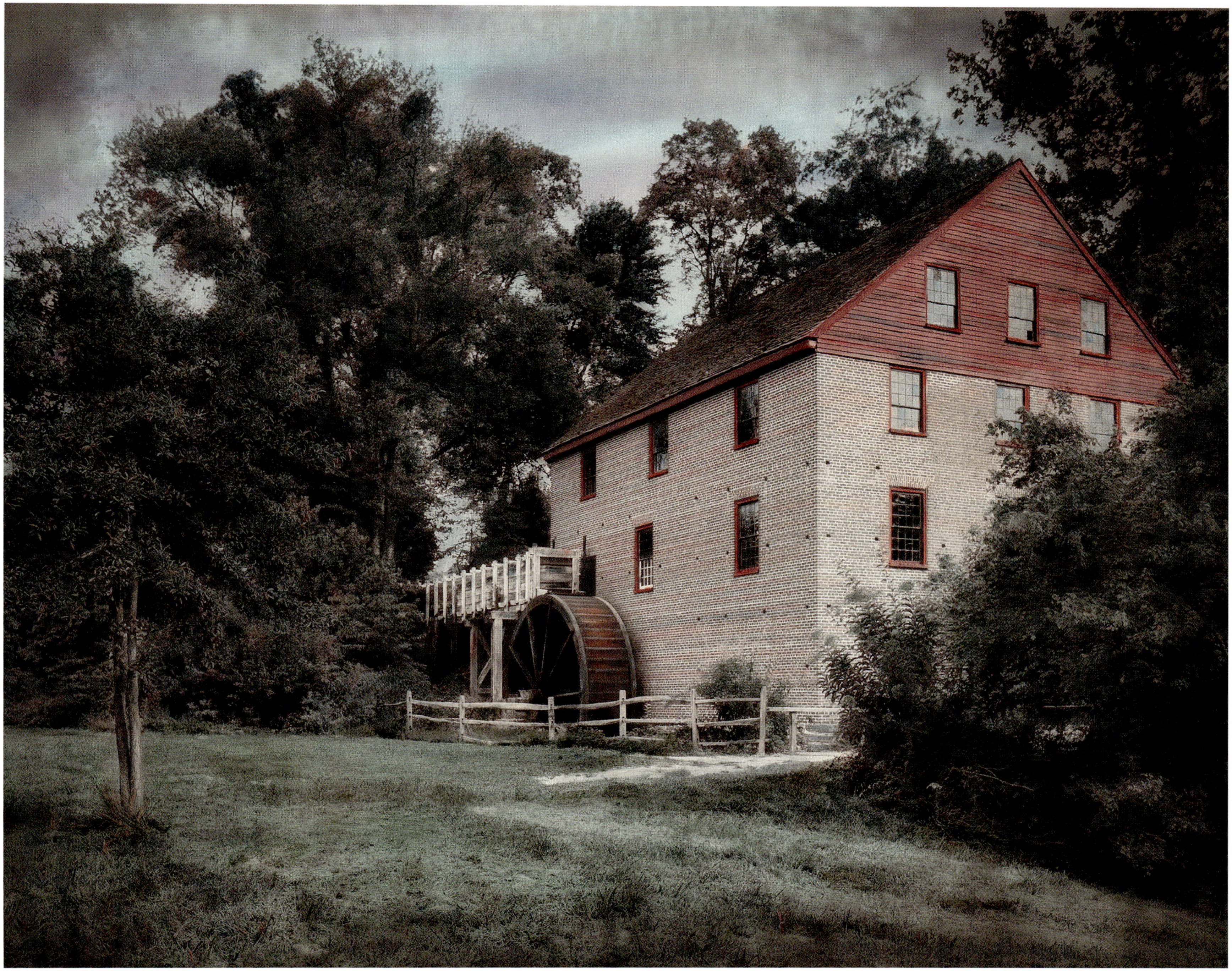

This image shows Colvin Run Mill on the author's first visit about forty-five years ago. Interestingly, the exterior bricks are of a smaller-than-standard size, so any brick restoration work requires that the new bricks be made individually by hand.

Tingler's Mill

Paint Bank, Virginia, 1863 / 2004

SITUATED ALONG Potts Creek in a mountain valley near the Virginia–West Virginia state line, Tingler's Mill is part of the historic Paint Bank community, which dates to just after the American Revolution. The land the mill stands on was granted to a Revolutionary War hero in 1780.

The community's interesting name comes from the orange and red iron clay taken from the banks of Potts Creek. The Indigenous Cherokee people used the clay as war paint and to make their pottery with its distinctive red colors.

Due to geographic boundary changes over the years, the mill has been in two different states and five different counties without ever moving an inch! Beginning in the last half of the nineteenth century, Paint Bank emerged as a timber-cutting and mining community. By the early twentieth century, the population of Paint Bank had boomed past two thousand, making Tingler's Mill more vital than ever to the town.

The building is a two-and-a-half-story timber frame design with weatherboard siding. The roofline details include cornice decorations at the gabled ends. The siding is painted blue gray with burnt red trim, in keeping with Paint Bank's history. The flume and the twenty-two-foot-diameter waterwheel are located to the right of the main entrance to the mill. The site includes a restored buggy and other exterior and landscaping appointments that add to the experience of visiting the mill.

The concept and design are based on Oliver Evans's automated milling technology to reduce the labor intensity, as grain is handled, ground, cooled, sifted, and packaged with mechanized equipment. As a result, the interior is busy with belt-driven roller mill machinery, wheat cleaners, wooden elevators, and bolters/sifters. Over the years, the mill has processed corn, wheat, and oats, as well as gunpowder.

Beginning in 2004, the mill was completely restored and reconstructed. It is presently operational and runs occasionally, and it is maintained in immaculate condition for tours and visitors.

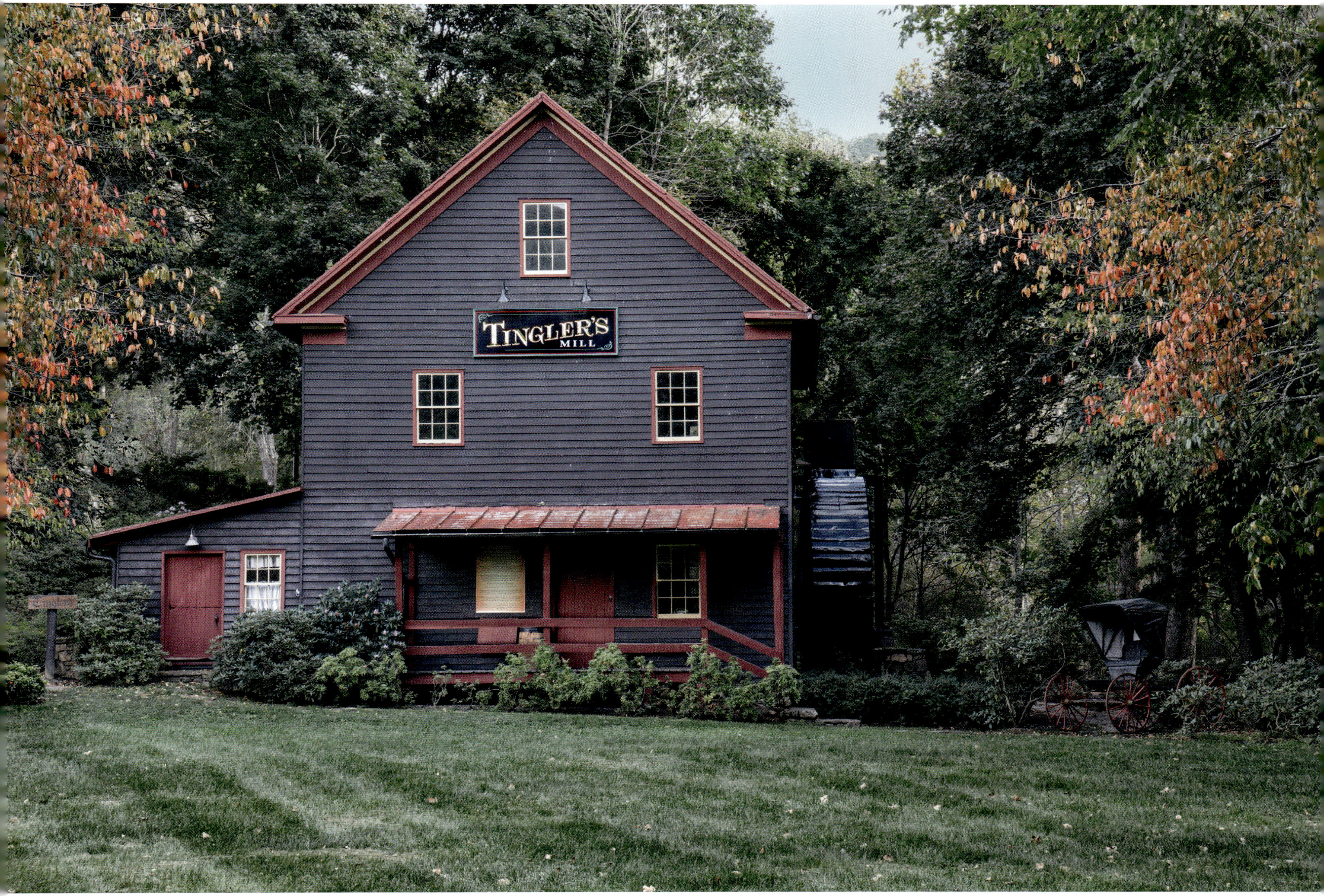

Tingler's Mill is a stately and visually delightful watermill in a remote but scenic mountainous Appalachian location.

George Washington's Gristmill

Mount Vernon, Virginia, 1770 / 1932 / Recent Restoration

BY 1769, George Washington had decided to build a gristmill on Dogue Run, about half a mile from an older 1730s mill his father had operated. The lumber for the mill was cut nearby on the Mount Vernon estate, and the millstones were floated down the Potomac River from near Great Falls.

A sixteen-foot wooden pitchback waterwheel was installed on the lowest level inside the mill building for better protection from the weather. The mill was designed with two sets of stones: French buhr stones for grinding wheat, and German "Cullin" stones (made from a dark blue-gray volcanic rock) for corn, barley, oats, rye, and other grasses. Washington specified machinery for the mill that was an early application of Oliver Evans's automated milling system. Evans sent his brothers to Mount Vernon for the machinery's installation.

Since the 1760s, Washington had been transitioning his estate from tobacco cultivation to growing corn and wheat, as the latter were more consistent cash crops than tobacco, which was dependent on British markets. The mill was used to grind grain from the Mount Vernon estate and to provide milling services for area farmers.

Water was delivered to the mill through a cut canal that was almost two miles in length. Enslaved workers dug the canal, which was seven to eight feet deep and six feet wide. Maintenance of this long water delivery system proved to be an ongoing problem for the mill.

After Washington's death in 1799, the operation and condition of the mill gradually deteriorated, and by the 1850s the structure had essentially been torn down. Beginning in 1932, the bicentennial of Washington's birth, the Commonwealth of Virginia built a replica, which operated until 1997. The mill has recently been redone and is again open to visitors, as is a reconstructed whiskey distillery. The two structures are located only feet apart, and both take water from the same canal.

In Washington's day, the gristmill and the distillery were operated by skilled enslaved millers and brewers, along with some hired workers. The two operations were of commercial scale, and both were profitable for Washington. The mill could produce about three thousand pounds of cornmeal and five thousand pounds of wheat flour a day.

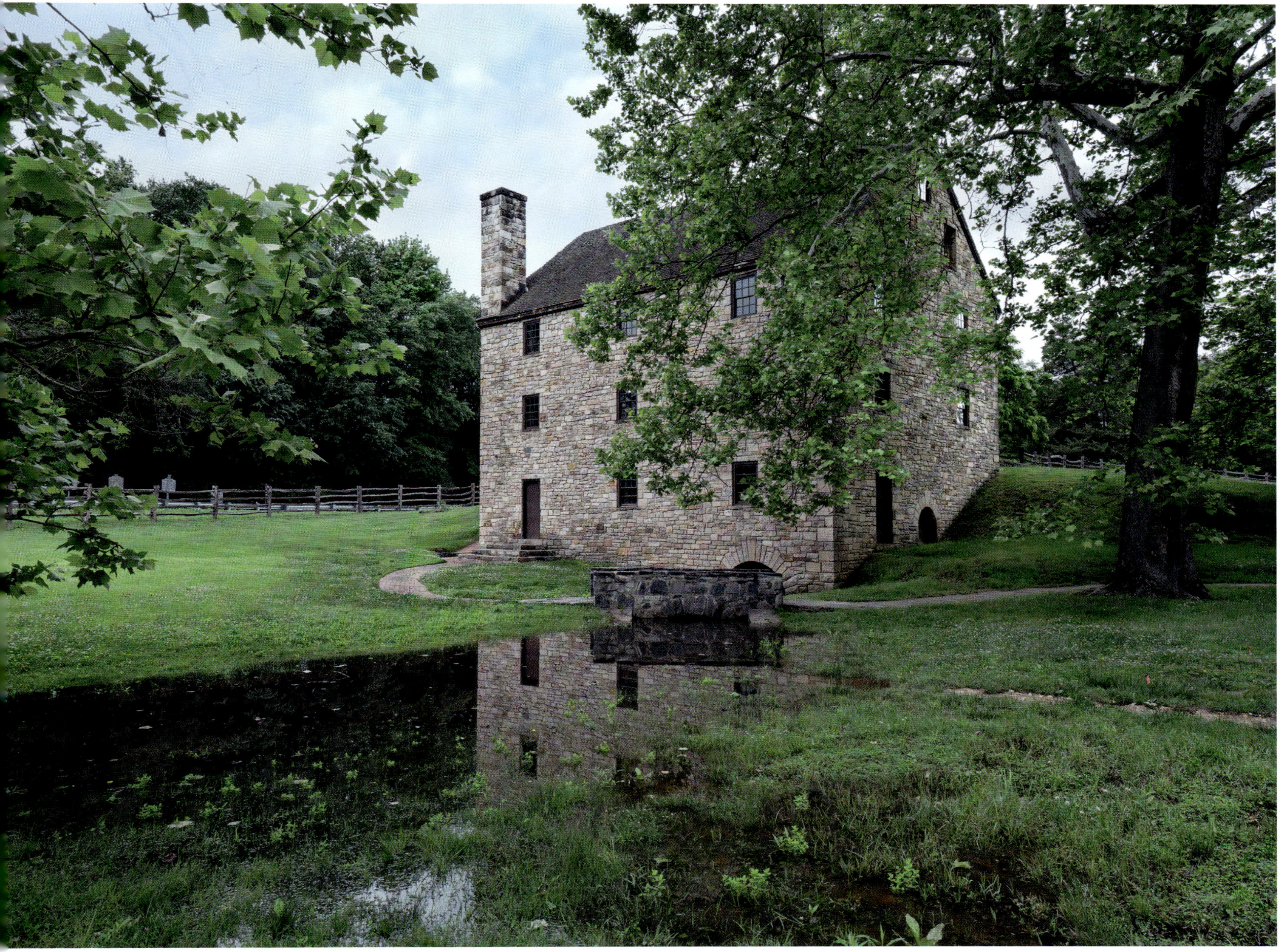

The restored Colonial Revival–style gristmill at Mount Vernon is shown here after heavy rains had flooded the grounds next to the creek, providing an almost mirrorlike reflection.

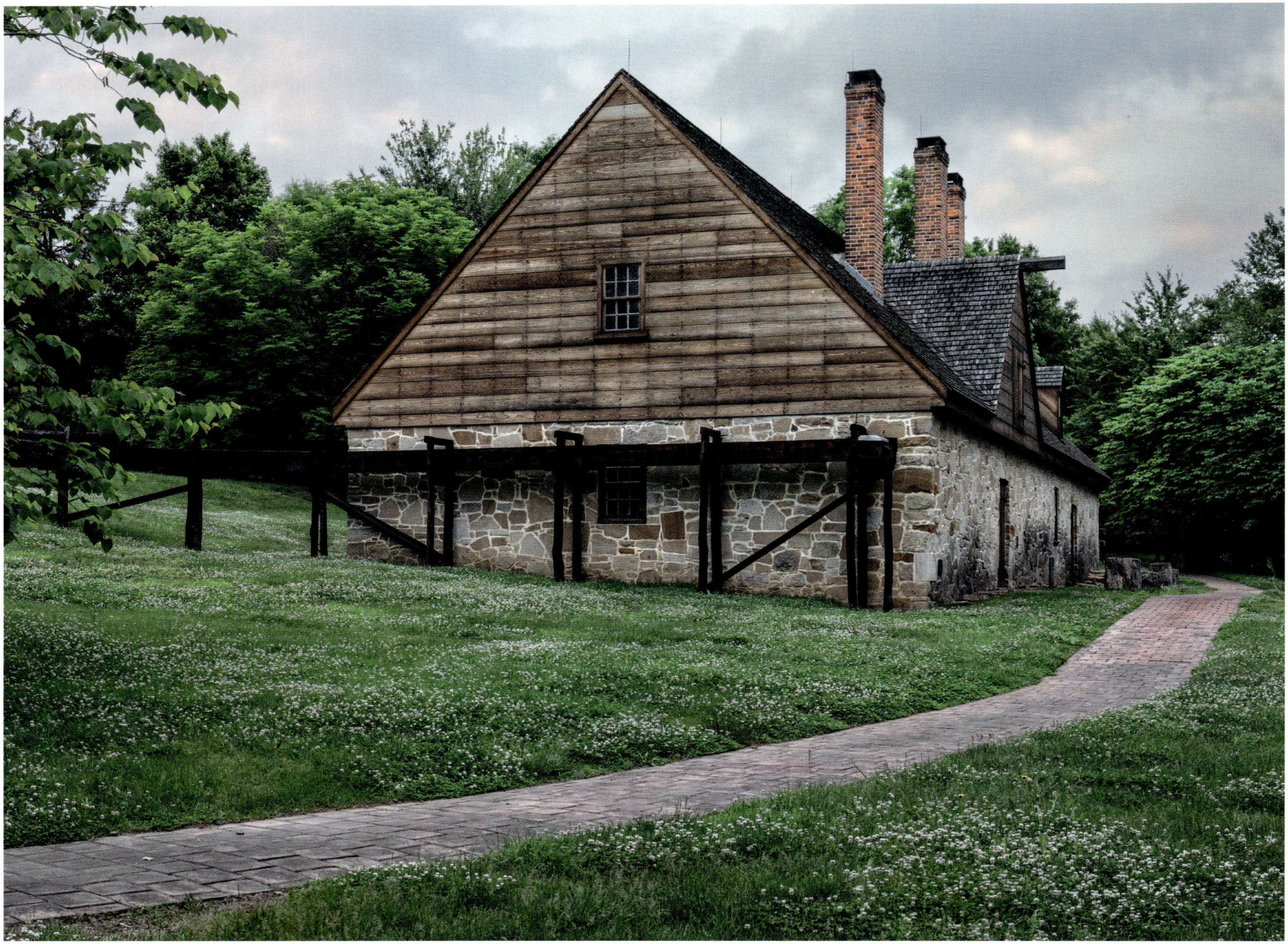

George Washington's original whiskey distillery became operational in about 1798. In 1814, a fire consumed the structure, and it was not rebuilt. This replica was constructed as accurately as possible from historical records.

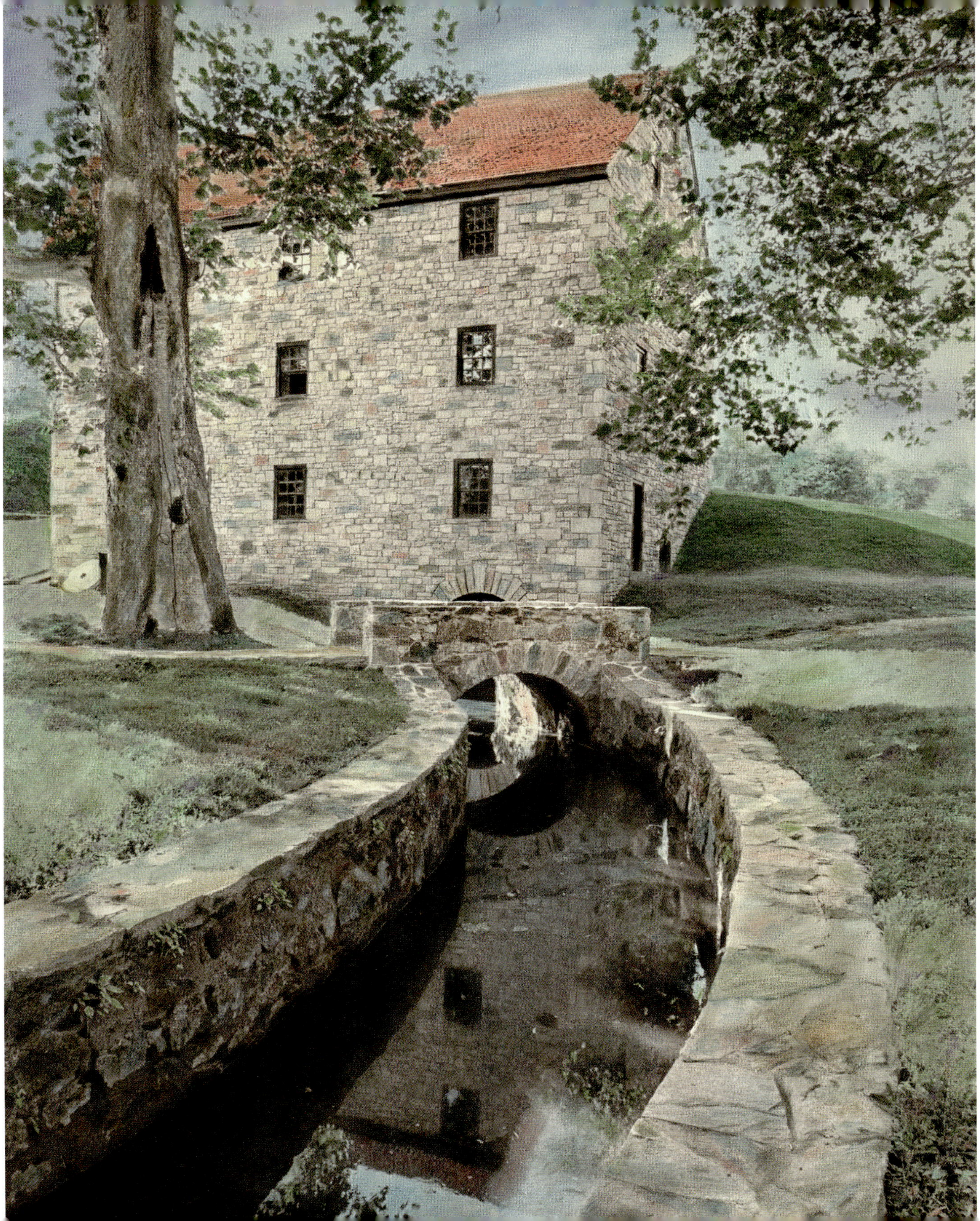

This is a hand-tinted image of the George Washington gristmill restoration as it looked in the late 1970s. The original mill was a long walk or horseback ride from the main area of the Mount Vernon estate.

Woodson's Mill

Lowesville, Virginia, 1794 / 1984–1986

WOODSON'S MILL, in the mountains of western Virginia at Lowesville, is one of the largest gristmills in the South. The mill is fully operational and has two active overshot wheels. At one time, a third waterwheel powered a sawmill and an ice plant, and a fourth wheel was used for some now-unknown application. Miraculously, the mill was not damaged during the Civil War.

For many years after the war, the mill was the hub of an active community. The site also included a stable, a foundry, a steam-powered sawmill, a machine shop, and a 650-acre apple orchard. For more than sixty years beginning around 1900, Dr. Julian Woodson owned and operated the mill. Woodson ran his family practice from an office on the main floor of the mill.

After Woodson passed in 1963, the mill suffered the fate of countless other Southern watermills. It had been deteriorating for years even before his death, and through the 1960s and 1970s, its future looked bleak. During this period, unskilled workers attempted to operate the mill and further damaged the gearing and machinery. Offers came in to buy the milling equipment, but fortunately everything was retained.

In 1984, work began in earnest to reclaim the mill and make it operational again. The first phase was to restore the drivetrain, clean the millstones, and clear the interior of assorted junk that had collected over time. Outside, at least one wall was in danger of collapse, and the water delivery raceway from the millpond needed to be reestablished for the two waterwheels. By 1986, the mill was up and running.

This story has a wonderful conclusion, thanks to the creative vision of a number of individuals and the expertise of Deep Roots Milling. Deep Roots has brought new life to the old mill through the experience, proficiency, and lineage of several skilled and committed millers, including the grandson of Julian Woodson.

The mill offers a full range of milling products on its online store. A large 1929 Arts and Crafts–style bungalow on the property is also available for rent; income from the house provides another source of revenue for maintaining the site.

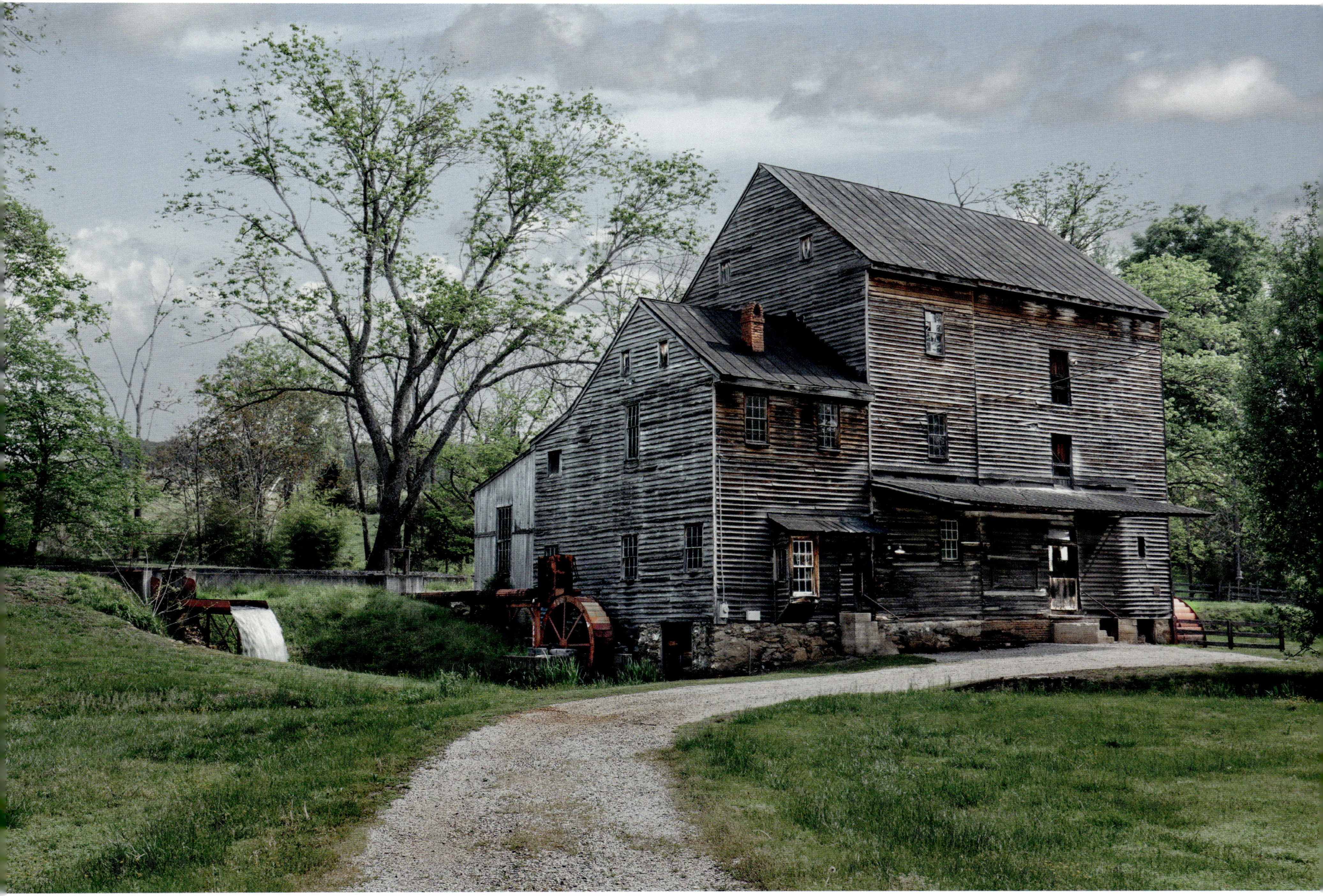

Woodson's Mill is one of the largest gristmills in the South. At one time, it operated with as many as four waterwheels. Today the mill has two operational waterwheels at opposite ends of the structure.

White's Mill

Abingdon, Virginia, 1790 / Mid-1800s / Ongoing Restorations

A GRISTMILL has operated on this site in southwest Virginia since about 1790. Most of the current mill dates to the mid- to late 1800s.

Structurally, White's Mill is a four-story wooden design with two major working floors. It is supported on a stone foundation. The waterwheel is a twenty-four-foot Fitz metal overshot design.

Inside, the mill was laid out and set up with Oliver Evans's ingenious technology of elevators, buckets, and lifts for automated operation. Many mills with this automated technology no longer use it because production volumes are just too low, or because some of the automation would need restoration and cleaning and require costly and time-consuming maintenance.

As an automated operation, White's Mill included bolting machinery for sifting the products into different grades and sizes. Traditional millstones were replaced years ago with roller mills.

As work at the mill flourished, a small community began to develop along Toole Creek in the area around the mill, with local farmers depending on the mill for meal and grits and other products. During the Civil War, the Union army raided the mill and knocked the waterwheel off its underpinning. They planned to burn the mill, but as the story goes, a big thunderstorm miraculously blew in and saved the mill from the torch. After the war, it continued to serve the community for more than a century.

By the 1980s, business had dwindled and the mill was falling into disrepair, and it finally closed in 1989. A decade later, in 1999, the White's Mill Foundation was formed to preserve and protect the mill. The foundation acquired the mill, mill store, and miller's house and undertook a total restoration.

As of 2024, the restoration project was well along, with plans outlining a complete overhaul of the site. As a part of those plans, the mill has received a much-needed new roof, new windows and siding, and substantial structural work. The large brick chimney is original. The siding has received mixed reviews from folks in the area, but it certainly protects the old structure. Plans are to restore and update the miller's house across the road in the hope that it could eventually be home to a miller.

The mill remains an important part of Abingdon's historic community. The mill grounds are host to local festivals, and the store offers assorted local arts and crafts. Mill products can be purchased at the store and in several markets in Abingdon.

The mill's name is sometimes a source of confusion, as "White's Mill" is a fairly common name for watermills, much like "Miller's Mill," "Red Mill," and "the Old Mill." For example, a famous watermill in Ohio is also named White's Mill. The White's Mill in Abingdon is the oldest operational gristmill in southwest Virginia and is in an especially historic region of the state.

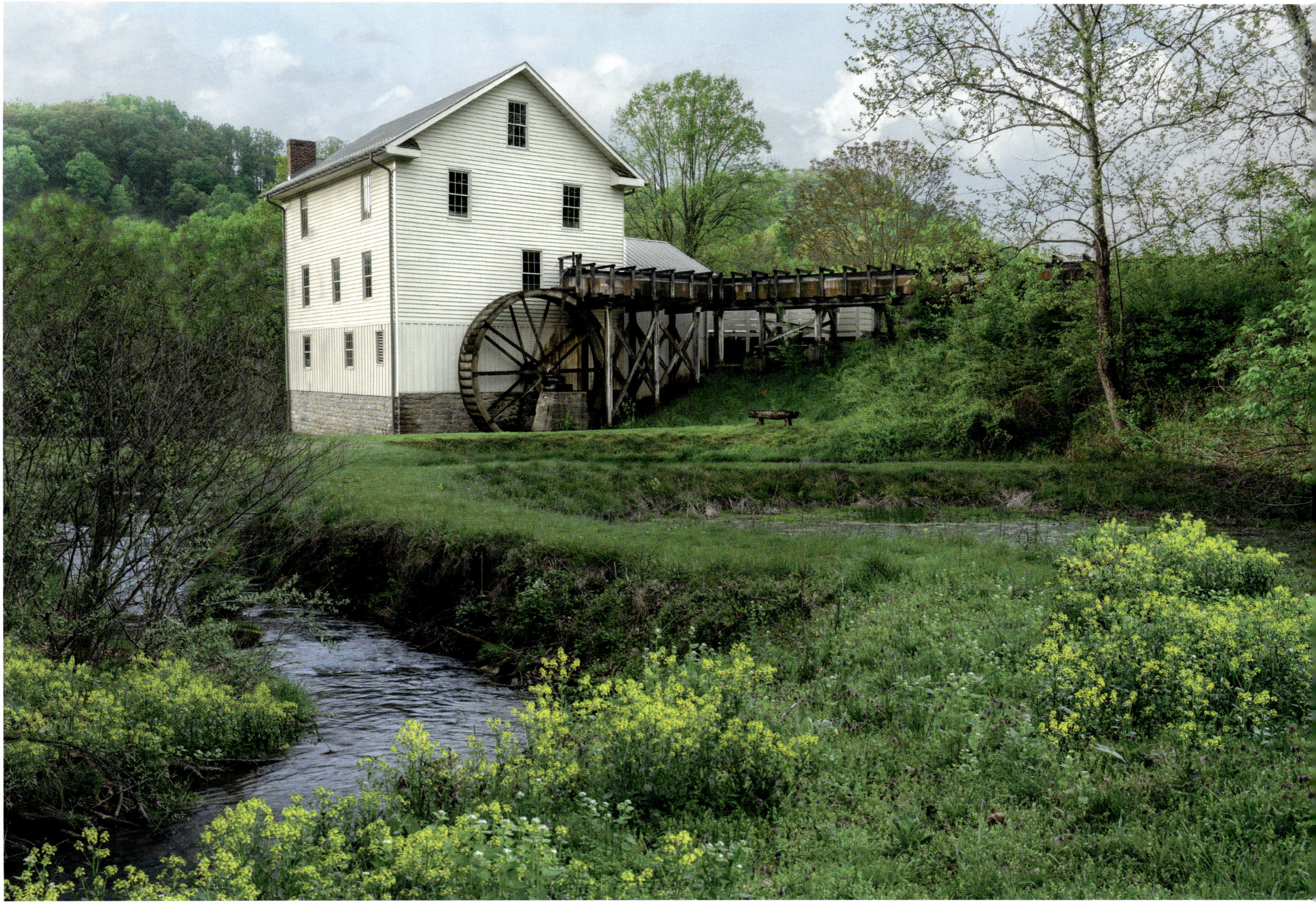

This recent image of White's Mill shows the new siding and flume.
The chimney at the rear of the mill is original. The white siding has received mixed reviews from some historians and preservationists.

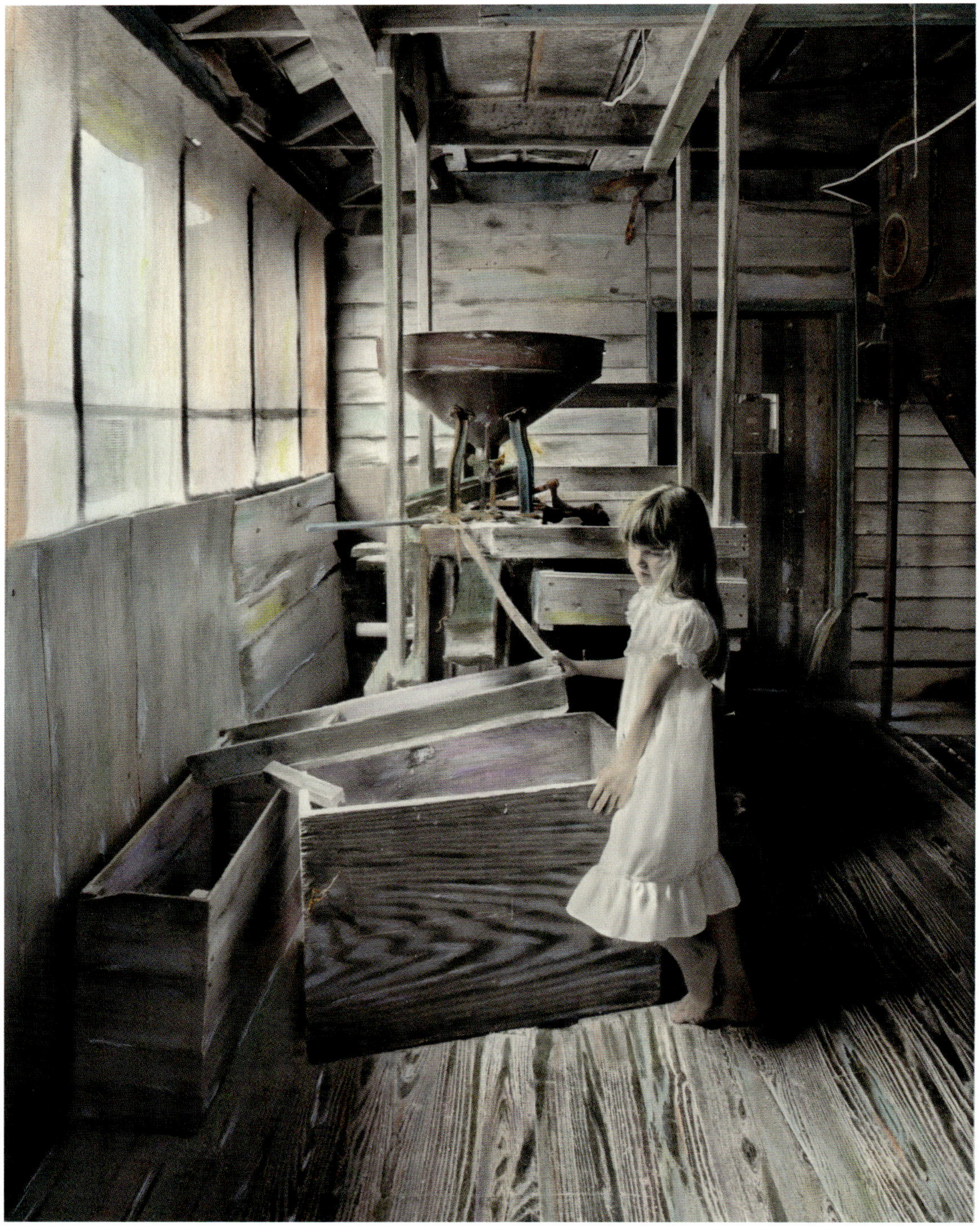

This 1985 image of the author's daughter, Kathy Boyd, at White's Mill shows the general condition of the interior at that time.

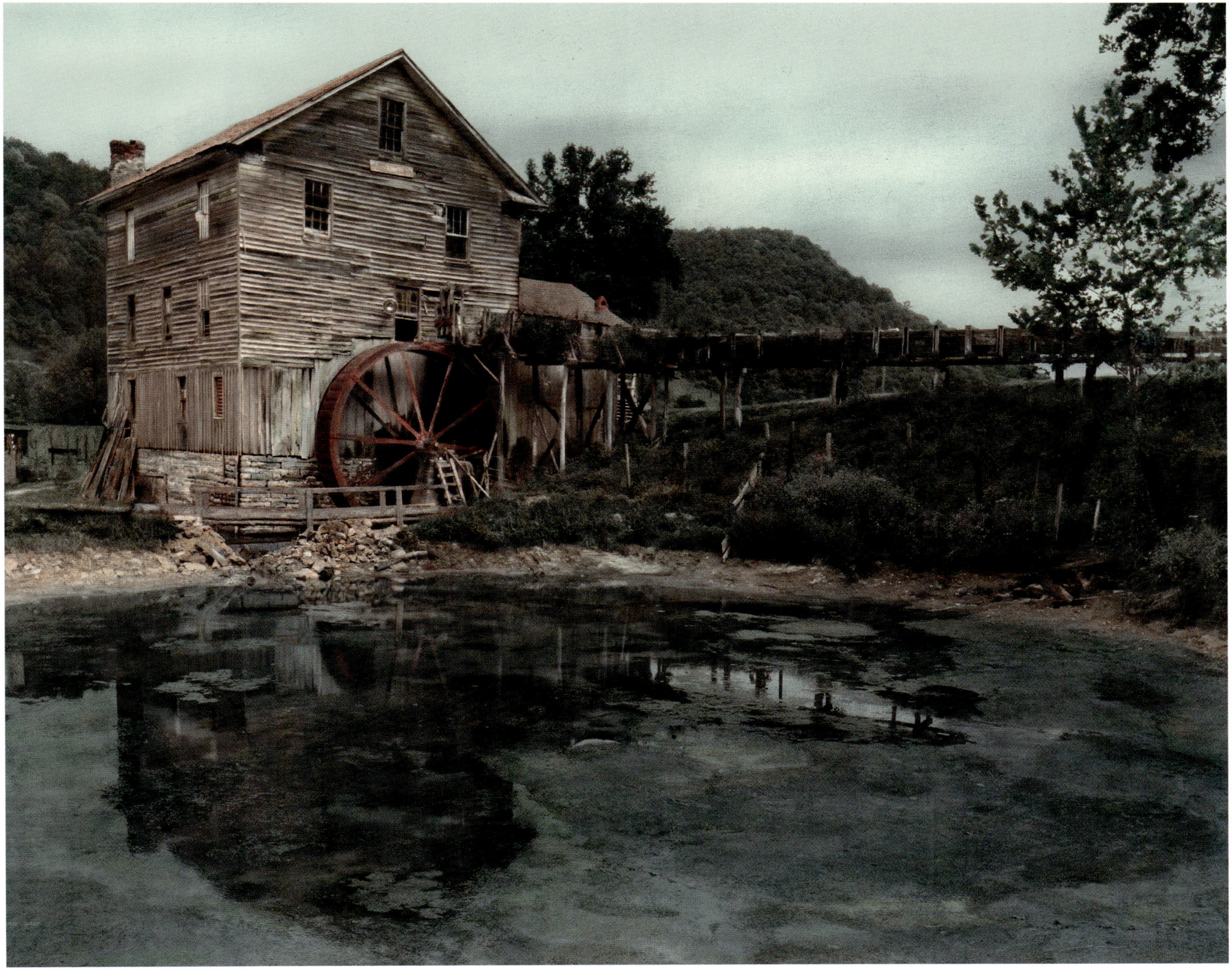

This is a hand-tinted image of White's Mill in 1985, well before the recent restoration work. The mill was operating the day this photograph was taken. The image shows the original, unpainted wooden siding, which some historians prefer to the new white siding.

Cook's Old Mill

Greenville, West Virginia, 1857 / 1893 / Modern Restoration Ongoing

GREENVILLE, WEST VIRGINIA, is classic Appalachia in an area with a colorful past that has now evolved into a beautifully peaceful community. The current mill on Laurel Creek was built in 1857 on the original foundation and site of a mill dating to the late 1700s. The property surrounding this mill makes up an interesting historic park that includes the dam and millpond, an iron forge building, a log cabin, and the miller's home. A number of millstones and assorted pieces of milling equipment are displayed around the site. According to records, the mill was constructed from massive hand-hewn mortise and tenon posts and beams.

Little is known about the earlier mill or its machinery. As for the 1857 mill, a waterwheel was installed when the mill was built, but it was replaced with a double turbine system in the late 1800s. The millstones were swapped for roller mills in about 1893, and roller mills are still in place today.

As with many historic watermill sites, this mill has passed through a number of owners over the years. Three generations of Cooks built, ran, and rebuilt the mill, starting with Valentine Cook, who in 1773 arrived in what was known at the time as the Indian Creek community. The last of the Cook family relocated to Kansas in 1887.

In the closing years of the nineteenth century, the McNeer family took over operations at the mill and installed the twin turbines and the roller mills. Beginning in 1906, a succession of owners ran the mill, until grinding ended in 1964.

Between 1964 and 1987, the milling machinery was removed, and the owner of the building, Harold Labelle, used it as a furniture store. Labelle disappeared in 1972, but it wasn't until 2006 that his remains were identified, confirming he had been murdered. Following his disappearance, his son, Dave Labelle, continued the family business.

In 1987, ownership passed to Jim and Nan Wells, who completed major repairs and restoration at the site. They worked on the millpond and the dam, stabilized the mill foundation, and fixed the roof. They researched and acquired milling machinery from mills across the area, constructed the adjacent metal forge facility, and relocated a log house to the property. The Wells are credited with saving the mill, which they were able to do in part by using resources they obtained from the West Virginia Department of Culture and History.

In 2002, Fred and Barbara Ziegler purchased the site, and through their vision, the mill has become a welcoming showplace for anyone with an interest in watermills and Southern history. The Zieglers have continued the work of the Wells, with further upgrades to the mill, the forge, and the miller's house.

The Laurel Creek Forge structure measures about twenty-four feet by one hundred feet. The Wells completed the forge in the late 1980s with the goal of beginning a blacksmith business, which historically was a fairly common practice at many Southern watermill sites. The forge was set up to include a number of early twentieth-century metalworking and woodworking items. The Zieglers are currently updating the structure with new windows and doors and are introducing more old machinery over time. The Zieglers would like to develop a fully functional blacksmith and cabinetry shop and encourage local artists to demonstrate technologies from one hundred years ago.

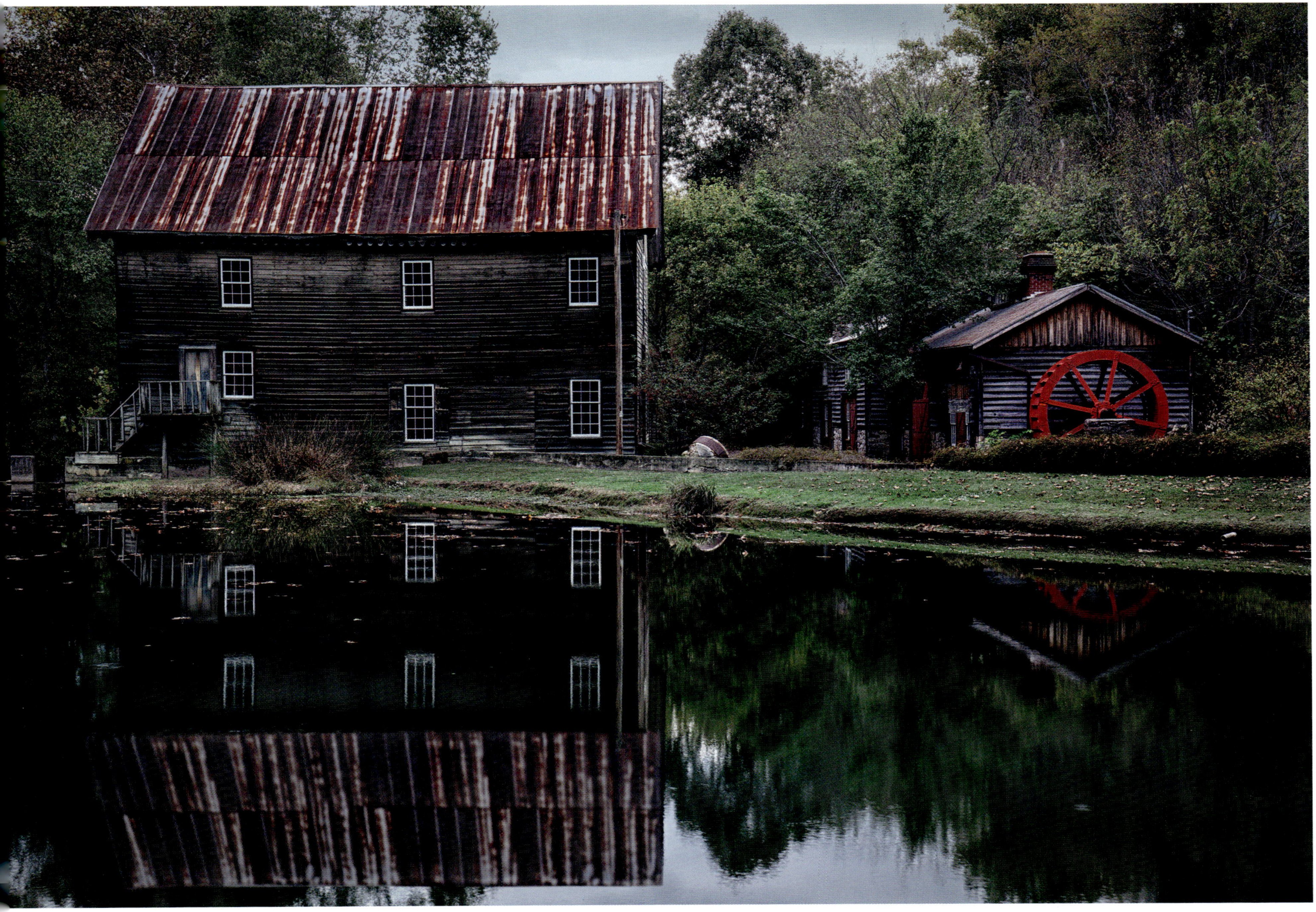

This picturesque, reflective millpond is fed by a large spring about a quarter mile upstream from Cook's Old Mill. The water from the spring meanders through a system of underground caverns before emerging into the pond.

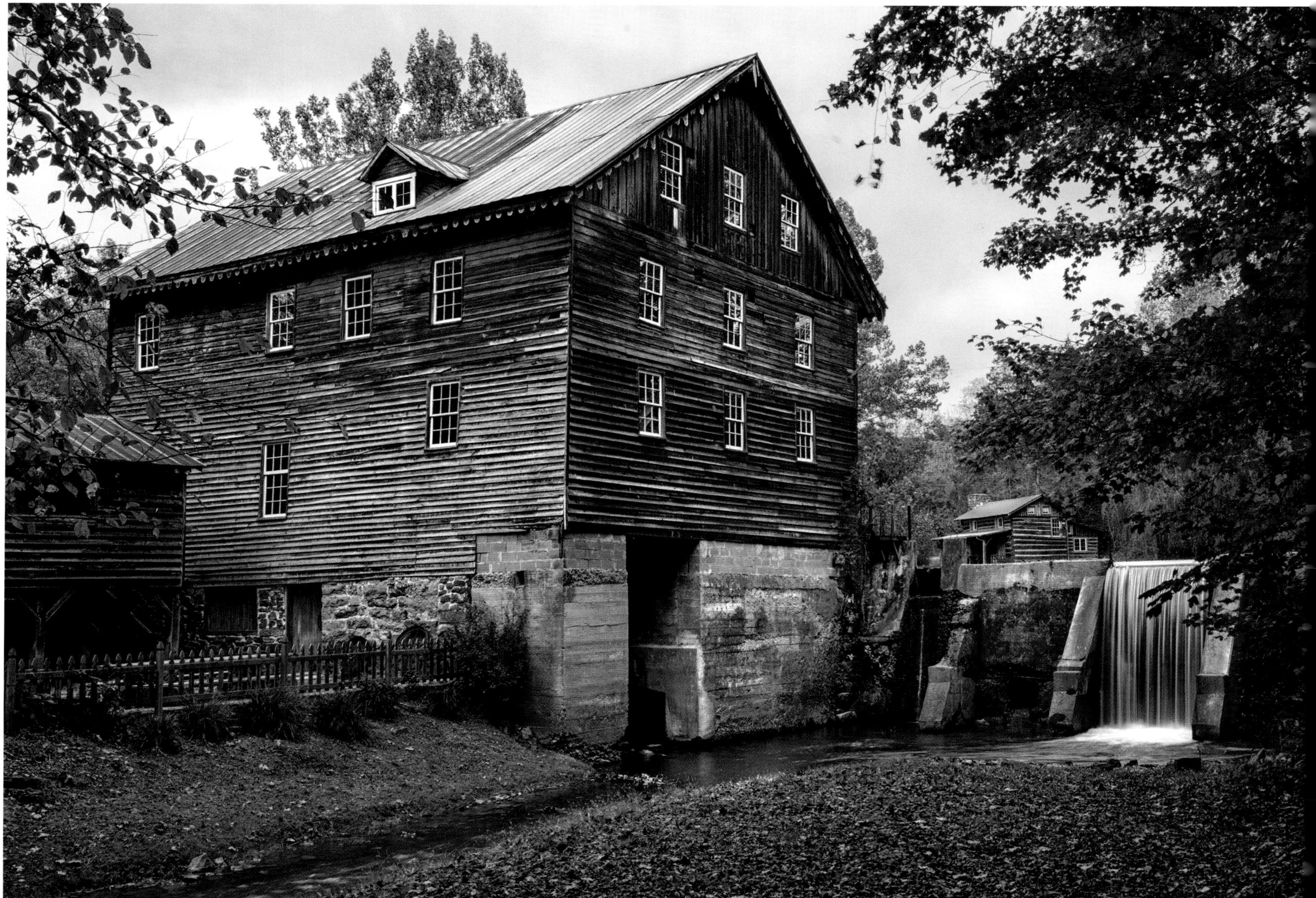

This image highlights the restored mill foundation and the concrete dam. A log cabin can also be seen in the background, across the millpond.

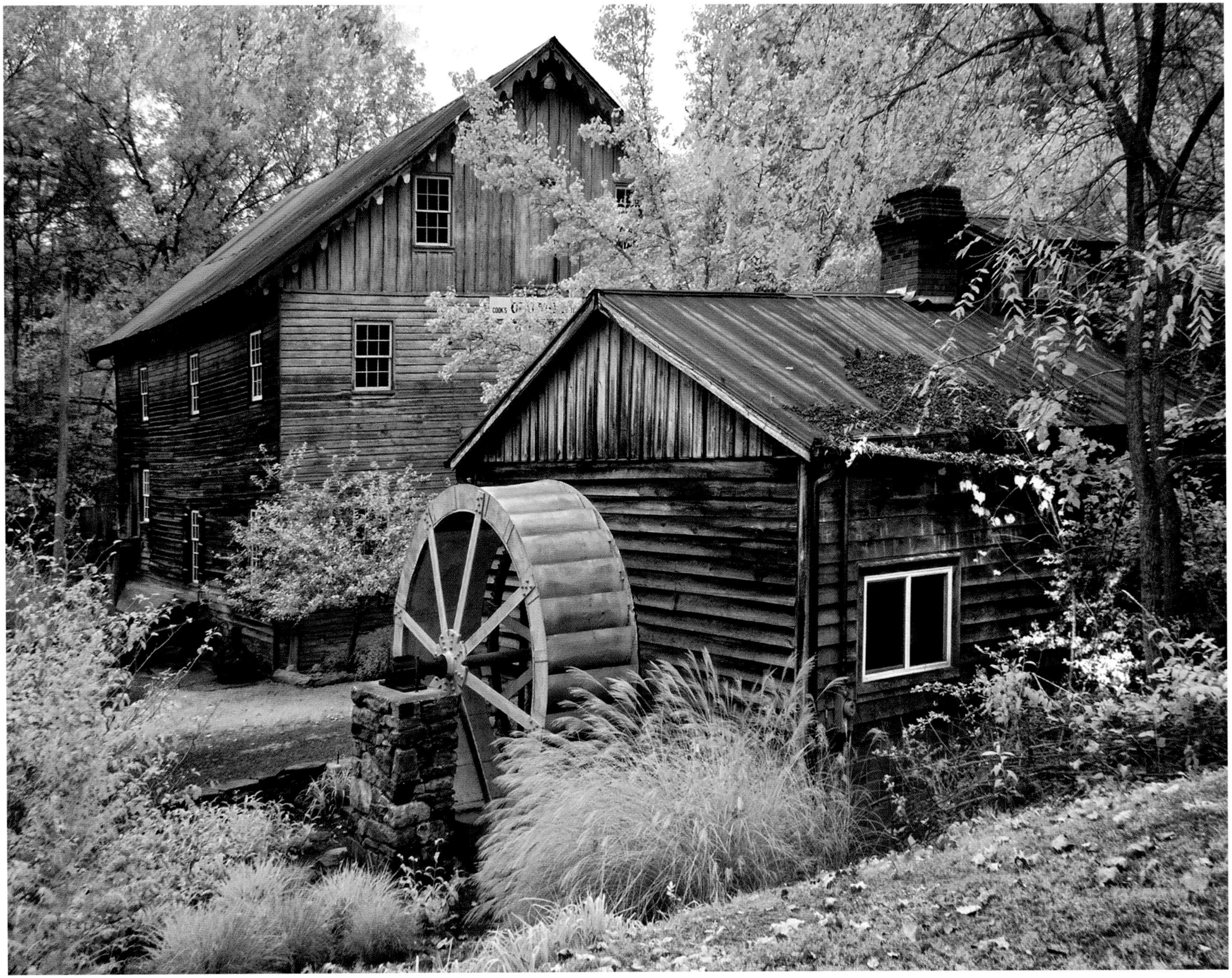

Laurel Creek Forge was added to the site in the late 1980s to attract artists and encourage historic blacksmithing demonstrations. The handsome waterwheel is cosmetic.

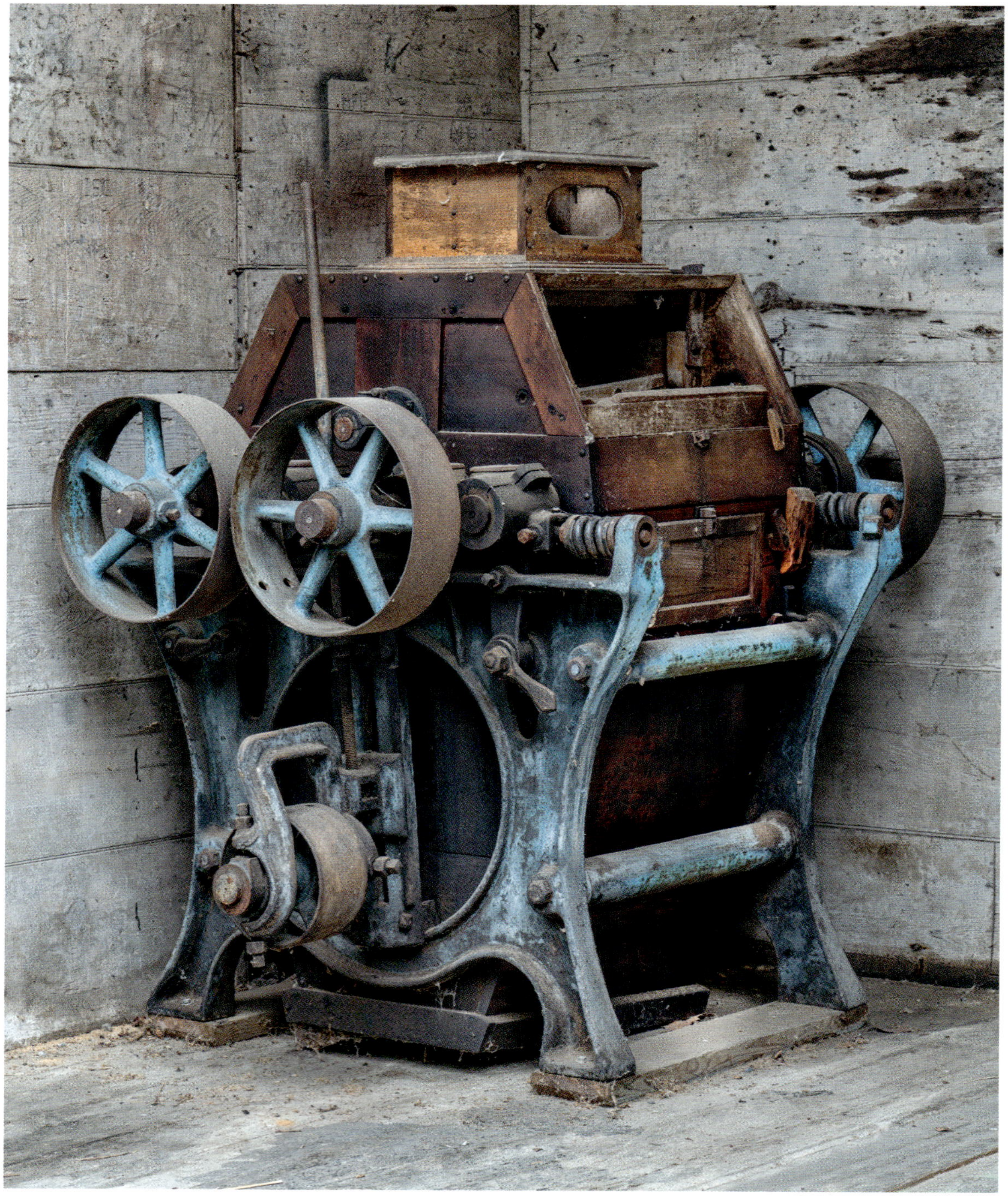

This colorful antique roller mill is prominently displayed at the mill's main entrance.

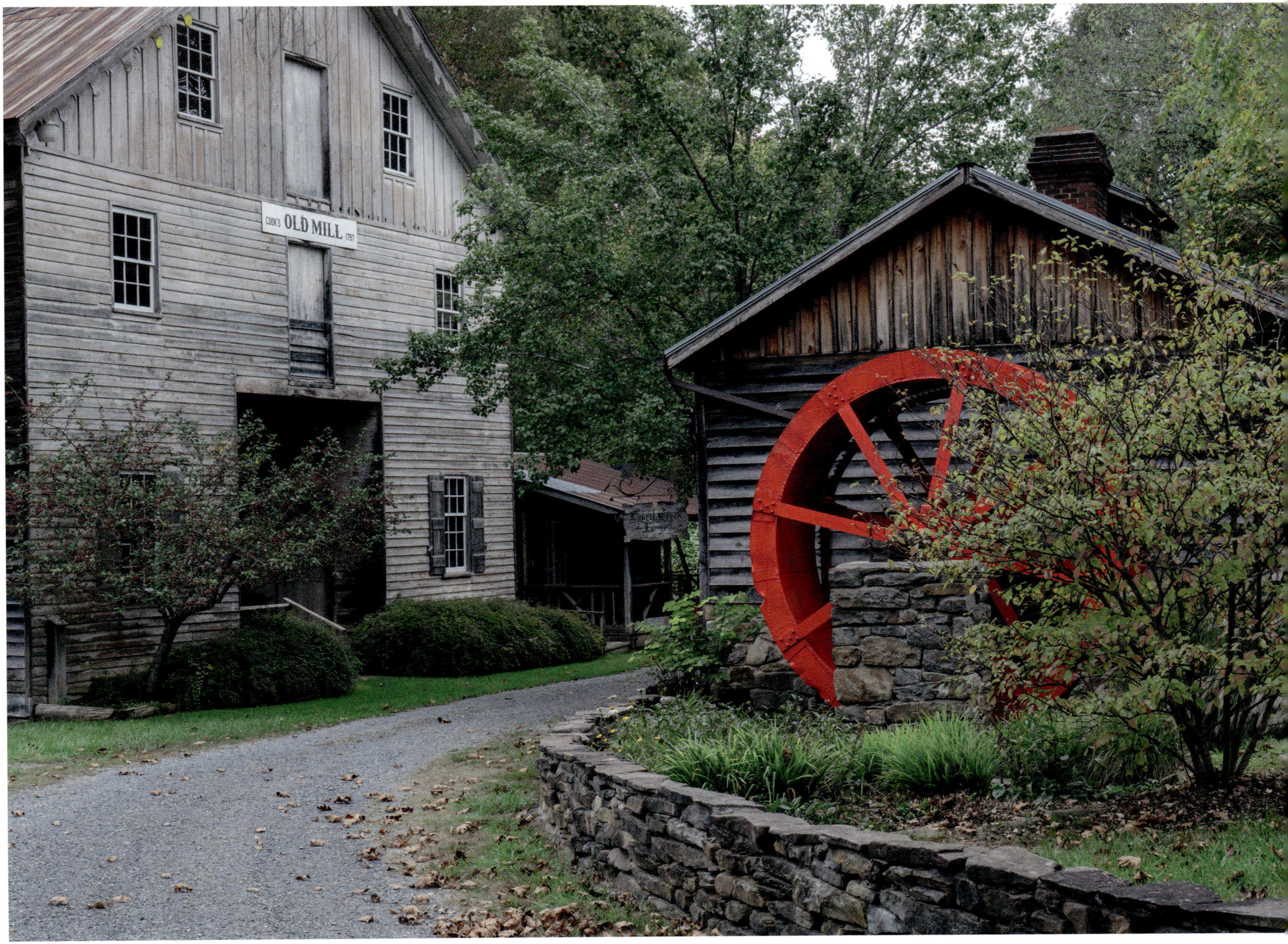

The pathway between Cook's Old Mill and Laurel Creek Forge is part of a popular walking trail through the property and along Laurel Creek.

Fitz-Tuscarora Iron Works

Martinsburg, West Virginia, 1851

THIS HISTORICALLY remarkable mill was built along Tuscarora Creek in 1851 on property that had been known as an oil mill site. The oil mill had operated there since 1843 to press oil from flax seed. Flax seed oil was an important lubricant at the time. Samuel Fitz took over the site from the family of Daniel Kennedy, who had passed at a young age.

Fitz was from Hanover, Pennsylvania, and he and his son and grandson would soon become world-famous for their refinement of the metal overshot waterwheel. Estimates are that at least a few thousand Fitz waterwheels were installed across the South, with seven hundred going to Virginia alone.

The Fitz family already operated a foundry in Hanover that made metal hubs for wooden waterwheels and repaired and restored all kinds of wheels. In Martinsburg, Samuel Fitz fabricated an all-metal waterwheel, which he thought was the first of its kind in the world. (In actuality, a few all-metal wheels were already in use at the time in the United States, and metal wheels had been operating in England and France for a number of years.)

Fitz carefully curved his wheel buckets, sometimes called cups or blades, to optimize performance. A keen marketer, he then advertised the wheels as being 90 percent efficient. History generally regards the Fitz waterwheels as the finest gravity waterwheels ever made.

For most of the last half of the nineteenth century, the Martinsburg and Hanover businesses operated independently, but in 1896, they merged to form the I-X-L Water Wheel Company. (Note that "I-X-L" is pronounced "I excel!") Then in 1902, the company took on the more familiar name of Fitz Water Wheel Company.

The ironworks building in Martinsburg was constructed of limestone and is 140 feet long and 40 feet wide, with two levels that are each 12 feet tall. A large blacksmith shop and an office building became part of the campus. Only the main limestone ironworks mill remains along Tuscarora Creek today. Just beyond the ironworks and in the creek are the remains of several concrete and stone structures from what were at one time a dam, waterwheel underpinnings, and a bridge.

The company continued to operate until just after the Fitz grandson died in 1965, and the Martinsburg mill officially closed in 1967. Other businesses have occupied the building since then, but as of the early 2020s it had been vacant for at least twenty years. At one time, local plans were to demolish the structure, but efforts are now well underway to preserve the mill as part of the Downtown Martinsburg Historic District.

In 2018, Larry DeMarco II and Shawn Wolford purchased the property. They are currently in the process of cleaning and stabilizing the structure. As can be seen in the accompanying image, a new roof has been installed, and windows and doorways have been secured. One goal is to bring a waterwheel back to the site and use it to generate electricity.

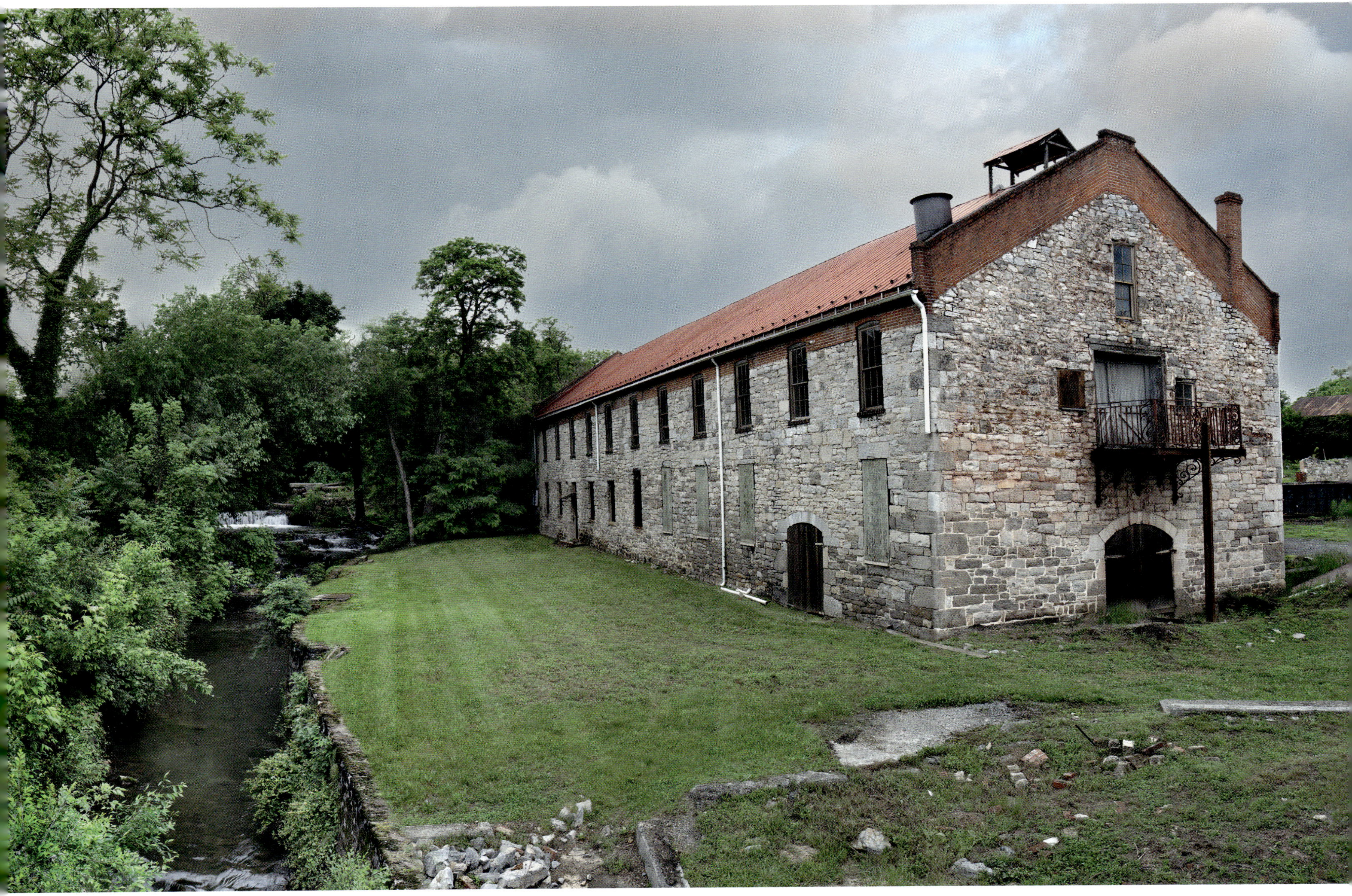

The Fitz-Tuscarora Iron Works remains an impressive limestone structure along Tuscarora Creek in the Downtown Martinsburg Historic District. The exterior has benefited from some cosmetic and structural rework. Plans call for the reestablishment of a waterwheel that would be located to the left and near the rear of the structure.

Glade Creek Grist Mill

Babcock State Park, West Virginia, 1975

THE GLADE CREEK GRIST MILL is located high in the West Virginia mountains at Babcock State Park, about twenty miles from the famous New River Gorge Bridge. The mill is sometimes described as having a put-together or "patchwork" appearance, and for good reason: It was assembled in the park in 1975–76 using parts from three older watermills in the area—Stony Creek Grist Mill, Spring Run Grist Mill, and Onego Grist Mill. In a way, Glade Creek preserves all three of those old mills.

The story is that the Glade Creek mill was fashioned to look something like the old Cooper's Mill that had been located nearby. The mill is set into a picturesque rocky mountainside alongside Glade Creek, just above a series of lovely stone-lined cascades. The mill is one of the most photographed, posted, and printed watermill scenes in the world. Millions have admired its image on notecards and T-shirts and in calendars, books, and framed prints.

People come from as far away as Europe and Asia to see the mill and nearby New River Gorge. Most anyone who visits Babcock State Park must have driven there deliberately, because the location is remote and not always easily found. Still, the consensus among visitors is unanimous—the mill is always worth the drive, no matter the time of year.

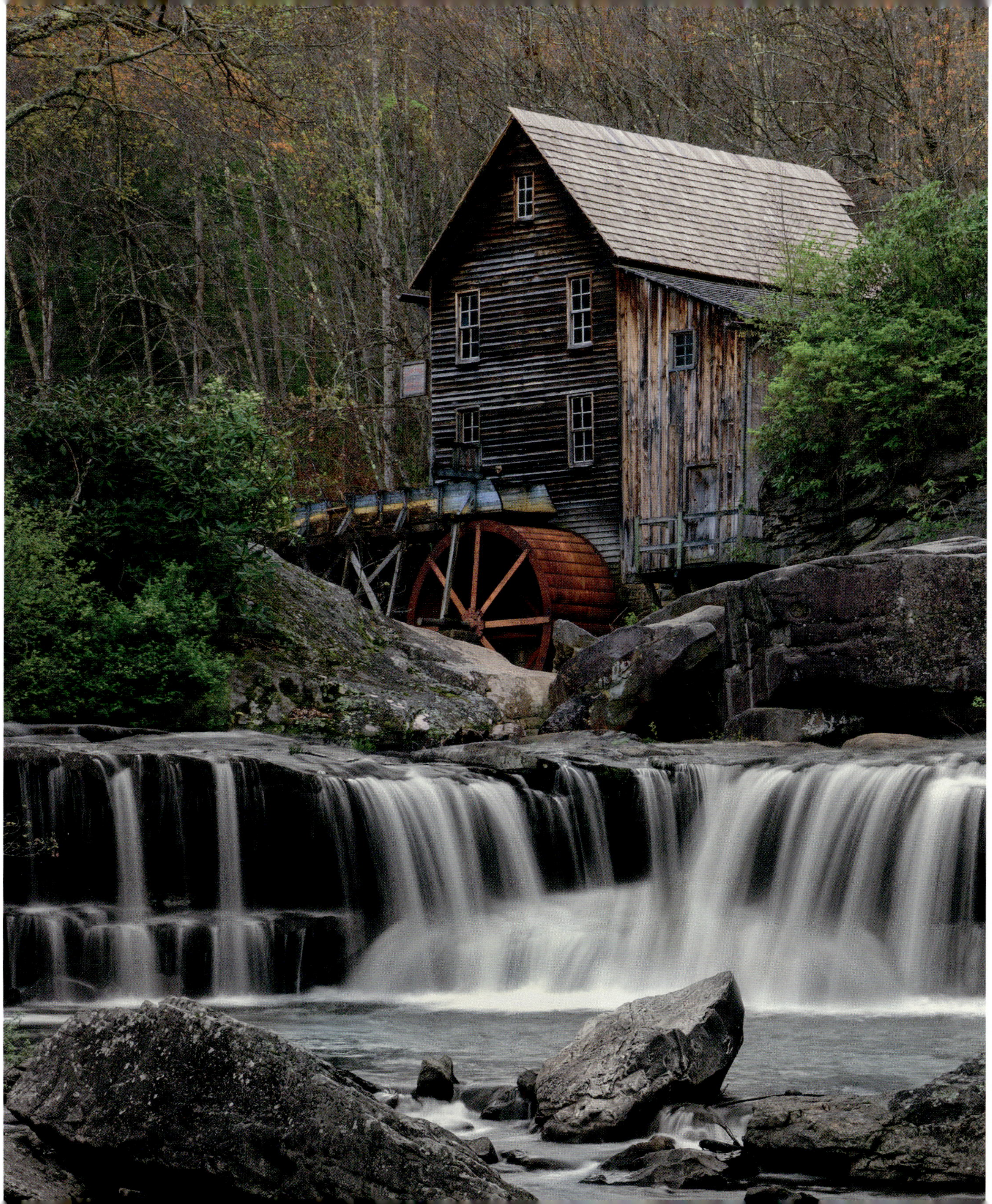

The Glade Creek Grist Mill attracts artists, most notably photographers, from all over the world, especially in the fall, when the leaves change to brilliant reds and golds. As can be seen in this image, the mill is incredibly beautiful in its design and setting.

Jackson's Mill

Weston, West Virginia, 1837

JACKSON'S MILL was the home of Confederate General Thomas "Stonewall" Jackson. Six-year-old Thomas and his younger sister, Laura, came to the mill in 1830 to live with their step-grandmother, Elizabeth Jackson, and their uncle Cummins Jackson due to their mother's declining health (their father had died several years earlier). Thomas stayed at Jackson Mill until 1842, when he left to attend the United States Military Academy at West Point. In the years leading up to the Civil War, the young Jacksons used the site to teach several enslaved residents about reading and writing, an illegal practice at the time.

The mill is situated along the rushing waters of the West Fork River near Freeman's Creek. The first sawmill/gristmill on the site was built in 1809. The 1837 mill is a three-story wooden structure with a stone foundation and a wood shingle roof. The mill is powered by two early turbines underneath it. Much of the interior is still as it was in the years leading up to the Civil War. The site was listed on the National Register of Historic Places in 1972.

The mill was deeded to the State of West Virginia in 1923, and the state has since been responsible for maintaining the site. In recent years, it has often been called the Jackson's Mill Center for Lifelong Learning. It also serves as a state 4-H camp. The campus area around the mill is affiliated with the West Virginia University Extension Service.

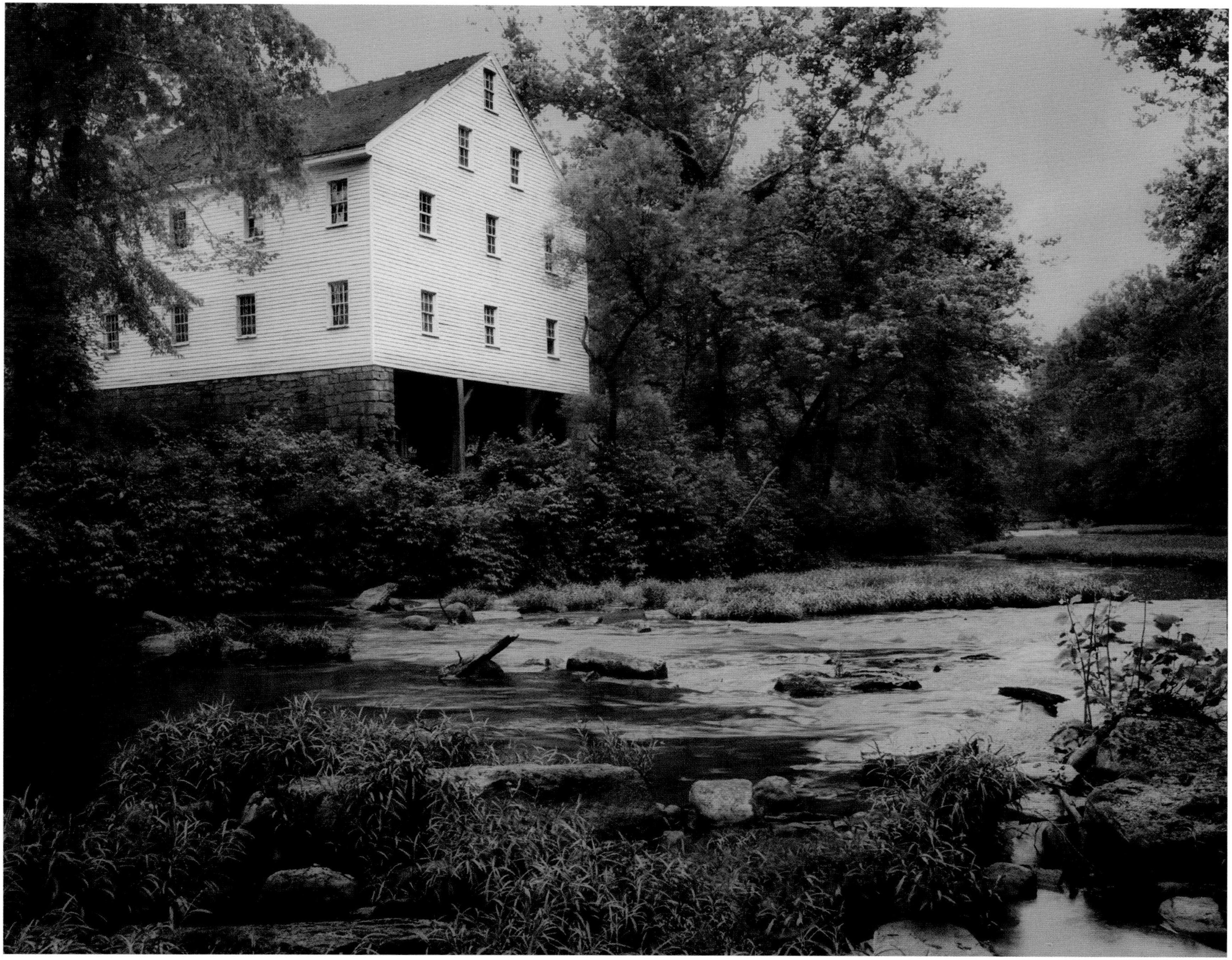

Jackson's Mill is set well above the water level on the west side of the West Fork River, away from the rushing currents that could wash out a mill during high water.

Yellow Spring Mill

Yellow Spring, West Virginia, Circa 1896

YELLOW SPRING MILL in Yellow Spring, West Virginia, is appropriately painted mostly yellow, with bright blue highlights and features. In fact, yellow paint is a popular color for homes and other buildings in this small panhandle community.

The mill ceased operations in 1990 but is cosmetically maintained in good condition, and the landscaping is neat and orderly. The main section of the mill structure is tall, with four stories above ground level. An additional one-story wing was used for storage at one time. In later years, the mill appears to have served as a feed mill and store.

The mill was powered by a large metal waterwheel that is still in place on the back side of the main building. Two millponds are situated above the mill, and a small, picturesque pond or basin is just below the wheel.

An earlier mill likely stood on the site, as some references mention operations as far back as the 1840s, before the time of this mill. Yellow Spring is unincorporated and located along the nearby Cacapon River. The community's name is sometimes written incorrectly as "Yellow Springs," with a final "s," or as "Yellowspring" (i.e., as one word). The entire rural community seems historic and well sustained and without the usual rundown and abandoned buildings.

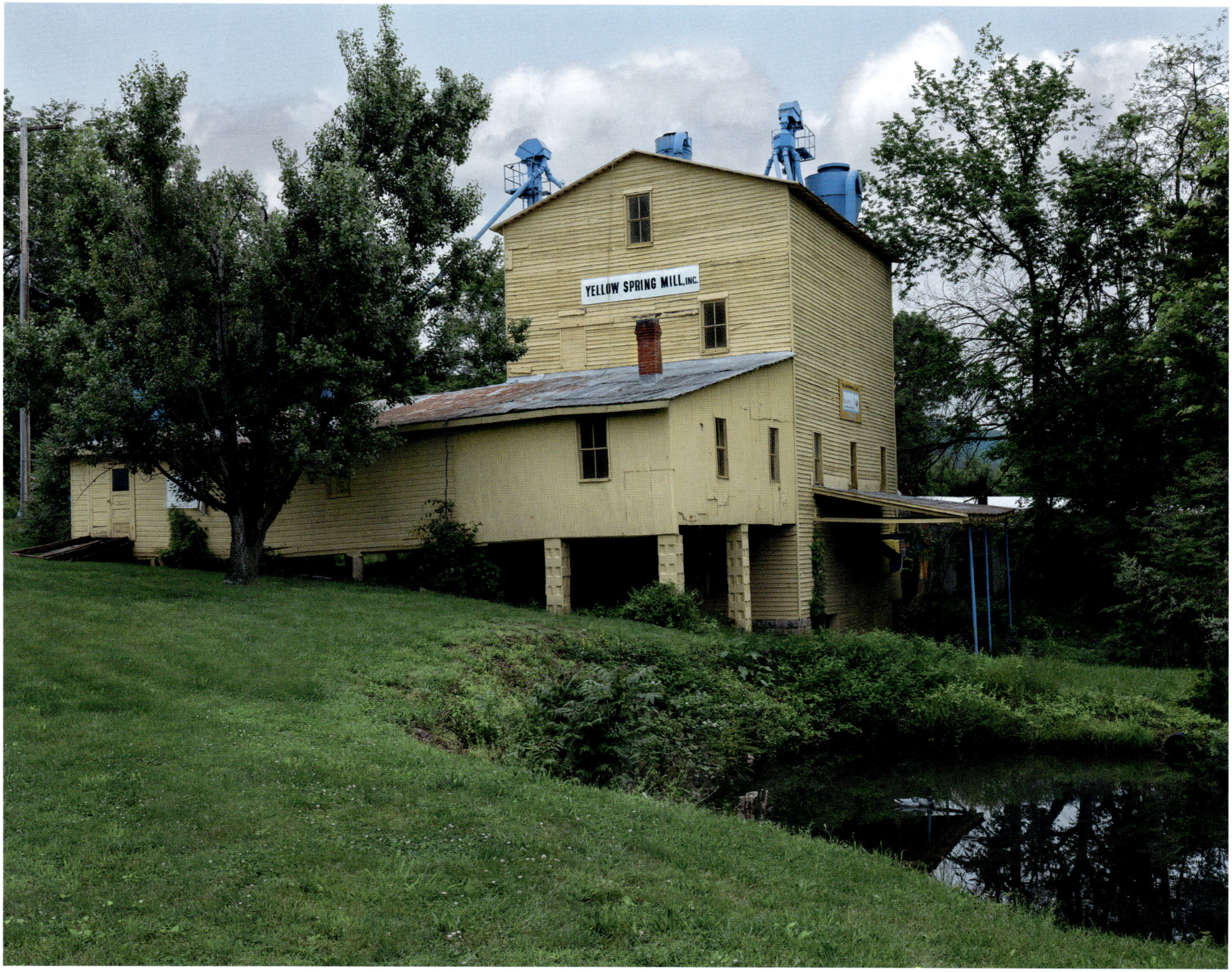

Yellow Spring Mill is unique in its color and design. Interestingly, yellow and blue paint colors and combinations were never popular choices for watermills in the South, or anywhere else for that matter. The large metal waterwheel is just barely visible beneath and beyond the shed at lower right.

McClung's Mill

Zenith, West Virginia, Original Construction Date Not Known

MCCLUNG'S MILL is a lonely and intriguing historic structure in an incredibly beautiful region of West Virginia. Although the original construction date seems to be missing from documentation, H. O. McClung and Homer Miller are known to have relocated the mill from Craig County to the current site in Zenith in the early 1900s. The mill is a one-and-a-half-story wood-framed structure on a concrete-block basement.

McClung operated the mill until the 1960s. For most of the time that he ran it, the mill was used to grind corn and animal feed with a roller mill system. (As clarification for anyone researching "McClung's Mill," from 1871 into the 1920s the McClung family owned and operated another area mill that was also known as McClung's Mill.)

After McClung's retirement, the structure quickly fell into disrepair, and within a decade the mill was in peril of being lost. Alarmed by the condition of the property, the Monroe County Historical Society leased the mill from McClung in 1971 and spent about $6,000 plus many hours of volunteer work to construct a new foundation beneath the mill, restore the massive twenty-six-foot overshot waterwheel, and rework the drainage at the mill and dam and around the property.

In 1973, the waterwheel turned again for the first time in over a decade, driven by water from Dropping Lick Creek. The creek originates as a spring from a cave just above the mill and flows through a large pipe that serves as a flume to the overshot wheel.

The mill is not in operation currently, but the site is still well maintained and includes one small outbuilding. Although well off the beaten path for most travelers, Zenith is a neat, quiet little community along a one-lane asphalt road. The area has the appearance and feel of a ghost town, with only a few residents and a number of deserted buildings, including a volunteer fire station, a gas station, and a general merchandise store. There is also the usual array of old vehicles, a common sight along rural roads.

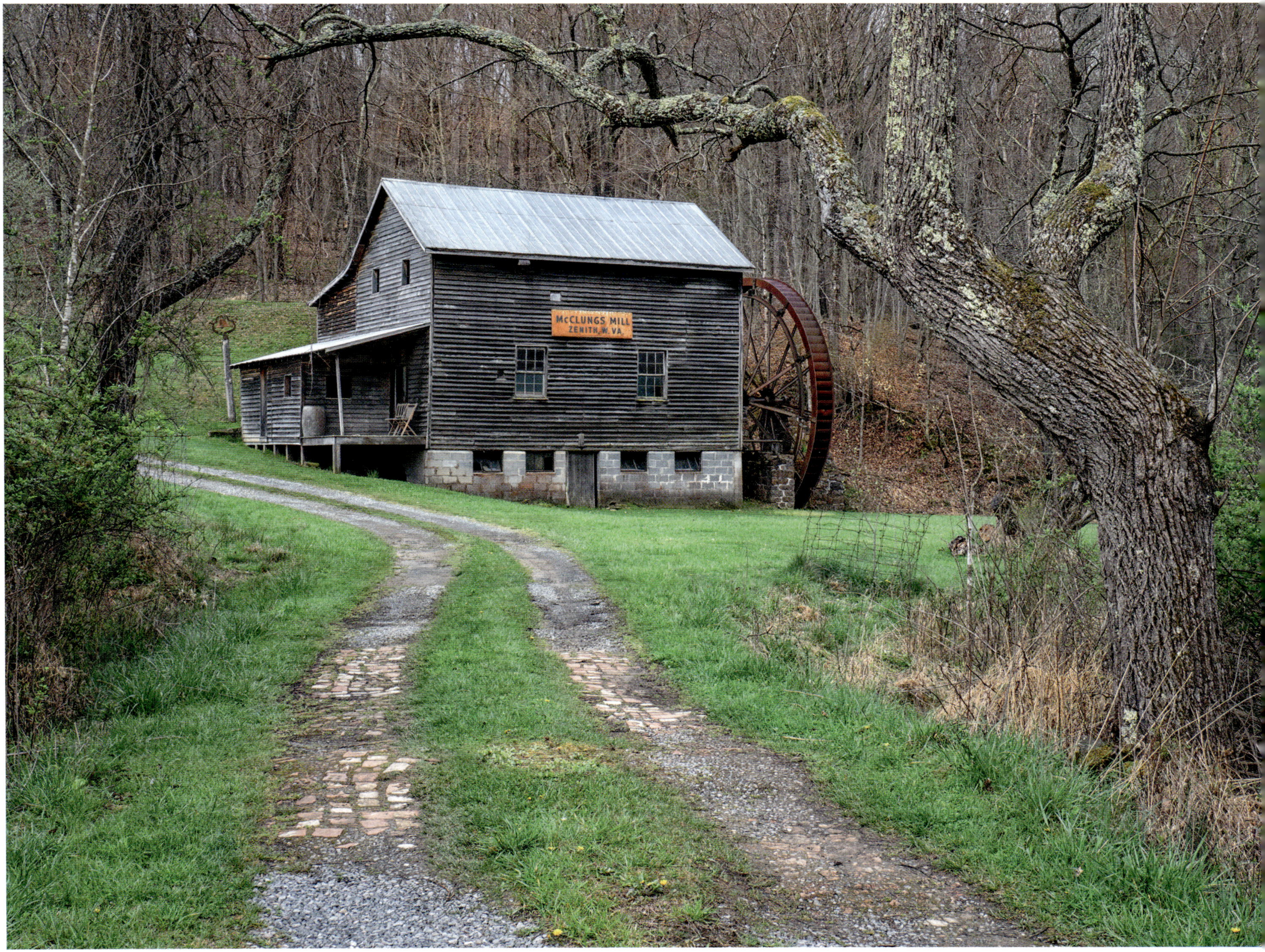

McClung's Mill is located in Zenith, West Virginia, in an isolated area of the picturesque Appalachian Mountains.

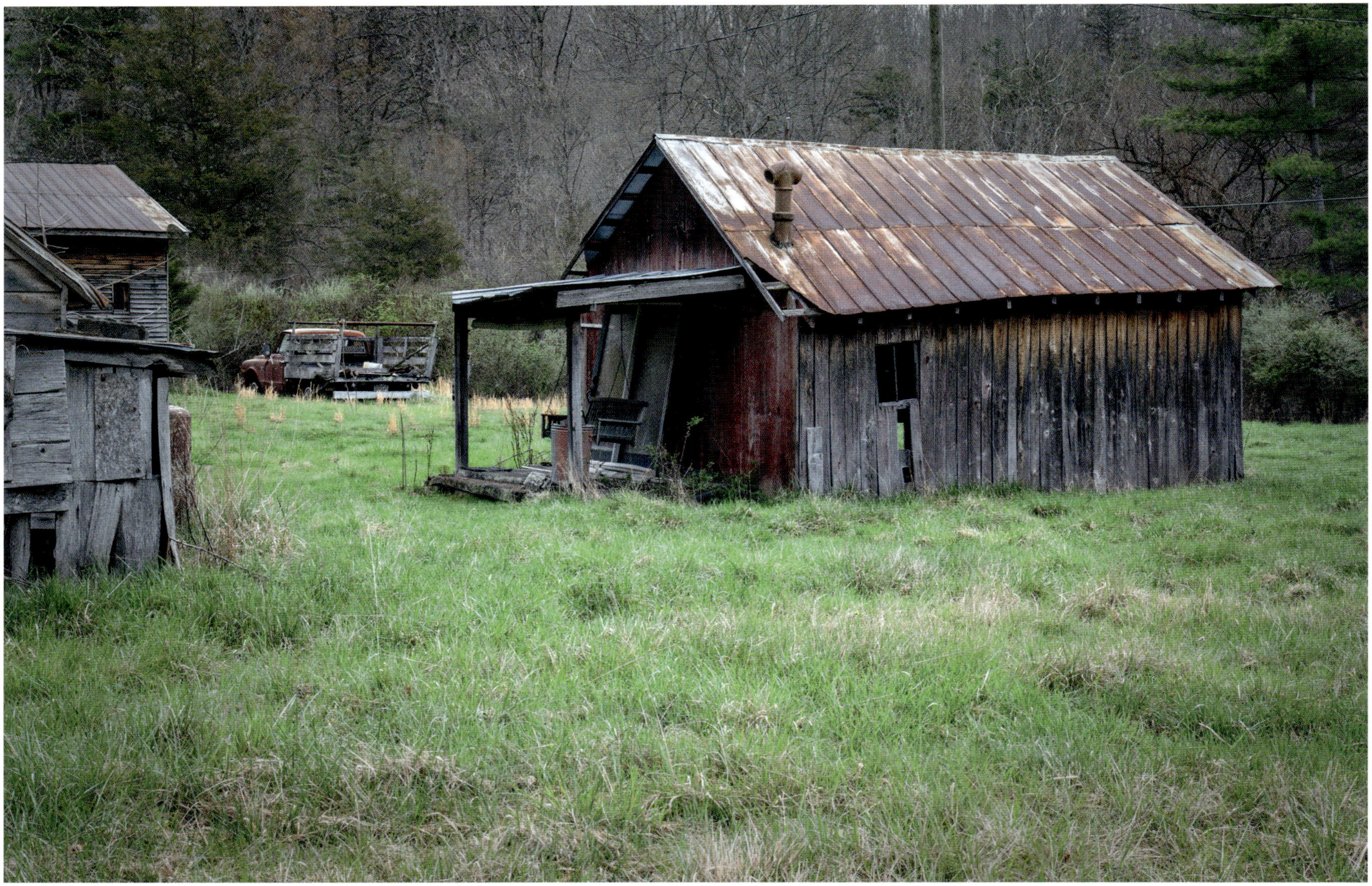

The town of Zenith is beautifully scenic, but the area is sparsely inhabited and has the look and feel of a ghost town. The town consists mostly of empty structures and abandoned vehicles.

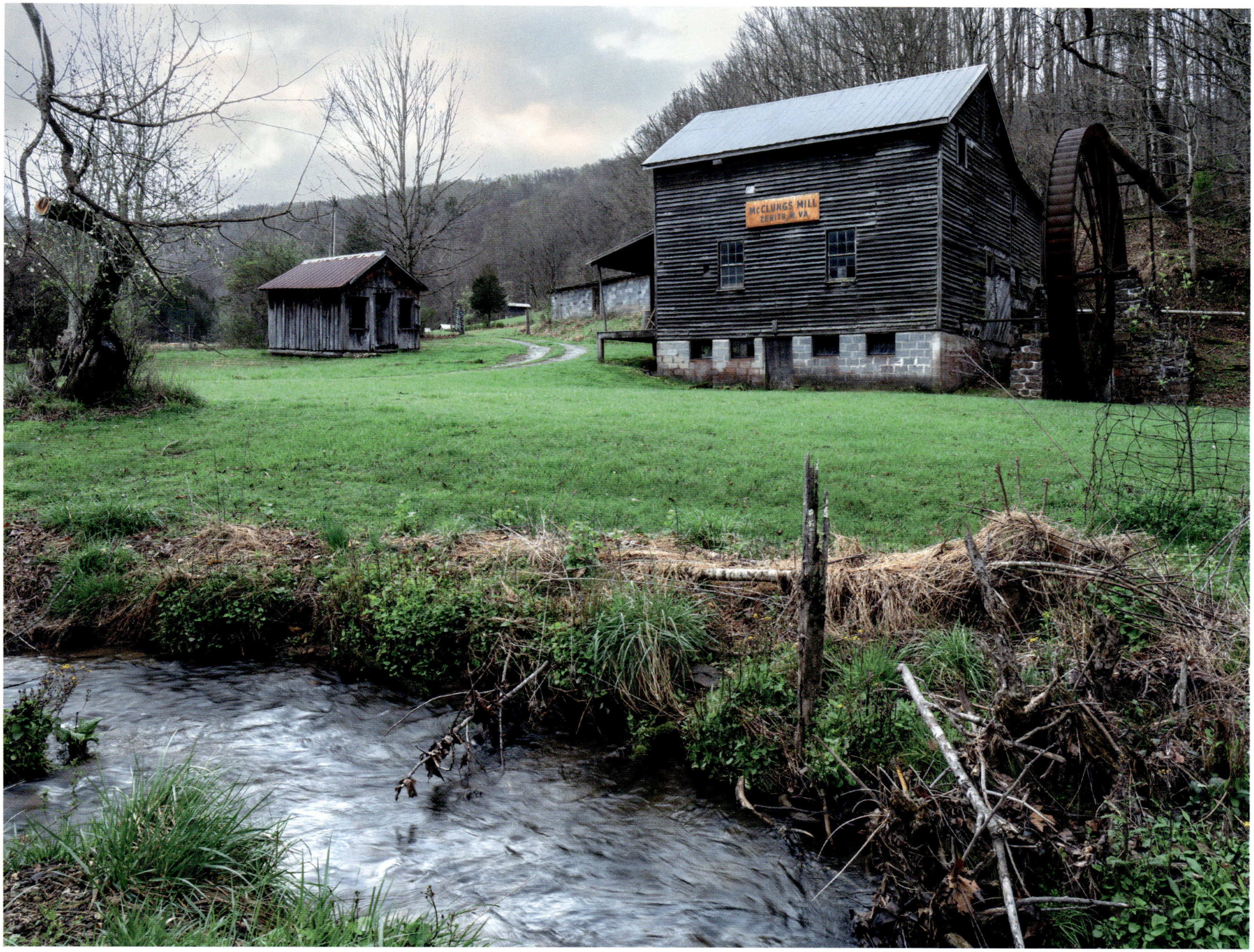

The setting around McClung's Mill is quiet and tranquil. The site is well maintained, and both the mill and the adjacent outbuilding seem in good repair.

Bunker Hill Flour Mill

Berkeley County, West Virginia, Circa 1800 / 1889

BY THE MID-NINETEENTH CENTURY, as many as thirteen watermills, including Bunker Hill Flour Mill, were churning the waters of Mill Creek in northeast West Virginia, just south of Martinsburg. Some of these mills had been there for more than a century by the time the Civil War broke out; tragically, most were destroyed during the conflict.

Fortunately, Bunker Hill Flour Mill survived the war and is still with us today. This is a tall, classically styled, and imposing white limestone block structure. The main floors of the mill cover three stories, with an additional one-and-a-half stories in the loft area of the high gabled roof. The first mill on the site dated to about 1735. Some references show that the current mill was constructed circa 1800, with a major rebuild being completed in 1889 after a fire gutted the interior two years earlier.

Bunker Hill was a large commercial-scale mill with a pair of low and wide tandem waterwheels mounted together on the back side of the building. The smaller wheel is said to be just under fifteen feet in diameter, with the other wheel slightly larger.

The mill mostly ground wheat flour, along with some corn, rye, barley, and animal feeds. Records show that the mill did not run at its full rated capacity very often. It could grind three hundred bushels of wheat in a day, but the output had to be reground up to three times to get the desired product quality!

As happened at a number of Southern mills, a diesel engine was added in 1920 to supplement grinding in periods of dry weather. In later years, a Midget Marvel roller mill was brought in. The Midget Marvel was a self-contained machine that handled the entire milling process in a single unit.

Although this was a commercial mill, the regrinding and the lack of convenient railroad access created ongoing financial difficulties, and the mill changed owners several times before ceasing commercial operations in 1964. In the 1980s, a feed business began operating from the mill building. That business closed in 1995.

Today the mill is part of the Mill Creek Historic District along the creek valley in the panhandle of eastern West Virginia. The mill's interior houses an interesting assortment of nineteenth- and twentieth-century watermill equipment. Most of the machinery is thought to still be operational. Of the thirteen mills that were once located on Mill Creek, only Bunker Hill Flour Mill remains.

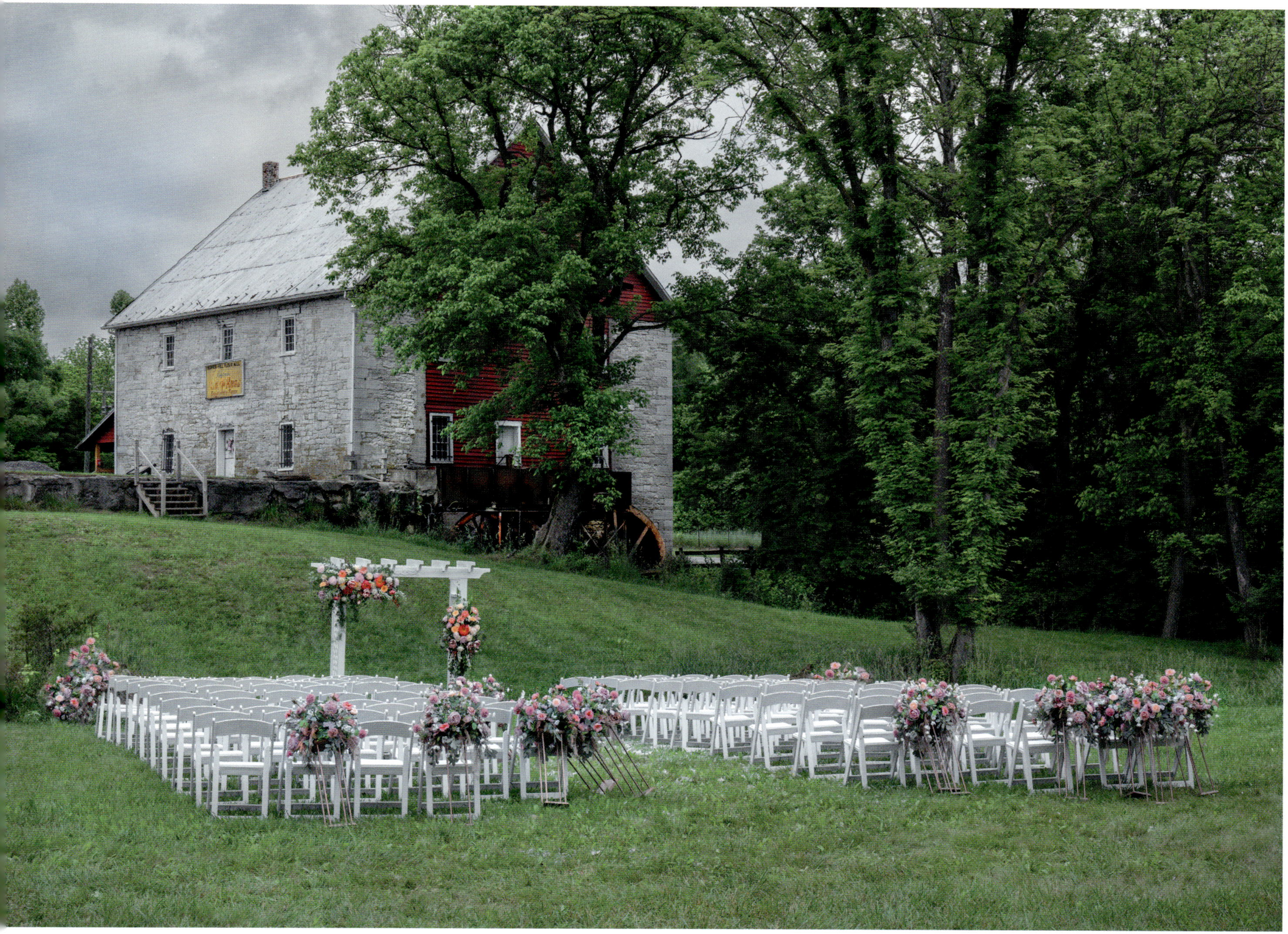

Bunker Hill Flour Mill has become a popular venue for weddings, receptions, and other special events, something it has in common with a number of other watermill sites. The income from hosting these events provides capital to help maintain the property.

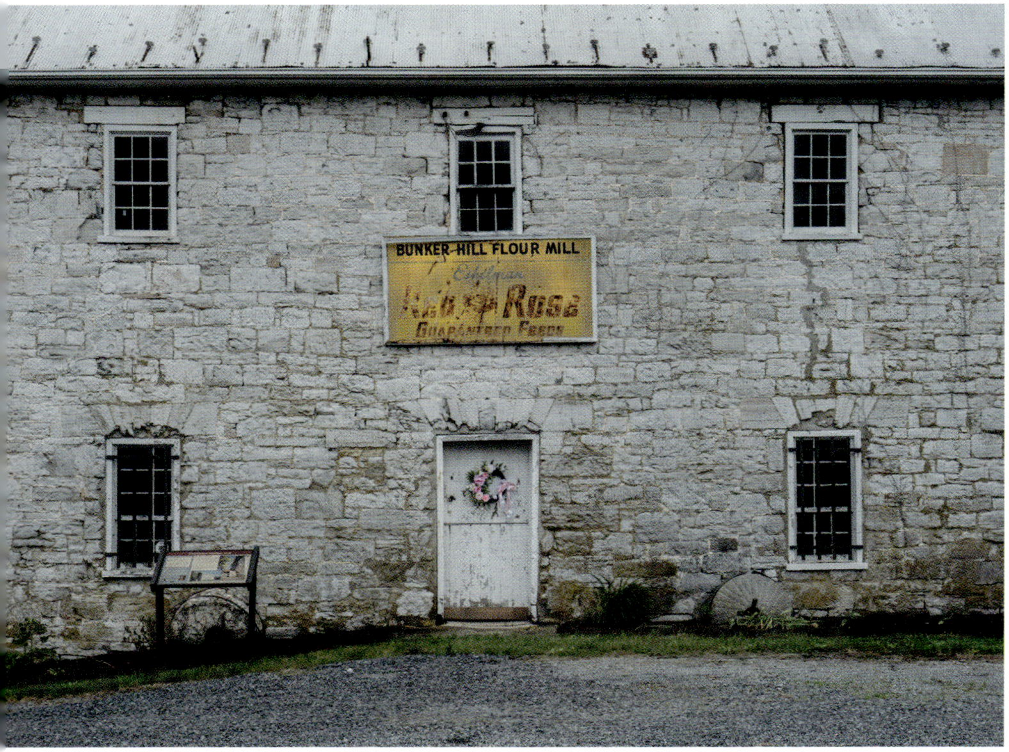

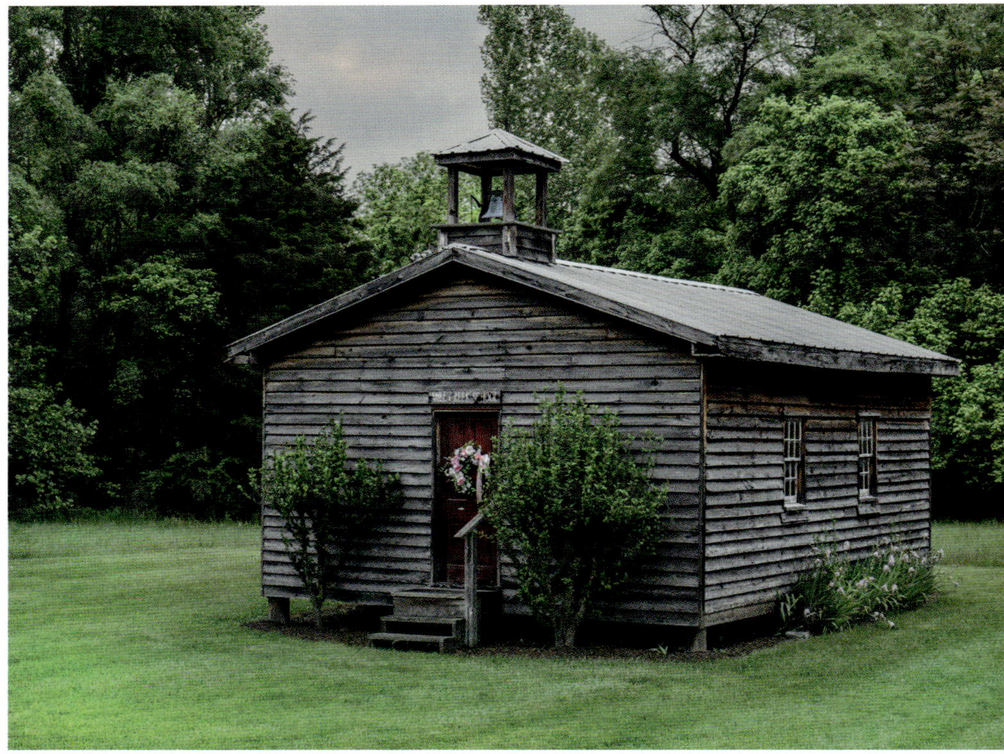

The yellow sign is for Eshelman Red Rose, a popular brand of livestock feed across the region. The brand was founded in 1842 and continued until the business was sold in the mid-1970s. The trademark continued to be used for a number of years afterward.

The little schoolhouse in the Mill Creek Historic District sits just across the road from the watermill. The building can serve as a prep and support area for weddings and other events.

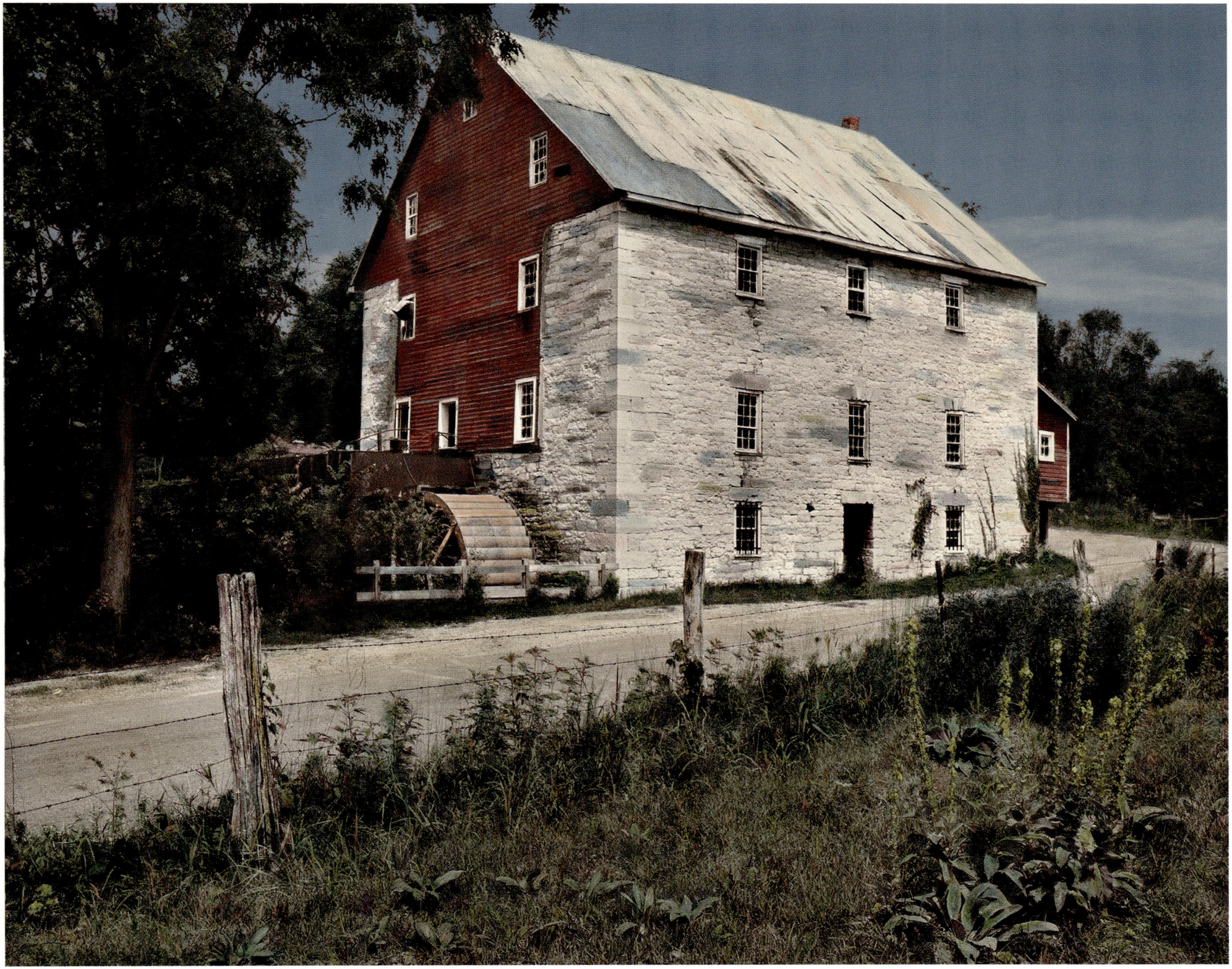

Bunker Hill Flour Mill features tandem, or dual, steel waterwheels, but only one waterwheel is visible in this vintage hand-tinted image. The mill is thought to incorporate some elements of a 1730s mill that predated the current mill at the site.

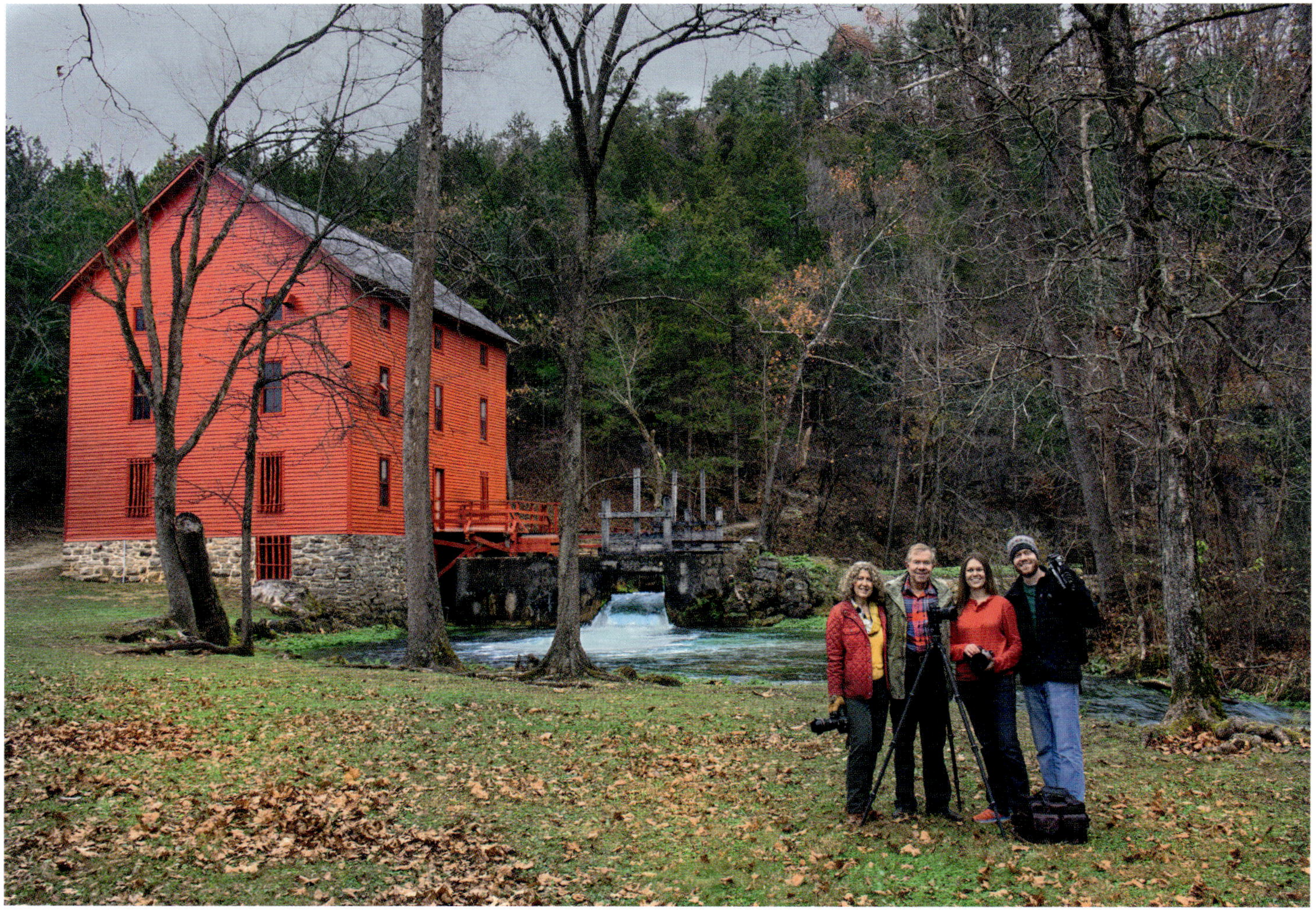

The temperature was in the mid-20s when the author and his family posed for this self-portrait at Alley Spring Mill. From left, Dori, Ken, Kathy, and Kevin Boyd.

Acknowledgments

A book such as *Historic Watermills of the South* represents a family commitment. I would first like to thank my wife, Dori, for her years of encouragement, patience, and perspective on this project. She has traveled with me, reviewed and critiqued images, and guided me through much of the planning. My children, Kathy and Kevin, have been a part of this project for years, tagging along to mill sites as children, helping out in the darkroom, and offering suggestions on the digital files.

Of special note, my longtime friend Carl Schinasi must be recognized for his invaluable guidance, editorial reviews, and expertise. Carl has been with me on all my book projects, and I am not sure any of them would have been possible without his contributions.

The author wishes to acknowledge the historic and technical expertise and the thoughtful comments and suggestions of Dr. John Lovett and Chuck Ketchie. Their passion for the South's old watermills is a true delight.

In addition, Arthur Johnson must be recognized for the untold hours he expended in his thorough review of the text to make it more concise, accurate, and consistent—thank you.

The book also required that a publisher see value in the project. From the outset, Dr. Nathaniel Holly had the vision for a project that would preserve the history of the watermill in the South. Jon Davies joined Dr. Holly as the project editor to carefully guide the preparation of this book. Now their vision can be shared with future generations.

Selected Bibliography

Beals, Fred, ed. Mill Lists. Society for the Preservation of Old Mills, late 1970s–2018.

Boyd, Ken. *Historic Watermills of North America: A Visual Preservation*. University of Alabama Press, 2020.

Boyd, Ken. "Kymulga Grist Mill: An Operating Mid-Nineteenth-Century Alabama Watermill." *Alabama Heritage*, no. 138 (Fall 2020): 32–43.

Brown, Jonathan. *Water Power and Watermills: An Historical Guide*. Crowood Press, 2011.

Dryden, Steve. *Peirce Mill: Two Hundred Years in the Nation's Capital*. Bergamot Books, 2009.

Ferguson, Eugene S. *Oliver Evans: Inventive Genius of the American Industrial Revolution*. Hagley Museum, 1980.

Fowler, Harriet. "Circling the Past." *Keeneland*, Winter 2016.

Fox, William. *The Mill*. New York Graphic Society, 1976.

Hasse, Larry. "Watermills in the South: Rural Institutions Working Against Modernization." *Agricultural History* 58, no. 3 (1984).

Hawkins, Leslie Erin. "'I Am History, Don't Destroy Please': Three Gristmills and Their Communities in Wake County, North Carolina." Master's thesis, North Carolina State University, 2008.

Hazen, Theodore R. "The Preservation of Historic Mills, Preservation in Brief." Pond Lily Mill Restorations, 2001. https://www.angelfire.com/journal/millrestoration/historicmills.html#anchor374190.

Hobbs, Alexis. "25 of the Most Beautiful Old Mills in America." *Country Living*, February 10, 2016. https://www.countryliving.com/life/travel/g3023/grist-mills.

Hodge, Nathanael. "The Decline of Wind and Water Mills." Mills Archive, Mills Archive Trust, June 3, 2016. https://new.millsarchive.org/2016/06/03/the-decline-of-wind-and-water-mills/.

Howard, Robert A. *Water Power: How It Works*. Eleutherian Mills–Hagley Foundation, 1979.

Howell, Charles, and Allan Keller. *The Mill at Philipsburg Manor, Upper Mills and a Brief History of Milling*. Sleepy Hollow Restorations, 1977.

Lakwete, Angela. "Antebellum Industrialization." *Encyclopedia of Alabama*. Last updated May 8, 2023. https://encyclopediaofalabama.org/article/antebellum-industrialization/.

Larkin, David. *Mill: The History and Future of Naturally Powered Buildings*. Universe, 2000.

Macaulay, David. *Mill*. Houghton Mifflin, 1983.

McLaughlin, Bob, and Dale Cook, eds. "Mill Lists with Sites." Document Library, Society for the Preservation of Old Mills. https://spoom.org/content.aspx?page_id=86&club_id=664666&item_id=60416.

Mill Pictures and Historic Information. https://millpictures.com/index.php.

Old Mill News. Published quarterly by the Society for the Preservation of Old Mills (SPOOM); primarily issues from the 1980s. https://spoom.clubexpress.com/content.aspx?page_id=22&club_id=664666&module_id=299533.

Parker, Barry, ed. "The Magic of Mills." *Country Highways* 1, no. 1 (1989).

Sass, Jon A. *The Versatile Millstone: Workhorse of Many Industries*. Society for the Preservation of Old Mills, 1984.

Sloane, Eric. "The Mills of Early America." *American Heritage* 6, no. 6 (October 1955).

Smith, Elmer L. *Grist Mills of Early America and Today*. Applied Arts Publishers, 1978.

Steelhammer, Rick. "Retired Teacher Documents WV's Water-Powered Mills." *Charleston Gazette-Mail*, May 25, 2017. https://www.wvgazettemail.com/life/retired-teacher-documents-wv-s-water-powered-mills/article_9474dc58-db51-5d7c-b3c1-585d93e3b3db.html.

Stephenson, Mary A. *Mills in Eighteenth Century Virginia, with Special Study of Mills near Williamsburg*. Colonial Williamsburg Foundation Library Research Report Series, no. 116. Colonial Williamsburg Foundation Library, July 1947. https://research.colonialwilliamsburg.org/DigitalLibrary/view/index.cfm?doc=ResearchReports%5CRR0116.xml&highlight.

Walton, Steven A. "Geography, Landscape and Mills." Medieval Technology and American History, In-Depth Articles, Penn State University. 2012. https://www.engr.psu.edu/mtah/articles/geography_landscape_mills.htm.

Washburn, Frank S. "The Power Resources of the South." *Annals of the American Academy of Political and Social Science* 35, no. 1 (January 1910): 81–98. https://www.jstor.org/stable/1011493.

Wells, Don, and Diane Wells. *The Grist Mills of the Mountain Communities*. Mountain Stewards, 2011, updated 2024. https://mountainstewards.org/wp-content/uploads/2020/02/The-Grist-Mills-of-the-Mountain-Communities-Complete.pdf.

Wikipedia. "List of Watermills in the United States." https://en.wikipedia.org/wiki/List_of_watermills_in_the_United_States.

Index

About the Author

Ken Boyd has been a passionate, enthusiastic, and award-winning photographic artist, researcher, and writer since the 1970s. Over the past half century, he has visited more than three hundred watermill sites across the United States, Canada, and Europe, with multiple visits to a number of these locations. His travels have taken him to beautiful and picturesque settings, rural farming areas, remote mountainous regions, deep forests, coastal wetlands, and even historic urban industrial sites.

Through the years, Ken's photographic process has evolved. It is often described as painterly and as having the quality of fine art. Many of his images are often mistaken for oil or watercolor paintings. His approach to photography involves a careful use of his different cameras. He maintains a keen vision for lighting and composition along with previsualizing the processing of his final images. His goal is always to capture a simple, clean, and orderly presentation for each watermill.

He has previously published several books and numerous feature articles on historic subjects. His most recent book before this one is *Historic Watermills of North America: A Visual Preservation* (University of Alabama Press, 2020).

In addition, Ken's books on historic locomotives have been distributed internationally for a number of years, and he is well known as an author on this subject. He currently teaches photography courses at Samford University. He also taught at the University of Alabama at Birmingham from 1985 until 2003 on topics ranging from traditional darkroom methods and vintage nineteenth-century photographic processes to creative digital techniques. Ken also leads tours to watermills and other historic sites.

Ken is a graduate of Jacksonville State University and the University of Alabama at Birmingham. He is retired from a career in environmental research with Southern Company. He and his wife, Dori, live in Hoover, Alabama, and enjoy travel, historic studies, and art.

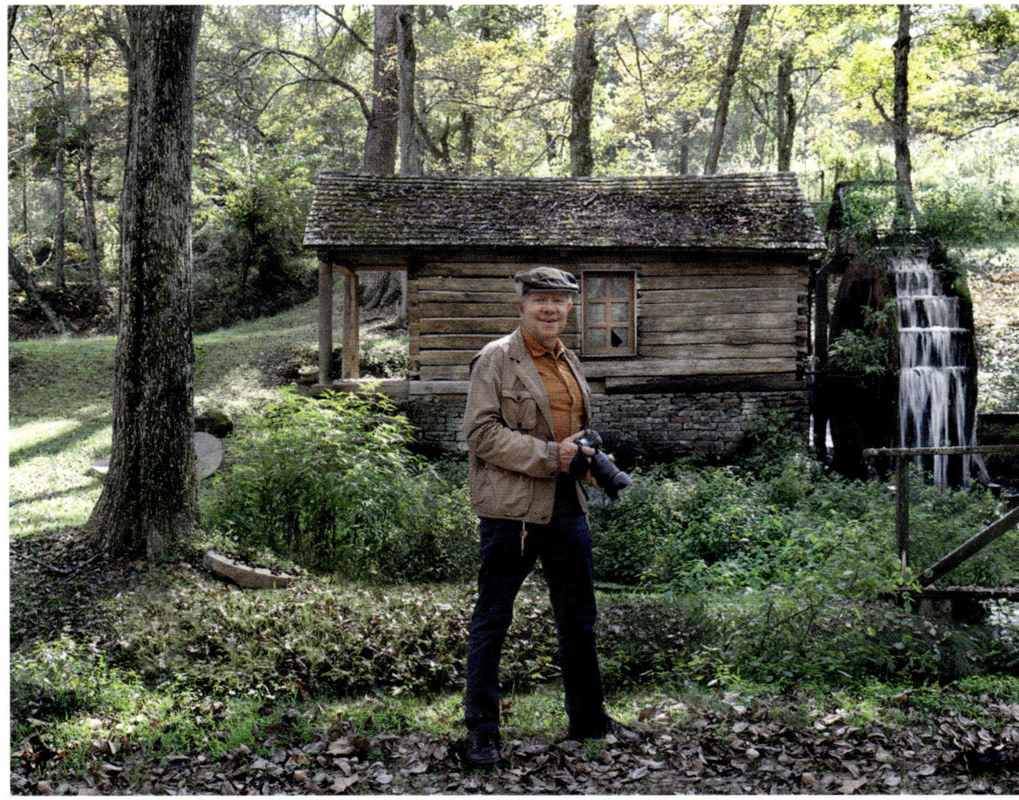

Author self-portrait, Reed Spring, Missouri.